Morrison Rogosa
Micah I. Krichevsky
Rita R. Colwell

Coding Microbiological Data for Computers

With 12 Figures

Sponsored by the Committee on Data for Science and
Technology of the International Council of Scientific Unions

Springer-Verlag New York Berlin Heidelberg
London Paris Tokyo

Morrison Rogosa
National Institute of Dental Research
National Institutes of Health
Bethesda, Maryland 20892, USA

Rita R. Colwell
Department of Microbiology
University of Maryland
College Park, Maryland 20742, USA

Micah I. Krichevsky
National Institute of Dental Research
National Institutes of Health
Bethesda, Maryland 20892, USA

Series Editor:
Mortimer P. Starr
Department of Bacteriology
University of California
Davis, California 95616, USA

Library of Congress Cataloging-in-Publication Data
Rogosa, Morrison.
  Coding microbiological data for computers.
  (Springer series in microbiology)
  Bibliography: p.
  1. Microbiology—Data processing. 2. Information
storage and retrieval systems—Microbiology—Code
numbers. 3. Information storage and retrieval
systems—Microbiology—Code words. I. Krichevsky,
Micah I. II. Colwell, Rita R., 1934–
III. Title. IV. Series.
QR69.D35R64 1986   576'.028'5   86-20348
ISBN 0-387-96417-7

© 1986 by Springer-Verlag New York Inc.
All rights reserved. No part of this book may be translated or reproduced in any form without written permission from Springer-Verlag, 175 Fifth Avenue, New York, 10010, USA.
The use of general descriptive names, trade names, trademarks, etc. in this publication, even if the former are not especially identified, is not to be taken as a sign that such names, as understood by the Trade Marks and Merchandise Marks Act, may accordingly be used freely by anyone.

Printed and bound by R.R. Donnelley and Sons, Harrisonburg, Virginia.
Printed in the United States of America.

9 8 7 6 5 4 3 2 1

ISBN 0-387-96417-7 Springer-Verlag New York Berlin Heidelberg
ISBN 3-540-96417-7 Springer-Verlag Berlin Heidelberg New York

PREFACE

As the title suggests, this book presents an open ended system in which computer techniques facilitate encoding, entry, management, and analysis of microbiological data derived from the study of bacteria, algae, fungi, and protozoa. The system is not constrained by any taxonomic point of view or proprietary computer technology and is freely and easily adaptable to the special needs of special problems. This enables investigators to deal efficiently with unforeseen areas of their investigation or with new data generated by old or new technology.

We are sincerely grateful for the invaluable advice and encouragement of Professor V.B.D. Skerman. Many individuals contributed ideas and technical knowledge, and users of the system have contributed data and responsive criticism for more than a decade. Particularly, we acknowledge Lesley Jones for some editorial help; Keith E. McNeil for contributions to Section 7; Elwyn G. McIntyre for invaluable assistance with Section 8; Dr. B. Amdur for major contributions to Section 21; Dr. A.J. Wicken for assistance in expanding and updating Section 23; Dr. S. Feingold for some features of anaerobic bacteria in Section 24; Drs. J. MacLowry and C.S. Thornsberry for assistance with Sections 19/40 and Dr. C.S. Thornsberry for major contributions to Section 35; Graham Manderson and Dr. Horst W. Doelle for some of the enzyme list in Section 34; John J. Wilson, Lesley Jones, and Graham J. Moxey for contributing general ideas on coding questions; Dr. Raymond Johnson, Claude Jackson, and especially Frances A. Benedict for development of reference system coding; Dean Wagner, Sallie McLaughlin, and Eric Molitoris for putting final order into the *Salmonella* series; Dr. P.M. Daggett and Frank Simione for contributing data set features applicable to protozoa; Drs. S. Jong, D. King, H. Furuya, and C. Philpot for contributing descriptive statements on various fungi; Dr. L. Wayne and the International Working Group on Mycobacterial Taxonomy for using the system to handle masses of data in many studies and publications; Dr. R. Atlas for many extensive studies and publications on marine bacteria using the system; Vesta M. Jones and Sallie McLaughlin for extensive testing of the system; A. William Davis for the chemical synonym list; Cynthia A. Walczak for development of file maintenance procedures and table organization; Claude Jackson and Lewis M. Norton for development of data processing procedures; and Eric Molitoris for taking our rough draft in hand, critically reviewing it for both content and format, and turning the draft computer files into a final camera ready document.

We express our appreciation to the Committee on Data for Science and Technology (CODATA) and its officers for sponsoring the publication of this book, and especially Dr. Edgar J. Westrum, Jr., Editor, CODATA Publications. CODATA was established in 1966 by the International Council of Scientific Unions (ICSU) to promote and encourage the production and international distribution of scientific and technological data. Its initial emphasis was in physics and chemistry, but its scope has been broadened to data from the bio- and geosciences. CODATA is "especially concerned with data of interdisciplinary significance and with projects that promote international cooperation in the compilation and dissemination of scientific data."

We appreciate the truly significant contributions of Elaine J. Krichevsky, whose enduring devotion to the maintenance of this document over the years, logical insights, computer knowledge, editing skills, meticulousness, and intelligence contributed greatly to the completion of this book.

Respondents requiring general information about the system, directions for coding, coding sheets or forms, or who wish to offer suggestions for improvements, please contact:

>Microbial Systematics Section
>Westwood Building, Room 533
>National Institute of Dental Research
>National Institutes of Health
>Bethesda, Maryland 20892, USA

CONTENTS

INTRODUCTION		1
DIRECTIONS FOR CODING DATA		25
Coding Numerical Data		31
Recording Specimen and Associated Microbial Data		33
Directions for Coding Reference System Data		37
Example Set of Code Sheets		40
SECTION		
1:	GENERAL INFORMATION CODE SHEET	40
2:	SPECIFIC STRAIN INFORMATION	50
3:	INDIVIDUAL CELL MORPHOLOGY	53
4:	INDIVIDUAL VEGETATIVE CELL SIZE	
	Narrow Increment Range, Longest Axis	58
	Wide Increment Range, Longest Axis	58
	Wide Increment Range, Second Longest Axis	58
	Narrow Increment Range, Shortest Axis	58
	Wide Increment Range, Shortest Axis	58
	Miscellaneous Measurements	59
5:	INSOLUBLE INTRACELLULAR AND EXTRACELLULAR DEPOSITIONS	60
6:	ENDOSPORES AND CYSTS	61
7:	MYXOSPORES, SPOROCYSTS, FRUITING BODIES, SPORES, SPOROZOITES, TROPHOZOITES, AND GAMETOCYTES	63
8:	BRANCHING, HYPHAE, AND PRODUCTION OF ASEXUAL SPORES	
	Branching Hyphae and Septation	66
	Fragmentation and Spore Formation	67
	Hyphal Pigmentation	68
	Conidiophores	68
	Conidia	69
	Conidial Location and Arrangement	70
	Properties of Conidia and Chains of Conidia	71
	Sporangia and Sporangiospores	76
	Sporangial Location and Arrangement	76
	Properties of Sporangia	77
	Sporangiospores	80
	Zoospores	82
	Sexual Reproduction	83
9:	STALKS	84
10:	SHEATHS	85
11:	CAPSULES	86
12:	STAIN REACTIONS	87
13:	MOTILITY, FLAGELLATION AND EXTERNAL ORGANELLES	
	General Mode of Locomotion	88
	Characteristics of Flagella	88
	Protoplasmic Extrusions	90
	Cilia and Cirri	90
	Membranellae and External or Paroral Membrane-like Structures	92

	External Architecture	
	General	92
	Tentacles	94
	Shells and Lorica	94
	Pellicle, Mucous Envelope, Adhesive External Organelles, Attachments, etc.	95
14:	MODE OF CELL DIVISION	97
15:	ARRANGEMENT	99
16:	CULTURAL CONDITIONS, INHIBITORS, NUTRITION, GROWTH, LIFE CYCLES	
	Solid Media	101
	Liquid Media	103
	Growth Kinetics	104
	pH Limits of Growth	104
	Relationship to Oxygen and Carbon Dioxide	104
	Light Requirements for Photosynthesis	105
	Tolerance to Inhibitory Substances	106
	Nutritional Characteristics	
	Growth in Complex Media	110
	Growth on Mineral Salts Media	112
	Ingestion of Biological Material	112
	Bioassociations	
	Symbiosis or Parasitism	113
	Vectors	113
	Life Cycles	113
	Habitat	114
	Differential Media	114
17:	VEGETATIVE CELL TEMPERATURE RELATIONS	116
18:	SODIUM CHLORIDE & OTHER OSMOTIC AGENTS - TOLERANCE AND REQUIREMENTS	
	Sodium Chloride	118
	Other Osmotic Agents	119
19/40:	ANTIBIOTIC SENSITIVITY	
	Lists of Selected Concentrations	120
	Lists of Selected Disc Concentrations	128
20:	PIGMENTS AND ODORS	
	Pigments	137
	Odors	139
21:	CELL CONTENTS	
	Cell Bound Lipids, Fatty Acids and Sterols	140
	Cell Contents	
	Inorganic	144
	Hydrocarbons	144
	Proteins, Peptides, Amino Acids	145
22:	LYSIS	146
23:	CELL SURFACE (WALL OR MEMBRANE)	
	Cell Surface Appearance	147
	Cell Wall	147
	Surface Structure-Other Than Cell Walls	152
24:	METABOLIC REACTIONS	
	Catabolic End Products	153
	Extracellular Polysaccharides	155
	Complex Organic Substances	155
	Complex Basal Medium Catabolism	156
	Fructose Catabolism	156

Contents

	Glucose Catabolism	156
	Lactose Catabolism	156
	Lactic Acid Catabolism	156
	Pyruvic Acid Catabolism	156
	Threonine Catabolism	156
	Synthesis of Specific Organic Compounds	157
	Catabolism of Organic Compounds	
	Acids	157
	Alcohols	157
	Amino Acids and Peptones	157
	Carbohydrates	158
	Miscellaneous	158
	Inorganic Products or Substrates	
	Ammonia, Nitrate, Nitrite, Nitrogen	159
	Sulfur Compounds	159
	Oxygen, Carbon Dioxide, Hydrogen, Miscellaneous	159
	Dye Reduction and Indicator Tests	160
	Toxin Production and Hemolysis	161
25:	CARBOHYDRATE METABOLISM	
	Utilized-Oxidized-Reduced-Acid	162
	Gas-Sole C Source-Hydrolyzed	164
26:	ALCOHOL METABOLISM	
	Utilized-Oxidized-Reduced	167
	Acid-Gas-Sole C Source	170
27:	ALDEHYDE METABOLISM	
	Utilized-Oxidized-Reduced	173
	Acid-Gas-Sole C Source	173
28:	CARBOXYLIC ACID OR ESTER METABOLISM	
	Utilized-Sole C Source-Oxidized	174
	Reduced-Gas-Decarboxylated	178
29:	AMINO ACID METABOLISM	
	Utilized-Required-C Source-N Source	182
	C & N Source-Deaminated-Decarboxylated	184
	Oxidized-Reduced-Gas	185
30:	AMINE, AMIDE, LACTAM, PURINE, PYRIMIDINE METABOLISM	
	Utilized-Required-C Source-N Source	187
	C & N Source-Deaminated-Oxidized-Reduced	189
31:	HYDROCARBON AND KETONE METABOLISM	
	Sole C Source-Utilized-Oxidized	191
	Reduced-Acid-Gas	195
32:	FAT AND OIL METABOLISM	199
33:	PRESERVATION OF STRAINS	200
34:	METABOLIC PATHWAYS AND ENZYMES	
	Glycolysis (EMP Pathway)	202
	Entner-Doudoroff Pathway	202
	Hexose Monophosphate or Warburg-Dickens System	203
	Phosphoketolase Pathways	203
	Citric Acid Cycle	204
	Glyoxylate Cycle	204
	Autotrophic Carbon Dioxide Fixation	204
	Heterotrophic Carbon Dioxide Fixation	205
	General Reactions Involving Pyruvate	205
	Pyruvate to Alcohol Pathway	205
	Pyruvate to Butanediol Pathway	206

Pyruvate to Butyrate Pathway	206
Pyruvate to Isopropanol Pathway	206
Propionic Acid Fermentation	206
Polyol Dehydrogenases	206
Miscellaneous Enzymes	206
Nucleosidases & Related Enzymes	
Hydrolysing N-Glycosyl Compounds	208
Hydrolysing Thioether Compounds	209
Pentosyl Transferases	209
Transferring A Phosphorus - Kinases	211
Phosphotransferases - Phosphate Acceptor	212
Diphosphotransferases	213
Nucleotidyltransferases	213
Phosphoric Monoester Hydrolases	214
Phosphoric Diester Hydrolases	215
Triphosphoric Monoester Hydrolases	216
Exodeoxyribonucleases Producing 5'-Phosphomonoesters	216
Exoribonucleases Producing 5'-Phosphomonoesters	217
Exoribonucleases Producing Other Than 5'-Phosphomonoesters	217
Exonucleases (RNA or DNA) Producing 5'-Phosphomonoesters	217
Exonucleases (RNA or DNA) Producing Other Than 5'-Phosphomonoesters	217
Endodeoxyribonucleases Producing 5'-Phosphomonoesters	218
Endodeoxyribonucleases Producing Other Than 5'-Phosphomonoesters	218
Site-Specific Endodeoxyribonucleases (Sequence-Specific)	218
Site-Specific Endodeoxyribonucleases (Not Sequence-Specific)	222
Site-Specific Endodeoxyribonucleases For Altered Bases	222
Endoribonucleases Producing 5'-Phosphomonoesters	222
Endoribonucleases Producing Other Than 5'-Phosphomonoesters	223
Endonucleases (RNA or DNA) Producing 5'-Phosphomonoesters	223
Endonucleases (RNA or DNA) Producing Other Than 5'-Phosphomonoesters	224
Phosphatases	224
Carbon-Carbon Lyases	225
Phosphorus-Oxygen Lyases	225
Ligases Forming Aminoacyl-tRNA	225
Ligases Forming Phosphoric Ester Bonds	225
Enzymes Acting on Peptide Bonds	
Peptide Hydrolases	226
Alpha-Aminoacylpeptide Hydrolases	226
Dipeptide Hydrolases	227
Peptidyldipeptide Hydrolases	227
Serine Carboxypeptidases	228
Metallo-Carboxypeptidases	228
Serine Proteinases	228

Contents   xi

	Thiol Proteinases	229
	Microbial Carboxyl (Acid) Proteinases	229
	Metalloproteinases	230
	Proteinases of Unknown Catalytic Mechanism	231
35:	QUANTITATIVE ANTIBIOTIC SENSITIVITY	
	Disc Zone Diameters	232
	Minimal Inhibitory Concentrations	238
	Numerical Inhibitory Zone Diameters	240
36:	INTERNAL ORGANELLES	245
37:	NUCLEUS	249
38:	NUCLEIC ACIDS	252

REFERENCE SYSTEM

101:	*SALMONELLA*	
	O Antigens - Polyvalent Antisera - Bacto	254
	O Antigens - Polyvalent Antisera - CDC	254
	O Antigens - Polyvalent Antisera - Wellcome	255
	Spicer-Edwards Polyvalent H Antisera	255
	H Antigens - Wellcome	256
	List of Manufacturers' Specific Antisera	
	O Antigens	256
	H Antigens	258
	Pooled H Antigens	260
103:	*STREPTOCOCCUS*	
	Neufeld Quellung Reaction	261
	Lancefield Groups	261
105:	*STAPHYLOCOCCUS*	
	Staphylococcal Toxin	263
	Immunodiffusion	263
	Phage Typing - Lytic Groups	264
	Phage Types	264
106:	*ESCHERICHIA*	
	O Antigens - Bacto	265
	H Antigens - Polyvalent Antisera - Bacto	266
	H Antigens - Single Factors - Bacto	266
	OK Antigens - Polyvalent Antisera - Bacto	267
	Specific OK Antigens - Bacto	267
107:	*CAMPYLOBACTER*	
	Penner Method - Hemagglutination - CDC	270
	Lior Method - Polyvalent Antisera	272
	Lior Method - Single Factor Antisera	272
APPENDIX - SYNONYMS OF COMPOUNDS		275
REFERENCES		295

INTRODUCTION

In their document entitled "Method for Coding Data on Microbial Strains for Computers", Rogosa, Krichevsky, and Colwell (1971) proposed the use of computers in microbial identification and classification and recommended the establishment of an international data bank or network for microbiology. The argument offered was that the experience of microbiologists with computers was sufficient and computer technology adequate for current and forseeable demands. Indeed, models of computers available to major universities and research centers in many countries at that time had storage capacities, software programs, and programming versatilities capable of accommodating and handling masses of data. The most compelling argument was that the body of historical data and the rapidly accumulating data were too large, cumbersome, and complicated to be handled conveniently or efficiently by conventional (i.e., non-computerized) methods. Clearly, with passing time and experience, these arguments are even more convincing and the remarkable developments in computer technology make it sensible to employ computers for the management of large data sets from many microorganisms. A number of commercial ventures in computer aided identification and classification of microorganisms have proven successful. An international data bank or network is the next logical step.

A microbiological data network, as a centralized repository of comprehensive information, would be an invaluable resource. It should include historical, nomenclatural, classificatory, epidemiological, special purpose applications, and other available and relevant information on individual strains, mutants, clones, and groups of microoganisms (such as bacteria, rickettsia, chlamydia, viruses, algae, protozoa, fungi, and yeasts).

The great potential value of computer technology and of a data network outweighs the onerous task of transferring and coding currently available data in the proper format for entry into an international network of computer centers. The magnitude of daily output of information from thousands of laboratories throughout the world requires a collaborative effort to catalogue and file these new and older data for contemporary applications and future reference. Computer storage and analysis of microbial strain data are already underway at the American Type Culture Collection, the University of Brisbane, the National Institutes of Health in the United States of America, the Japanese Federation of Culture Collections, the National Collection of Type Cultures (U.K.), etc., and at university and foundation laboratories around the world.

A computer network of strain data would greatly facilitate the updating of major microbiological reference works such as Bergey's Manual of Determinative Bacteriology (Buchanan and Gibbons, 1974) and Index Bergeyana (Buchanan, Holt, and Lessel, 1966). Despite the clear need for a computerized data network for microbiology, no general solutions to the international information explosion have been proposed. In fact, the 8th edition of Bergey's Manual was published in 1974 only after a seventeen year interval during which international committees and other experts diligently but slowly updated the very large volumes of information by conventional techniques involving hand sorting, collating, eye inspection, etc. in the management and interpretation of the data.

Computer assistance would be of great value in filing literature references on the taxonomy of microorganisms at an international data center for rapid retrieval and consultation by any individual in any country at any time. The history, epidemiological significance, nomenclature, identification, and classification of microorganisms require the same diligence in literature search as other scientific activities. The lack of immediate access to information on microorganisms, as is available for chemical compounds, is an obstacle to identification and classification of microorganisms, whether for academic needs or for specific applications such as in medical diagnosis, industrial processes, or in sanitary control. The explosive developments in biotechnology, in which microorganisms are the source material for genetic engineering, has created yet another set of requirements for microbial strain data. Although lists of references are available from libraries at the National Institutes of Health, the National Library of Medicine, large university centers, and foundations, these are generally limited to recent references cited within the past decade and are not always easily accessible to many individuals. A comprehensive computer library of strain data and literature sources will provide a capacity to achieve ready and orderly retrieval of wanted specific information.

The classes of information acceptable for computer storage and the kinds of questions which can be put to a data center are, for most practical purposes, limitless. The RKC format (Rogosa, Krichevsky, and Colwell, 1971) for coding microbial strain data (the acronym was invented by Prof. V.B.D. Skerman) was published in the form of a questionnaire whose answers were applicable to computer entry, storage, retrieval, and analysis of data descriptive of individual strains of microbial taxa. Strain or clonal data encompassing morphology, physiology, biochemistry, genetics, serology, ecology, nomenclature, pathogenicity, etc., have already been included in our growing data base. We herein present an expanded and revised questionnaire based on the RKC format. It has been extensively expanded with respect to the bacteria and now includes protozoa (Daggett et al., 1980) some fungi (Philpot et al., 1982) and some algae (Van Valkenburg et al., 1977, and personal communication). This contemporary effort is based on ten years of further experience with the RKC format and the benefit of much help from the scientific community. Because this format for data entry is open-ended, it allows for the computer entry of new, contemporary, or past information to be accommodated in the system. Additional questions or statements about newly discovered features can be included at any time in current or newly established sections of the document. Present programs permit this and are constantly being improved. These programs have the advantage of being without taxonomic prejudice and of being adaptable to numerical taxonomy or other types of analysis.

There are invaluable computer functions beyond the handling of strain data. For example, Index Bergeyana (1966), or similar lists of strain information, provides a format suitable for computer storage by available keypunching operations, without major modification or rearrangement. Thus, synonyms, as listed in Index Bergeyana (1966) and elsewhere, can provide historically complete cross-indexed lists of nomenclatural epithets. Also, requests for information concerning nomenclature, location of strains with specific biochemical or other properties, or a combined request for any or all of these or other matters may be answered within seconds by a computer system. Computer linked terminals remote from the computer installation would provide ready and convenient access to computer memory storage. By any one of several internally programmed methods, organisms can be identified or classified. If insufficient differentiating data are provided, additional tests necessary for more positive identification can be given the investigator at any remote terminal station. Thus, a valuable investigator-computer interaction can now be achieved, constituting a feed-back control system for diagnostic purposes. Based on this powerful process, an effective probabilistic identification system is now available which uses the RKC format (Johnson, 1979).

The design of questions is generally independent of test method or taxon. Although there are occasional problems in comparing results of some present methods from different laboratories, we here assume that, regardless of how it was determined, the test result is acceptable for coding if the method is reproducible and reliable by the best technology currently available. Investigators have long known the technical weakness of certain tests and test methods and recognize the desirability and necessity of standardizing all methods used in microbiological studies. Thus, the focus of this document is on data handling by objective methods to promote the standardization of microbiological tests. At the very least, all tests and test methods are entered into the data bank and can be linked with answers to specific questions. If particular tests give conflicting results, either in the same laboratory at different times or among laboratories, such discrepancies can be detected and the efficiencies of tests and test methods can be computed. If discrepancies are the result of defective methods, suggestions for improved technology can be proposed and developed.

The system for data handling offered here is independent of any given philosophy for classification. There is no pre-evaluation or judgment as to what data may be stored. Consequently, there is no pre-selection, exclusion, or hierarchical sorting of input data.

All computers have certain restrictions because of hardware and/or software design. A standard data code, fixed location of related data, and specific ordering of data is therefore followed. In the present instance, information is coded in the RKC format, as described below, in order to be compatible with computer entry. With minimal and careful forethought, data in laboratory record books can, in many cases, be entered directly into the computer. Binary logic was chosen as the basic coding method so that all answers to questions could be accessed by the same methods. A YES answer is signified by the digit "1" and a NO by the digit "0".

Numeric data, as well as yes/no data, can be coded (see Directions for Coding). This numeric coding complements the binary coded attributes (e.g., allows coding of a number denoting a delay in becoming positive) through linkage in the computer. Numeric coding allows coding of data that is more naturally numeric to the user (e.g., minimum inhibitory concentration).

The list of criteria included in this document has been left open-ended. Additional attributes can be added to the list so that expanded versions can be produced at any time without causing major programming problems in the computer. As new data become available and is supplied by an investigator, updating of the data files is done. Examples of new information are records of new strains, additional test information on old strains, recent publications, isolation and/or accession of strains, etc. Lists of culture collections, literature references as in the U. S. National Library of Medicine, synonyms, other systems of taxonomy, whether in French, Russian, Japanese, or other languages can be entered into the microbiological data bank as it is presently designed.

The document was edited, written and arranged by computer using the text editor program, WYLBUR. IT SHOULD BE EXPLICITLY UNDERSTOOD THAT THE ENTIRE DOCUMENT DOES NOT HAVE TO BE FILLED OUT. The forms can be provided with a list of any combination of characteristics or expanded to include an index. However, an index is not included in this document, since the total number of pages would be quadrupled, at least. We have made allowance for institutions without computers or with only limited computer facilities so that card decks can be used. The text editor programming system employed permits modification of the form. For example, changes in card deck lists, test categories, etc. are possible. The form is inclusive by design, rather than selective, since data can be discarded or eliminated readily, but going back to the laboratory bench and reproducing data is far more difficult. Thus, if it appears that the same genes or gene effects are scored a number of times, the implied redundancy in the data can be eliminated when the situation warrants.

There is an element of arbitrariness in selection of genus and species names for storing and search by computer. No significance is involved or implied since any tag that allows search and coding can be used. It is only that some label is necessary. Names are accepted as given to us. However, we have generally striven to abide by the International Microbiological Code, valid publication, the judgments of International Committees, and decisions of the

International Judicial Commission. Where Bergey's Manual (8th edition) meets the above criteria, we have taken cognizance of names included therein.

Special purpose forms have been prepared, published or put into use in the past (Quadling and Martin, 1968). These, and other programs for handling culture collection data (Simpson, 1968) have been accommodated within this document, which has been written to allow inclusion of most kinds of available or proposed microbiological information and documentation. Fixed field format is used to simplify programming to generate the query system, allowing less expensive and more efficient use of computer and programmer time. Existing software for sort and search routines can be utilized and open formats can be used, but they require larger computer facilities and more sophisticated programs. Since one of the objectives of our work was to standardize the form of data communication, the entire document was keyed to the 80 column Hollerith punch card. Thus, a simple card sorter can be used to handle the data, if necessary.

Difficulties in interpretation of the questionnaire by individuals whose mother tongue is not English may require that the form be translated into other languages. Translation can be done when conditions and need dictate. However, this may not be an obstacle because the document is already in use in non-English speaking countries.

In conclusion, classification, the grouping together and ordering of objects with characteristics in common, can be done in a standardized fashion, regardless of the objectives of the classification. In any case, the data should be accumulated and assembled so as to be available to all interested microbiologists. Modern taxonomic analyses involving several thousands of strains, for which several hundred features have individually been scored, can not be scanned efficiently without computer assistance, no matter how expert the taxonomist. With this in mind, we propose the following compilation of microbiological criteria as an aid to definition of microbial taxa.

The Data Structure

In the original RKC document (Rogosa, Krichevsky, and Colwell, 1971), the method for coding data comprised the following parts: general information; strain source and related information; and strain characteristics.

At that time, we recognized the tentative and somewhat experimental nature of some of the coding procedures. Our experience over the past ten years, together with significant input from the scientific community, has demonstrated the need for a revision and expansion of the method.

The individual questions and format of this document are designed to facilitate the use of computer technology for the retrieval and analysis of data derived from the study of individual strains of bacteria and other, generally unicellular, microorganisms. We believe that the manner in which the questions must be answered helps to eliminate subjective or unclear impressions of organisms. Rather, objectivity and maximal clarity are more likely to be achieved. At the same time, the data can be retrieved, sorted, and analyzed by powerful logical, statistical, or mathematical programs in taxonomic, or other studies. Indeed, some studies may be impractical or impossible without the type of technology suggested here. We are aware of the considerable work involved in coding and proof-reading the coded data on appropriate forms. However, we cannot overemphasize the importance of verifying the accuracy of coded data by meticulous proofreading at every step. After all, if data are recorded carelessly or incorrectly, in laboratory note books, cards, or in any format, such data would be considered not trustworthy and would be rejected. The advantages and potentialities of computer technology are so great that the work involved in coding data should not be considered onerous even when it appears to be.

The data structure is illustrated in Fig.1. All data are categorized and labeled on entry into the computer by a system of six digit numbers. Each number denotes a single statement. The statements fall into one of two categories: a statement discriptive of the dataset as a whole or a statement descriptive of a single strain. Thus, general information describing the data set as a whole is entered under the lowest numbers. Next is historical information on each strain, followed by phenotypic attributes of each strain coded in binary logic (including a span of numbers for applying binary logic to any strain-associated data of the the user's choosing), and, finally, attributes which can be coded numerically. The numerical item numbers represent the identical attributes as those in the binary sequence of code numbers. Thus, an attribute can be entered into the computer as binary or numerical information. More details of the data structure are presented in the "DIRECTIONS FOR CODING DATA".

Individual Statement Design

The individual statements are descriptors of individual characteristics or features of individual microbial strains. The statements are expressed positively, i.e., "Cells are spherical" and are never expressed negatively, i.e., as "Cells are not spherical". Each statement concerns a single attribute; thus, we have avoided composite statements like..."Cells are spherical, have gas vacuoles, and deposit sulfur outside the cell". This is really three statements about three different attributes; therefore, the statements in this document are: (1) "Cells are spherical"; (2) "Gas vacuoles (aerosomes) in the cell"; and (3) "Sulfur deposited outside the cell". Each statement is individually coded. Although the descriptive statements are not expressed grammatically as questions, there is an implicit question in each statement; thus, "Cells are spherical" really asks... Are the cells spherical? And the expected answer is yes or no. We shall explain how these responses are to be coded in a following section of this document entitled "Directions for Coding Data."

Introduction

FIGURE 1

RKC code number conventions
===

```
000000-----|
           |---Reserved for free text describing the
           |   data set and/or individual strains
002012-----|

002013-----|
           |---Reserved for permanent assignment to
           |   specific phenotypic attributes
           |   (binary coding)
089,999----|

090,000----|
           |---Reserved for temporary assignment to
           |   specific phenotypic attributes pending
           |   editing and reassignment to permanent
           |   numbers (binary coding)
097,999----|

098,000----|
           |---Reserved for free use by each submitter
           |   of data (binary coding)
099,999----|

100,000----|
           |---Reserved for reference system attributes
           |   (serotyping, homologies, etc.)
           |   (binary coding)
199,999----|

200,000----|
           |---Unassigned, reserved for future
           |   expansion (binary coding)
499,999----|

500,000----|
           |---Reserved for numerical coding in the
           |   same sequence as the binary coding above
999,999----|
```

Because of the implicitly questioning nature of the descriptive statements, users of the RKC Coding System have often referred to the individually coded feature items as "questions" and groups of questions have been referred to as "question sets". The term "data sets" indicates the data comprising the responses to "question sets".

We have practiced the principle of assigning unique code numbers to individual feature items because it is critical to assign code numbers unambiguously. Thus, individual items of data, once coded under a valid number, will always be associated with that number and no other. Editorial changes in the language of a statement are restricted to clarification of the meaning of the statement. If language changes in a statement are of such a magnitude as to introduce a different meaning or concept, then the statement must be assigned a different code number. The system design requires acceptance of this rule by contributors of data in order to achieve orderly and unambiguous management of data sets.

In any standardized coding system, it is essential to understand the nature of the test being performed as well as the logic of the system in which the test result is coded and recorded. For example, a microbiologist speaking of "gamma" hemolysis means that there is an absence of alpha or beta hemolysis of red blood cells. In keeping with this logic, the microbiologist records in the RKC format what was specifically observed:

024198: Sheep blood hemolysis is alpha.
024199: Sheep blood hemolysis is beta.

If both of these are coded negatively, there is "gamma" hemolysis. Similar statements about horse, human, and rabbit blood have the code numbers 024200 - 024205.

Meaningful responses in the RKC format are the strict corollary of specific knowledge. In recording whether a substrate was utilized, fermented, etc., the microbiologist must know whether the substrate used in the test had a D, L, or DL molecular configuration because the statements in this document most often use specific language asking for this very information. We have increased the accuracy of the coded statements about enzymes by using the "Key to numbering and classification of enzymes" in Enzyme Nomenclature, Recommendations (1979) of the Nomenclature Committee of the International Union of Biochemistry and by brief appropriate definitions of various terms, or occasional explanatory notes.

The fact that each item receives a unique code number lends specificity to the coding method. For example, the RKC code contains three separate statements concerning the detection of the ability to degrade hydrogen peroxide:

034109: Catalase (1.11.1.6) is present (hydrogen
        peroxide is decomposed by an enzyme containing
        a heme or porphorin structural group).
034110: Catalase-pseudo (an enzyme not containing a

> heme structural unit) decomposes hydrogen
> peroxide to water and oxygen.
024164: Hydrogen peroxide is decomposed.

Such specific answers to specific questions have the multiple virtues of ensuring maximal clarity and the simplification of coding, editing, and storage of the data. Also facilitated are the writing of computer programs and general computer management and analysis of the data. Thus, expenditures of time and money may be kept within practicable or feasible limits even for large projects.

Cell Morphology Statements

Most terms defining and quantitating individual cell morphology of microorganisms relate to plane surfaces and shapes having the two dimensions of length and width (breadth) only. Thickness is the third spatial dimension; one of the definitions given by Webster's Unabridged Dictionary (2nd and 3rd eds.) is that thickness is "the smallest of three dimensions". Applications of this definition may produce confusing results. The cell "thickness" of a cell in a chain of lactobacilli would be recorded as the dimension (diameter) of the cylinder perpendicular to the direction of division (i.e., along the chain). In contrast, the "thickness" of a *Simonsiella* cell would be in the direction of cell division since the height of the curved disc-shaped cell is considerably less than the diameter. For some geometric shapes such definitions are without real meaning. In a circle or square, length = width. Also, cylinders and solid cubic square figures may be constructed so that all dimensions are equal.

It is relatively easy to measure any individual cross-sectional dimension of simple undifferentiated single cells particularly where the cells are immotile or in fixed preparations. Indeed, the dimensions of spherical cells are expressed in the single dimension of what is observed to be the cell's diameter (as if the cells were circles) and the dimensions of rod-like cells are given in only the dimensions of length and width [See Bergey's Manual of Determinative Bacteriology, 8th ed., 1974].

For computer coding purposes it is not necessary or advantageous to state which dimension is length, or width, or thickness. The original RKC document (1971) managed this situation for bacteria by including in Section 4 ten descriptive statements as to the dimensional range in micrometers of the *longest* axis of each cell (004001-004010); similarly the ten items 004011-004020 are concerned with the dimensional ranges of the *shortest* axis of each cell. Daggett et al. (1980) and Philpot et al. (1982) employed this convention in the encoding of cell size data descriptive of selective groups of protozoa and fungi, respectively. Van Valkenburg et al. (1977, and personal communication) adopted the terms "longest" and "shortest" axes for individual cell size of certain algae (photosynthetic aerobic nanoplankton) as previously used for encoding bacterial, protozoan, and fungal data; however, they extended the usefulness of the coding system for their purposes by interpolating descriptor statements referring to the "Second longest axis of each cell is between X and Y micrometers". By this device Van Valkenburg et al. (1977, and personal communication) have created a descriptor for a third dimension.

For many of microbiological forms, third dimensional observations are difficult, inexact, and meaningless unless the definition of each dimension includes unambiguously observable geometric or anatomic reference points; in some cases the experience of the scientific community may be helpful in establishing generally accepted workable conventions. Generally, it serves no useful purpose to define third dimensional morphological aspects of simple bacteria because of the lack of acceptable points of reference or orientation. With micro-organisms having more complicated cell differentiation than the simple bacteria, more sophisticated techniques are required to encode all the morphological descriptors. The perspective of the observer of higher botanical structures and animals is generally obvious because of orientation of organisms with respect to the center of the earth, significant and directing anatomical reference points and structures (i.e., points of attachment or insertion, oral, anal, anterior, posterior, dorsal, ventral aspects of the organism, internal and external organ differentiation, etc.)

Biologists have difficulty in framing standard definitions of shapes, forms, or structures. Different sciences may use different methods of study and may tend to emphasize the study of certain features as compared to others. Sometimes more than one term has been used to describe the same structure. In such instances, equivalent known common terms are listed parenthetically in this document. A descriptive term often has varying meanings in different disciplines of microbiology although it may be clearly understood within each discipline. The term "spore" as applied to structures of bacteria, protozoa, fungi, and algae is often very different in many basic ways among these groups of organisms. Nevertheless, the use of a single question about a feature with a multiplicity of meanings is still possible. Although certain words may be used in common among disciplines, confusion is avoided by the specific meanings assigned specific words within a given discipline. We need only know the general kind of organism that the encoded information is describing (i.e. protozoa, bacteria, fungi, or algae). After all, this is what happens normally.

Morphological features have often been described imprecisely. For example, descriptions like "more or less round" have been used. To solve such a dilemma, the answers to two questions must be encoded: (1) is the structure spherical? or (2) is the structure elipsoidal? If the shape of the structure is variable, positive answers to both questions are encoded.

Botanists since Linnaeus (1753) have attempted to establish a more standardized vocabulary of descriptive terminology of plane surfaces and shapes. A committee of the Systematics Association of Great Britain published charts [see Taxon 1960, 1962a, 1962b] containing drawings of many geometric forms and shapes and associated coded names for the individual drawings of shapes in the charts. None of the charts were devised for use with microorganisms or descriptions of third dimensional perceptions.

The description of geometric forms of motile or live free floating cells or colonies is inherently difficult. We think that definitions would help and we provide many here. Protozoologists and algologists accept the direction of swimming as the anterior reference point for swimming forms. We have adopted this convention and have also accepted definitions of terms for the algae as defined by Bold and Wynne (1978). For morphological terms descriptive of protozoa, we have accepted definitions by Corliss (1979).

Differentiation of shapes and their exact obverse is especially troublesome. From pictures in charts or texts we have consulted, it appears that some of the plane shapes are redundant or so close to one another as to add confusion. In some cases, one figure blends into another with a small change in curvature or the ratio of length to width. Some examples of many that could be cited are: semilunate = plane convex; arcuate = bow-shaped = lunate; ovoid = oblong = ovate; discoid = flattened spheres. To improve clarity, we have combined a number of such features mentioned in the literature into appropriate broader based categories.

Section 3 of the original RKC code (1971) contained 23 descriptive items pertaining to the morphology of individual cells of bacteria. The descriptors were expressed in general terms such as: "003001: Cells are spherical.", etc. For the past ten years these have been very satisfactory for encoding morphological features of bacteria. Daggett et al. (1980) added 120 morphological feature items applicable to protozoa. Philpot et al. (1982) used already existing code numbers in Section 3 to encode for morphological features of selected groups of fungi and Van Valkenburg et al. (1977, and personal communication) have added 43 items especially applicable to the encoding of the morphological description of certain algae (nanoplankton).

Although Sections 3 and 4 deal more specifically with Individual Cell Morphology and Individual Vegetative Cell Size, respectively, there are at least 13 other sections which contain some morphological connotations. An examination of the titles of Sections 5-15, and of the specific items in Sections 36 and 37, is sufficient evidence of this. However, for logical reasons, and for convenience of encoding, we believe that the separation of these sections into broadly related categories is appropriate.

Gram Stain Statements

The Gram stain in its many modifications is one of the most widely used bacteriological procedures. Even the most earnest advocates of Adansonian theory view the Gram stain with some special emphasis. Soon after its initial use by

Gram in 1884 to visualize bacteria in tissue sections, it was realized that the Gram staining method divided bacteria into two very large groups of genera and species. Organisms which retained the stain after treatment with ethanol were termed Gram-positive and those which were decolorized by this or some other solvent treatment were Gram-negative. Gram-variable staining reactions also occur; often these are variably associated with the age of the culture, its acidity, or some other special conditions. This differentiation by Gram-staining has become a primary taxonomic property.

The reactions occuring in the Gram stain are complicated and are not clearly understood; over the years one explanation of the mechanism of the procedure tends to be superseded by a more current one attempting to explain deficiencies in previous schemes; no explanation seems fully satisfactory. In the Gram-stain procedure the insoluble crystal violet-iodine complex that is formed is extracted by alcohol from Gram-negative cells, but not from Gram-positive cells. It has been suggested that the relatively thick and dense walls of the latter are dehydrated by the alcohol or other solvent so that the wall pores are closed or diminished in size, thus inhibiting the escape by dissolution of the stain-iodine complex. In contrast, solvents more easily penetrate and dissolve the stain-iodine complex from the outer lipopolysaccharide and adjacent thin peptidoglycan layers of Gram-negative cell walls. Almost certainly the Gram stain reaction is not directly related to the chemistry of the bacterial cell wall because Gram-positive organisms, like yeast and many fungi, have a thick cell wall but are of different chemical compositions. To some extent, the physical structure, rather than the chemical composition of the cell wall, probably determines the Gram-stain reaction.

Even though the mechanism of the Gram-stain reaction is largely unexplained, it is now known that the reaction very often correlates well with cell wall composition and structure and generally with some other properties of bacterial cells. In brief summary, Gram-positive bacteria have a thick, dense cell wall having what appears to be a single layer, as revealed by appropriate electron microscopic techniques, and consisting of a thick peptidoglycan layer comprising as much as 90% of the wall, and also including glycerol or ribitol teichoic acids as the chief antigenic determinants; lipopolysaccharides are absent. The cell wall of Gram-negative bacteria appears multilayered and relatively thin; there is only a thin peptidoglycan layer comprising only 5-20% of the wall; teichoic acids are absent; and lipopolysacharides are a major component and are the principle antigenic determinants. However, Beveridge and Davies (1983) have re-examined the Gram reaction by a variety of physical and electron microsopic techniques and confirmed that the crystal violet - iodine complex is formed via a metathetical anion exchange. Their electron microscopic observations suggest that although some crystal violet coordinates with the wall polymers of Gram-positive

Introduction

and Gram-negative cells, the major site of deposition is in the cytoplasm.

In a more general phylogenetic sense, microorganisms fall into three major groups or proposed kingdoms: 1) the eukaryotes consisting of yeasts, fungi, algae, and protozoa, 2) prokaryotes, i.e., the so-called eubacteria or "true" bacteria; and 3) the archaebacteria. Until recently, the archaebacteria (which are procaryotic) were considered as members of the procaryotes; archaebacteria are known to comprise three major phenotypes, methanogenic, extreme halophilic, and thermoacidophilic (Fox et al., 1980, Woese 1982). The proponents of the creation of a separate kingdom for archaebacteria justify this as follows: 1) enormous genotypic differences as demonstrated by sequencing of the oligonucleotides of 16 S ribosomal RNA (rRNA); 2) uniqueness of the cell walls (where they exist) - none contain typical peptidoglycan (murein) (Kandler, 1982) - none contain muramic acid - archaebacterial cell walls contain mainly protein subunits (Jones et al., 1977, Weiss, 1974) or polysaccharides (Kandler and Konig, 1978) or are wall-less like the mycoplasmas (Darland et al., 1970) - lipids and membranes are unique - the translation apparatus as exemplified by tRNA is unique in containing no ribothymine in the so-called TψC loop (the RNA polymerase subunit structure is of several unique types (Zillig et al., 1978, Sturm et al., 1980) - a number of biochemical idiosyncracies are found (Wolfe and Higgins, 1979) and more are being revealed with further study.

Just as in the eubacteria, the archaebacteria are also split into Gram-positive and Gram-negative groups. *Methanosarcina barkeri* and the rod-shaped members of the genera *Methanobacterium* and *Methanobrevibacter* are Gram-positive. Although the cells in the genus *Halococcus* have a thick cell wall as in *Methanosarcina*, *Halococcus* is Gram-negative and thus a purely physical explanation of the mechanism of the Gram-stain may be inadequate. The other methanogenic organisms are Gram-negative, i.e. *Methanococcus*, *Methanomicrobium*, *Methanogenium*, *Methanospirillum*, *Methanothrix*, etc. The family of extreme halophiles, i.e., *Halobacteriaceae* and the thermoacidophiles *Thermoplasma* and *Sulfolobus* are also Gram-negative.

Unfortunately, some bacteria which would be expected to stain Gram-positive because of their cell wall structure and other properties, decolorize more readily in the Gram staining technique than we would anticipate; such a Gram-negative staining result is discordant with their taxonomic status as ideal Gram positive organisms. An ideal Gram-positive organism would have the cell wall structures and chemistry characteristic of generally recognizable Gram-positive groups of bacteria and would also give a Gram-positive staining result consistent with its ideal Gram-positive nature.

In order to accommodate such anomalies when the organism does not behave ideally, Wiegel (1981) has proposed the elimination of the terms "Gram positive, negative, and variable" and their replacement by the terms "Gram reaction positive, negative, and variable" (this refers to the Gram staining technique only) and "Gram type positive, negative, and zero" (this is a concept referring to the taxonomic properties of the cell). The terms "Gram type positive, negative" would refer to taxonomically relevant groups of eubacteria. The group of microorganisms named "*Archaebacterium*", as proposed by Woese and Fox (1977), has a cell wall chemotype different from that of other bacteria and would be described by the term "Gram type zero".

In keeping with this idea, Wiegel and Quandt (1982) have demonstrated polymyxin B dependent formation of protrusions (blebs) of lipopolysaccharides on the cell wall of all tested species of Gram-negative organisms. KOH solubility and some enzyme tests did not always correlate well with Gram reactions or types in other studies (Halebian et al. 1981; Carlone et al. 1982).

There are some logical difficulties in item definition and coding of Gram-stain information for computer entry. The terminology is an inversion of what really happens in the staining procedure. For example, by Gram-negative we mean that a positive act was achieved, namely that the stain-iodine complex was dissolved from the cell by alcohol, acetone, ether, or combinations of these (depending on which modification of the Gram-stain is used) and that a second contrasting stain was retained by the cell and helped visualize it in ordinary light microscopy; contrastingly, the initial stain-iodine complex is *not* extracted by solvents from Gram-positive cells, the secondary contrasting stain is *not* retained, and the cell is visualized by means of the initial stain-iodine complex. The Gram-variable feature could be encoded in either of two ways: 1) both Gram-negative and Gram-positive attributes could be scored as positive (1 in our binary notation); or 2) a separate item (Gram-variable) could be scored as such. We have chosen the latter as the preferred method because of other possible states (e.g., cells do not stain by Gram methods, variable or uneven staining within a cell, young cells are Gram-positive whereas other cells are Gram-negative, etc.). In keeping with the logic that each coded item should be defined to signify a single attribute, we have coded Gram-stain information as a series of usually mutually exclusive attributes as follows: 012001, Cells stain by Gram methods; 012021, Gram positive; 012022, Gram negative; 012023, Gram variable; and a series of six other coded statements each dealing with one of the remaining aspects mentioned in the parentheses above.

The Gram stain procedure, itself, even though it comprises a number of discrete operational steps, is a *primary* observation and each possible result, as we have already shown is coded separately. Contrastingly, Wiegel's (1981) proposal to use the terms "Gram type positive, negative, and zero" does not refer to *primary* observations of the Gram stain (and indeed does not refer to the Gram stain at all) but rather is derived from consideration of multiple *associative* taxonomic properties of the cell. We recognize the possible conceptual value of the taxonomic Gram type. However, directly coding a derived concept such as

Introduction

the Gram type blocks reconstitution of the original primary observations from which the type is constructed, thus obscuring variant patterns. The Gram type could be determined by similarity to patterns of individual item descriptors as follows:

TABLE 1. *Phenotypic characteristics and Gram types*

		Gram type		
		(+)	(-)	(0)
023001:	Cell wall present	1	1	1 or 0
023010:	Cell wall contains peptidoglycan (murein)	1	1	0
023040:	Teichoic acid in cell wall	1	0	0
023089:	Lipopolysaccharides in cell wall	0	1	0
023012:	Peptidoglycan >10% of the mass of the cell wall	1	0	0
023002:	... Cell wall appears as a layer homogeneous in density 100-800 angstroms thick ...	1	0	1 or 0
023003:	... Cell wall appears as a triple-layered structure ... 60-100 angstroms thick...	0	1	1 or 0
023004:	... Cell wall appears multilayered...	0	1	1 or 0
023013:	Backbone structure of cell wall peptidoglycan contains glucosamine and muramic acid residues	1	1	0
023108:	Cell wall protein appears to have "structural" role	0	0	1
038049:	Ribothymine in TψC loop in tRNA	0	0	1

Thus, from the above individual descriptors requiring data to give answers (0 = negative, 1 = positive), the concept of Gram type can be derived. Strains with one or a few aberrant traits can be accommodated on the basis of overall similarity to the "type" pattern of results. New and supporting statements can be framed and coded as new varieties of data are obtained.

Categories of Unlisted Data Subsets

Some kinds of microbial features are not listed herein. Other characteristics may be listed only as single feature statements and are not logically arranged in separate sections of related feature items. As a policy, we have excluded most complex pattern data which must be used in a relative or comparative sense. However, where single features can be extracted from a complex of pattern data, they often are included here as separate items in relevant sections. Types of such pattern data we have not treated in separate or unique subsets or sections are: genetic sequences, maps, or markers; patterns of cell components derived from liquid or gas chromatography or electrophoresis; general immunological diffusion patterns; etc. These can be coded in this system by use of the unassigned numbers.

Multivalued Tests

Microbiologists use certain tests which appear to be multivalued statements. Triple Sugar Iron Agar (TSI), Lysine Iron Agar (LIA) and Kligler's Iron Agar (KIA) are some examples of such multi-characteristic tests in which one medium, inoculated in two ways, gives multiple results. The microbiologist would conventionally record separate information on a number of items from the "one" TSI test: (1) acid slant (AS); (2) alkaline slant (KS); (3) acid butt (AB); (4) alkaline butt (KB); (5) no change in slant (NCS); (6) no change in butt (NCB); (7) gas production; and (8) hydrogen sulfide production. In the same way, statements about these attributes are coded individually in the RKC format; the first seven are included in the range of numbers from 016174 - 016180 and hydrogen sulfide production has the code number 024148.

If a question is of a multiple component type, such as Question 16 of Section 7, "007016: Myxospores are aggregated into sessile mucoid fruiting bodies without sporocyst formation.", it should only be answered as positive if each component is positive. To be positive, the aggregate of myxospores must be sessile, the aggregate must be mucoid and the fruiting bodies must give no evidence of sporocysts. If the fruiting bodies are dry instead of mucoid, but all other characteristics are positive, the answer is still negative.

In general, multivalued items should be re-stated and each clarifying statement should be coded and entered as if it was a separate item. The multistate nature is often obscured in the original statement; a statement, that a bacterial strain is a gram-positive rod really makes two

Introduction

statements and is coded in the RKC format by two statements, 001201: Gram positive; and 003008: Cells are rod-shaped.

Patterns of Statements in the Differentiation of Taxa

An isolated statement can be specific but it is also imperfect because it is only a bit, i.e., "a unit of information equivalent to the result of a choice between two alternatives." This choice in binary systems is yes or no...possessed or not possessed...and in the RKC coding system "1" is a positive response and "0" is a negative response. By itself, a descriptive statement contributes the least and most narrowly focused information possible to a field of knowledge.

But with the addition of each pertinent or related descriptive statement, new units of information accrue. As this occurs, there is more meaning in the accumulated statements and greater understanding of wider fields of knowledge. Let us take the example of hydrogen sulfide production. The RKC document has 19 coded statements about sulfur compounds: feature item 024148 states "Hydrogen sulfide is produced"; 6 other coded statements elicit information as to the substrates from which hydrogen sulfide is produced; and 12 other statements are concerned with oxidation or reduction reactions of thiosulfate, thionates, sulfate, sulfite, and sulfur. These 19 feature items are sufficient to contribute the significant units of information necessary to understand which key reactions, if not metabolic pathways, are present in the various sulfur metabolic cycles of 10 genera of the family *Chromatiaceae*, 5 genera of *Chlorobiaceae*, 2 genera of *Beggiatoaceae*, and 9 well-known and cultivable genera of chemolithotropic bacteria which have not been formally assigned to families. What is especially interesting here is that these genera of bacteria obtain the energy necessary for their growth, reproduction, and survival from the selective utilization of these sulfur compounds.

Similarly, there are 16 feature items in a subsection on reactions involving ammonia, nitrate, nitrite, and nitrogen. The information in this subsection is concerned with various reactions known to occur in the reduction of nitrate or nitrite, and in the utilization or production of ammonia. Obviously, such information is applicable to many groups of microorganisms.

A complete subset of data on sulfur metabolism or any other single aspect of metabolism is much richer in information than any informational statement within it. However, even this informational content is limited because other aspects of metabolism and the very many other characteristics (i.e., morphology, sporulation, staining reactions, mode of cell division, cultural conditions, etc.) have not yet been considered. In comparison to the total information and expression of phenetic characteristics controlled by the microbial genome, this small subset of information is very limited. Nevertheless, there are times when simple and limited data subsets can be powerful instruments of differentiation between microbial taxa. In extreme environments where external pressures are harsh and highly selective, minimal information will differentiate between taxa; in kinder environments allowing competitive survival, very much information is generally required for the differentiation of even a few taxa. An example of the first kind is one in which organisms have temperature optima near 70 C and utilize either sulfide, sulfite, or thiosulfate. After elimination of nearly all bacterial and algal taxa not behaving as above, perhaps 5 microbial taxa remain to be considered. On the other hand, differentiation of *Enterobacteriaceae* or *Mycobacteriaceae* requires extensive and complicated data sets to differentiate between taxa within the families or those in related groups of microorganisms.

From the Table of Contents, it is evident that descriptive data sets can be developed covering a comprehensive array of different microbial taxa and microbial attributes. Thus, 14 probabilistic matrices (Johnson, 1979), encompassing 14 groups of bacteria and 429 commonly encountered heterotrophic taxa, have been created and used successfully for the past 4 years. Although sets of feature items have been developed for protozoa (Daggett et al., 1980) some fungal taxa (Philpot et al., 1982), and some algae (Van Valkenburg et al., 1977, and personal communication), probabilistic identification matrices have not yet been constructed for these organisms. We welcome the collective wisdom and cooperative assistance of the scientific community in achieving this goal.

"Inverted" Coding

Although the coding logic in the RKC system is generally straightforward, there are certain apparent anomalies or coding inversions in which one could unwittingly violate the precept "say what you mean and mean what you say". In Section 16 there is a subsection entitled "Tolerance to inhibitory substances". Feature item 016289 states "Grows in media containing 0.02% sodium azide". In more general language, 125 additional statements say: Grows in media containing ....X% of some potential inhibiting chemical agent. If growth *occurs*, the result must be coded with a "1" (for positive) and the organism is *insensitive or resistant* to the inhibitory agent; if *no growth* occurs, this is coded as "0" (for negative) and the organism is *sensitive* because its growth is *inhibited* by the agent. Please note that 6 of the statements are in an inverted format, such as 016196: Growth is inhibited by mercuric chloride (disc) 1.0 micrograms. Here, if growth *occurs* the meaning of the coding is the opposite of the first example given above and "0" indicates *insensitivity or resistance*; if growth *does not occur* this is coded by "1" and indicates *sensitivity or inhibition*. Similarily, there are some apparently anomalous coding inversions and even some apparent ambiguity or some subjectivity in a small

Introduction

number (2.5%) of the statements in Section 19/40 dealing with certain aspects of "Antibiotic sensitivity".

These inversions were purposefully introduced into the coding to capture data as it is most commonly recorded in the laboratory. Examples of such apparently imprecise situations are: "019001: Sensitive to ampicillin concentration (disc) 2 micrograms." or "019429: Intermediate resistance to ampicillin (disc) 2 micrograms."

In examples 019001 and 019429 it is *not* the language which is imprecise. We simply may not know the *manner* in which *interpretations* of sensitivity and resistance are made and for our purposes we do not need to know this. The scientist or laboratory technician has made a judgment based on the presence or absence of growth in a region of medium adjacent to the disc. This may have been done based on experience or knowledge of diffusion properties of the antibiotic and growth characteristics of the test organism and without associating specific measured distances of clear zones to objective definitions of sensitivity, intermediate resistance, or resistance. Another pertinent example of this principle is: "025195: Acid produced from glucose." Acid production may have been determined in any number of ways. If we need to know more about the acid, i.e., which acid or acids are formed, or how acid formation was measured, etc., there exists a number of coded statements in the RKC format to elicit this information; if it becomes necessary, additional statements can be easily formulated and incorporated into the open-ended RKC system. We briefly recapitulate what was said earlier, that "The design of questions is generally independent of test method or taxon" and ... "that regardless of how it was determined, the test result is acceptable for coding if the method is reproducible and reliable within available knowledge and technology."

Attempts to define and quantify the results of disc agar diffusion susceptibility tests for clinical applications have as their purposes better guidance of the clinician in treatment of the patient and use in epidemiological studies of hospital-acquired infections, prevalence and location of resistant strains in the community, etc. These tests must be standardized and accurate to rationally compare results from different locations and circumstances. The methodology is based on the test described by Bauer et al., (1966) and the recommendations of the National Committee for Clinical Laboratory Standards (NCCLS) (1979) associating zone sizes of clear areas around discs with clinical sensitivity or resistance of the testing organisms to many commonly used antibiotics. We have reserved Section 35 of the RKC document for the feature items having clinical applications and dealing with the very important subject of "Quantitative Antibiotic Sensitivity (Disc Zone Diameters; Minimal Inhibitory Concentrations)".

Coding of Color

Another problem area is color perception. This is complicated by the fact that the color of an opaque object depends on the wave lengths *reflected* from it into the eye while that of a transparent object depends on the wave lengths *transmitted* through it.

The achromatic or neutral colors are white, black, or the many possible grays. Unlike the chromatic colors, their quality does not depend on wave length or combinations of wave lengths. The achromatic colors differ from each other only in the single dimension of relative lightness reflected from surfaces. A white surface reflects approximately 80% or more of incident visible light and a surface is perceived as black if it reflects *ca* 4-5% of incident light. The grays are intermediate. Visual perception of neutral colors is also associated with the intensity of the illuminating light. As illumination varies, the intensity of the light reflected by a surface of a given neutral color will vary accordingly.

Nevertheless, once the eye is adapted to ambient light, the general observer perceives neutral colors in good agreement with the reflectance of the surface on which they appear - a dark gray object tends to look dark gray in all sorts of light. This "constancy" effect was shown first by Katz (1935) and later studied by Helson (1943) and Wallach (1948, 1963) who postulated that a ratio effect could explain this "constancy" in which a gray sample and cardboard background proved resistant to changes in illumination. "Since any neutral surface reflects a constant fraction of the available illumination, the light intensities reflected by two different surfaces under the same illumination should stand in a constant ratio no matter how the illumination is changed." Wallach (1963) confirmed this "constancy" ratio experimentally.

Because the neutral or achromatic colors have the single dimension of relative lightness or darkness, reasonably accurate perception of them is relatively more simple than accurate perception of the chromatic colors which have the added dimensions of hue and saturation. The human eye is perceptive to hues in a limited range of spectral wave lengths (*ca* 400-700 mµ). In the classical theory, a pure color would be a single wave length and compound colors would be mixtures of these. Thus, Young in 1845, Maxwell in 1860, and von Helmholtz in 1865 proposed that three wave lengths in the red, green, and blue bands of the spectrum (the primary colors) could effect all color matches. This theory fits the facts of color matching with pairs of small spots of light and it was tacitly assumed to apply to all conditions of color sensation. An "opponent-colors theory" involving four primary hues was proposed by Hurvich and Jameson (1957) and seemed to account more neatly than the classical theory for color mixture phenomena. However, in color vision of complete images under natural conditions, the classical theory is not in accord with the facts and the opponent-colors theory does not really explain the psychophysics of color perception. More recent studies by

Land (1959 a,b,c) have shown that in the real world: (1) the spectroscopic peak is not sharp, pigments have broad reflection characteristics, and "... each pigment reflects some energy from wave lengths across the visible spectrum"; (2) colors over the entire visible spectrum arise not from a choice of wave length but rather from the interaction of longer and shorter wave lengths on each side of a "fulcrum" wave length and that within rather broad limits the specific wave lengths or the brightness of each make no difference in color perception; (3) different wave lengths of light can be mixed and separated without affecting one another; thus, "red light" (650 mµ) and "green light" (530 mµ) combine to form what the eye perceives as a "yellow hue" indistinguishable from a single yellow wave length of 590 mµ: (4) therefore, the "yellow" is in us, not in the light which remains unchanged by the mixing; the light rays *per se* are not color making but are "bearers of information the eye uses to assign appropriate colors to various objects in an image."

We can experience very many just noticeable differences (JND's) of color. The JND, also known as the difference threshold or discrimination threshold, is the smallest change in any physical stimulation that can be observed with the unaided senses. There are a number of models and systems to standardize or predict resultant hues from mixtures of colors. Most notable of these are the standards of Ridgeway (1912), Munsell (1929), and various modifications of the Munsell system as found in the ISCC-NBS method of designating colors (Kelly and Judd, 1955), A Universal Color Language (Kelly, 1965) and the National Bureau of Standards (NBS) special publication 440, "Color ... Universal Language and Dictionary of Names" (Kelly and Judd, 1976). There are more than 13000 color names in the NBS Special Publication 440.

The "Centroid Color Charts" (see NBS Special Publication 440) have been usefully applied to the textile and other industries. Special standardized illumination and viewing conditions must be employed for accurate matching of samples to the Centroid Color Charts. For viewing clear colored solutions, special apparatus is required. It is obvious from the context and discussion that almost all of the commercial application has been limited to opaque surface pigments (textiles, plastics, powders, metallic surfaces, etc.) and almost none to clear solutions. There are special problems in comparative color designations for opaque and clear samples, such as the following:

TABLE 2. Designations for opaque and clear samples

Opaque samples	Clear samples
White	Colorless
Pinkish white	Faint pink
Yellowish white	Faint yellow
Greenish white	Faint green
Bluish white	Faint blue
Purplish white	Faint purple

In using the Centroid Colors, a Centroid number is associated with an abbreviation of the ISCC-NBS color description. This number (1-267) is entered on a computer punch card or data processing card at each of six levels of accuracy. This has been done on a commercial basis. An example of the six levels of accuracy would be: Level 1, brown (Br); level 2, yellowish-brown (yBr); Level 3, light yellowish-brown (Ly Br). Level 3 accuracy is generally sufficient and has been very satisfactory in commercial applications. Levels 4-6 are much more complicated and are probably not practical for routine usage.

The Centroid Color Charts have seldom been applied to the study of biological pigments. For the possible great accuracy of Level 6 a whole color solid is theoretically divisible into 5 million very small blocks and even NBS Special Publication 440 suggests the spectrophotometer as the preferred instrument to produce the accuracy required at this level and to warn of the metamerism in which excessive color changes result from changes in illumination. If one is interested in the color of a thin section under the microscope whose thickness is not easily controlled, the hue name (Level 2) is usually adequate or Level 1 might suffice; the same might be said of any experimental parameter which is changing or variable during the period of observation, as is the case in biological studies.

There are some further difficulties with the ISCC-NBS method of designating colors as exemplified in NBS Special Publication 440. As an example, grass-green denotes perceptibly and widely different colors in different series of standards.

One person has suggested that in using the Centroid Color Charts, the complex name "dark grayish reddish brown" would "immediately call to mind an image of the color." We can not do this; we have not yet had an accurate visualization of this color description, rather we can visualize the very many colors this name description does *not* fit. We wonder what a spectroscopic analysis of a "dark grayish reddish brown" would show.

Color standards may fade and color standard systems may be too expensive, not universally available, or out of print. Also, matching of samples requires special artificial conditions of sample handling and illumination which may not be easy to achieve or reproduce with very many types of biological specimens as they are observed in the laboratory or in native habitats. In view of the difficulties in coding biological pigments, the "solution" adopted by the RKC Code is to record the "most recognizable" of colors; i.e., red, green, blue, yellow, brown, black, purple (mauve), orange. Mixtures (i.e., yellow-green, blue-green, etc.) can be coded by scoring two colors positive. This approach seems to work. Once it is explained, most investigators adapt well to this procedure. In some special cases, colors can be coded in a "098000" number series. This number series is created and defined by the investigator to suit his interest in any special study or project. For example, if an observer is interested in "pink" or "tan" hues, such features can be included. The

Introduction 23

"098000" series of numbers can also accommodate the coding of spectroscopic wave lengths associated with observed colors.

Reference System Code (RSC)

Until now, the RKC Coding Method has encoded method independent information and no reference strain is needed (except for test quality control). The original RKC Code had no mechanism for directly recording data pertaining to antigenic structures and reactions, bacteriophage types, time and intensity of reactions, behavior of purified proteins and enzymes in immunological or electrophoretic procedures, etc. Such data were recorded and managed separately, independent of the RKC System. In order to expand the range of accessible data, the following procedures were studied: antigen/antibody reactions; bacteriophage typing; bacteriocin typing; chromatographic patterns; electrophoretic banding; immunoelectrophoretic patterns; Ouchterlony and similar diffusion patterns; and nucleic acid hybridizations.

These procedures employ different techniques, assess different properties, and generate different data. However, they are identical in testing a facet of the comparative relationship betweeen some feature of a test organism and that of a "standard" or "reference strain". Data acquisition systems requiring the use of reference strains may be designated as "Reference Systems" (S. Lapage, Personal Communication).

Reference Systems may be unique to a particular laboratory because antisera, extracts, etc. are prepared in that laboratory by its own methods and from its own laboratory strains which that laboratory often uses as reference strains. An example of this are DNA or RNA extraction and testing procedures for homology studies. We present here a standardized format that facilitates design of coding data sets to record reference strain information.

The RSC is organized into sections by genus. For example, all code numbers and descriptors of *Salmonella* are in one section and features descriptive of *Streptococcus* are in another section. If the reference strain and the test strain belong to different genera, the RSC is cross-referenced. For example, a *Streptococcus* strain may be tested for sensitivity to a bacteriocin produced by a *Bacillus* strain. Organizing the RSC around genera enables the individual scientist (who is concerned with only a few genera in a given study) to locate expeditiously the extant RSC numbers and descriptors for a genus.

The types of reference system coding presented here are incomplete. For example, toxin-antibody and bacteriophage reactions are covered for the genus *Staphylococcus* only. The serological statements pertain to conventional serological characteristics of the genera *Salmonella*, *Streptococcus*, *Campylobacter*, and the species *Escherichia coli*. Obviously, these are a limited group of taxa. They were included here because these taxa are of great hygienic, medical, and epidemiological importance and because readily available commercial preparations are widely used in the determination of the reactions in most of these taxa.

What we present here are examples of what can be done with reference system coding and is not meant to be exhaustive. Particularly, we have omitted areas in states of rapid flux, such as monoclonal antibodies and identification by nucleic acid hybridizations, probes and sequences. These are examples of areas of relative newness and ferment in which coding problems are being seriously investigated internationally in such endeavors as the CODATA/IUIS Hybridoma Data Bank (Bussard et al, 1985) and various gene sequence data banks.

Specific directions for coding reference system data are given in the "Directions for Coding Data."

DIRECTIONS FOR CODING DATA

[NOTE 1: In the following discussion, computer manipulations of data encoded in this format are mentioned to facilitate the explanation. The programs to do so are available from one of the authors (MIK) but are *not* required to use this coding format.]

This compendium of microbial feature descriptors is designed to employ computer technology in the storage, retrieval and analysis of data descriptive of individual strains. A set of filled out code sheets of the type now used is appended as an example; it has been encoded for descriptors of the genus *Salmonella* and should be consulted as a model for the understanding of the following coding directions.

The original coding system (Rogosa, Krichevsky, and Colwell, 1971) is henceforth given the notation RKC1; the expanded and improved coding system presented herein is referred to as RKC2 (See Table 3 for specific differences).

The RKC1 document uses five-digit numbers to code more than 10,000 characteristics (= features = properties = attributes = qualities = traits = items) of bacteria. The items are arranged logically in 34 Sections. The RKC2 Code uses six-digit numbers; items are arranged logically in 38 Sections with expansions to include protozoa, some fungi, and some algae. The six digit expansion was made to accommodate features such as serology, phage typing, bacteriocin sensitivities, nucleic acid homologies, etc. The names of genera of bacteria should be abbreviated by a code of two to four letters as outlined by Rogosa et al. (1986). As far as we are aware, there is no equivalent system of abbreviations for any other group of microorganisms. Section 1, identical for RKC1 and RKC2, is reserved for certain general information which is submitted on a special code sheet, designated as Code Sheet for Strain Characteristics Section 1 - General Information (CSSC-1-GI). In the RKC1 document there are three additional CODE SHEETS: CSSC-I, CSSC-II, and CSSC-III. In RKC2, CODE SHEETS CSSC-I, CSSC-II, and CSSC-III are replaced by five newly designed CODE SHEETS: CSSC-05, CSSC-06, CSSC-07, CSSC-08/09, and CSSC-10. These CODE SHEETS improve the informational content and facilitate data handling. At the end of the Directions for Coding Data, filled out examples of all forms of the RKC2 coding system are presented.

Whereas Section 1 (entered on CSSC-1-GI) asks for certain necessary and optional *general* information, Section 2 is designed to elicit specific *strain* information.

Section 2 asks questions about the history of the strain, whether certain types of studies have been done (historically or by the respondent), whether the strain is a type strain or has been designated as a neotype, monotype, lectotype, holotype, or syntype strain of the species or subspecies, and if the strain has been utilized for a specific purpose (e.g., production of an item of commerce, an antimicrobial standard, etc.). Feature items for Section 2 are in the range of code numbers 001992 - 002037. Informa-

TABLE 3. Item numbers and code sheet column positions for selected strain information in RKC1 and RKC2 formats (The full text of the items is given in the following document. The punctuation in the abbreviations is for compatibility with computer conventions which are confused by internal blank spaces.)

RKC1

ITEM	ABBREVIATION	CODE SHEET	COLUMN RANGE
02001	STRAIN	ALL	1-10
02002	MUTANT	I	11
02003	PARENT-STRAIN	I	12-21
02004	SPECIFIC-EPITHET	I	22-46
02005	AUTHORITY	I	47-70
02006	BERGEY	II	11
02007	GC	II	12-13
02008	CULTURE-COLLECTION	II	14-19
02009	DATE-DEPOSITED	II	20-23
02010	PLACE-OF-ISOLATION	II	24
02011	DATE-ISOLATED	II	25-28
02012	SOURCE	II	29-70

RKC2

ITEM	ABBREVIATION	CODE SHEET	COLUMN RANGE
002001	STRAIN	ALL	1-10
002002	MUTANT	05	11
002003	PARENT.STRAIN	05	12-21
002004	SPECIES	05	22-46
001997	DEPOSITOR	05	47-70
002005	COMMENTS	06	11-70
002006	BERGEYANA	07	11
002007	GC	07	12-13
002008	COLLECTION	07	14-19
001996	DAY.DEPOSITED	07	20-21
001995	MONTH.DEPOSITED	07	22-23
002009	YEAR.DEPOSITED	07	24-25
002010	PLACE.ISOLATED	07	26
001994	TIME.ISOLATED	07	27-30
001993	DAY.ISOLATED	07	31-32
001992	MONTH.ISOLATED	07	33-34
002011	YEAR.ISOLATED	07	35-36
001998	SOURCE.LOCATION	07	37-30
002012	SOURCE	10	11-70

Directions for coding data

tion on items 001992 - 002012 is supplied by the respondent on appropriate code sheets. Information on items 002013 - 002037 is coded on CSSC-III for RKC1, or an improved version, CSSC-08/09 for RKC2.

The remainder of this RKC document (Sections 3-38) is arranged logically in groups of related feature items. For example, Section 3 is concerned with Individual Cell Morphology, Section 4 with Individual Vegetative Cell Size, and Section 37 with the Nucleus of the Cell. Certain of the larger sections, such as Section 8 (Branching and Production of Asexual Spores), Section 13 (External Organelles, Motility, and Flagellation), Section 16 (Cultural Conditions, Inhibitors, Nutrition, Growth, Life Cycles), Section 24 (Metabolic Reactions), and Section 34 (Metabolic Pathways and Enzymes) have sub-headings under which related feature items are listed in pertinent groups.

With few exceptions, we think we have achieved a satisfactory and usable logical arrangement. However, inherent complexities and imperfect knowledge may result in some uncertainty with respect to individual statements. For example, Sections 6, 7, and, 8 are generally concerned with anatomic descriptions and arrangement of various spores, cysts, conidia, etc. These are structures involved in stages of reproduction of the cell and are often a necessary and unique part of the Mode of Cell Division (Section 14), or Sexual or Asexual Reproduction (see subsections of Sections 8 or 16).

Obviously there will be questions which have not been asked. The reason for this is practical. In the face of the remarkable versatility of microorganisms, no one can anticipate all possible descriptive statements. Furthermore, it is impossible to anticipate the very many and diverse interests and purposes of the entire scientific community. For example, within the Section on Metabolic Reactions, there may very well be questions on enzyme functions missing. In these cases, we enlist your assistance to contribute questions or suggestions for expansion. Please forward comments, via letter correspondence, to the authors in care of the Microbial Systematics Section, Westwood Building, Room 533, NIDR, NIH, Bethesda, MD 20892, USA.

The responses for each logical group of strains are encoded on a set of code sheets; these constitute a self-contained DATA SET. The logic underlying the formation of the data set is chosen freely by the user for personal convenience. Normally, it is convenient to restrict all responses in a given set to a single genus. Thus, a new set of code sheets would be needed for each additional genus or grouping that a respondent wishes to include. The limitation of each response to a single genus or grouping does not imply any taxonomic prejudgement. It only serves to provide a nominal boundary condition for ease of coding and cross-indexing of the data set. Section 1 (GENERAL INFORMATION CODE SHEET) IS FILLED OUT IN ALL CASES (at least with the name of the submitter of data and the logical name of the data set).

In certain cases the genus or other taxonomic group will be uncertain or not have been determined. The data being supplied may be an arbitrary set of strains (e.g., a series of isolates from soil, water or clinical patients, etc.). The information should still be supplied as described below with the following simple exceptions. An arbitrary descriptive name may be supplied in place of the genus name if grouping of organisms is desired. In this case, a new set of data sheets ("Form" in the conventions we use) must be used for each group designated. Alternatively, the genus name may be left blank and the information supplied in one batch.

When entering information onto the code sheets, please use pencil, marking heavily, for those sections requiring coded answers. PRINT CLEARLY. Correct any errors by erasing, *not* by crossing out and/or writing over the error.

For accurate data entry and proof reading the code sheets should be neatly prepared.

In general, alphabetic characters should be printed as capital block letters and other characters should be clearly shown and distinguished. In particular, the following usage will eliminate confusion between certain alphabetic and numeric characters:

1) Use L for the alphabetic L

   and I for the numeric 1 (one)

2) use O for the alphabetic O (oh)

   and $\emptyset$ for the numeric 0 (zero)

3) Use S for the alphabetic S

   and 5 for the numeric 5 (five)

4) Use U for the alphabetic U

   and V for the alphabetic V

5) Use Z for the alphabetic Z

   and 2 for the numeric 2 (two)

In addition, when a field is to be duplicated, use a "Z" as shown in the example set of code sheets.

In Section 1: General Information, answer all questions where applicable. Use only the spaces provided. Separate all words and/or names with a blank space. Where specific word or number answers are called for, place one letter or number in one space. If a number is called for, use leading zeros (e.g., a "1" is coded):

|0|0|1|

Do not exceed the spaces provided on a line. If necessary, abbreviate or round off to the nearest whole number. Always use the English alphabet and Arabic numerals. All other

Directions for coding data

questions (those having only 1 space available) are to be answered "yes" or "no" by using a "1" for "yes" and a "0" for "no". However, if you do not know the answer for a given space, leave that space blank. Ignore Question 4 on the title page of the Questionnaire itself; it is primarily for internal bookkeeping purposes.

If one person from one address, submits a set of answers for several groups of organisms simultaneously, in one package, Questions 001001 through 001010 (on the front side of CSSC-1-GI and 001011 through 001031 on the back) need only be answered once.

In Section 2: Strain Information, use the special code sheets as follows. Answer Questions 002001 through 002004 and 001997 on CSSC-05. If more than 20 strains are being coded, use additional sheets. Serially number each code sheet used in the space provided (PAGE NUMBER) in the upper left corner. Answer each question in the columns on the code sheets as denoted in parentheses after each question in the text.

After filling in CSSC-05 for all strains, answer Question 002005 on CSSC-06. Again, enter the page numbers in the upper left corner, beginning with the integer following the last number used on the CSSC-05 series for the previous 5 questions.

The same procedure is used for CSSC-07 and CSSC-10. This structured approach is advocated (but not intrinsically required by the computer) to facilitate record keeping on data flow as part of data processing quality control. If data are not being submitted as called for on a given CODE SHEET (e.g., no COMMENTS, CSSC-06), those code sheets should be omitted from the data set. However, the data manager should be informed in writing to avoid confusion as to whether the sheets were intentionally omitted.

The upper right corner of all CODE SHEETS (except for Section 1) is for identifying information which is filled out under the same rules as for Section 1. In the spaces provided, the following information MUST be entered on each code sheet: 1) The respondent's surname; 2) The genus or other microbial group; and 3) The date. Items 1,2, and 3 should be the same as the corresponding answers in Section 1. The spaces for Form Number (i.e., data set number) should be left blank.

CSSC-08/09 is to be used as follows. It is to be used for Section 2, Questions 002013 through 002023, and all following sections. The features to be coded are decided on first. Enter these vertically from left to right on lines 1-6 on as many copies of CSSC-08/09 as are required to hold them. (There is space for 60 features per sheet. However, columns may be left blank to improve readability for later proof-reading.) Enter one feature identification number over each column (Fig. 2).

The feature identification number as entered is organized into a 6 digit vertical column: digit 1 is a leading "0"; digits 2 and 3 convey the Section; digits 4, 5, and 6 represent the feature number within the Section.

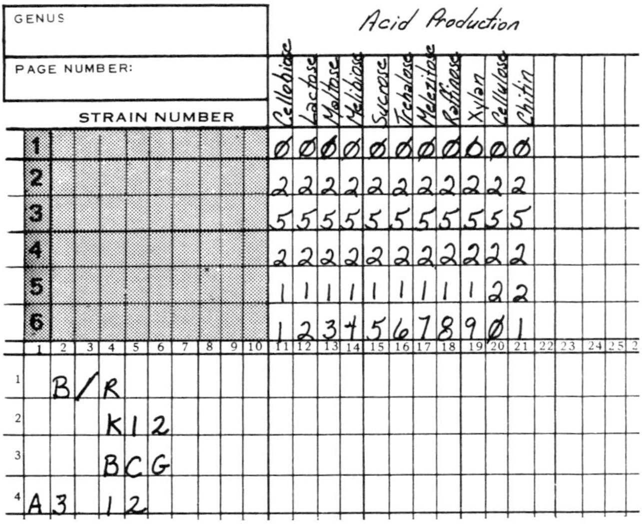

Figure 2. An example of coding features for binary data on CODE SHEET 08/09. Only the upper left hand corner is shown.

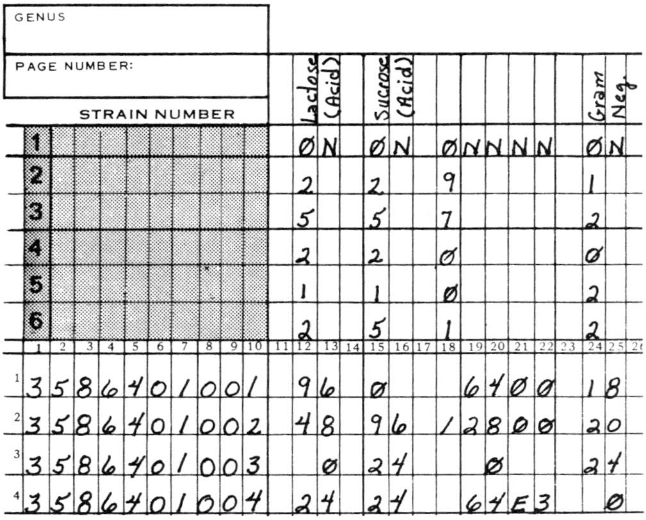

Figure 3. An example of coding features for numerical data on CODE SHEET 08/09. Only the upper left hand corner is shown.

Directions for coding data 31

Feature item 146 of Section 19 would be entered as follows:

|0|
|1|
|9|
|1|
|4|
|6|

Feature item 3 of Section 4 would be entered as follows:

|0|
|0|
|4|
|0|
|0|
|3|

The items may be entered in any order. The computer will rearrange the features, in order, automatically. The blank margin above the spaces may be used for the convenience of the respondent (e.g., for mnemonics of the features). The coding structure yields a table of strains by features.

Place the strain designation or identifier (as many as 10 characters, any letters, numbers, punctuation, i.e., ., :, -, /, etc.) in columns 1 through 10 on CODE SHEET-08/09, one strain to each row. Use of blanks or zeros is optional in this case. The information for as many as 20 strains may be entered for as many as 60 features on each code sheet. Thus, answers to 190 features regarding 105 strains would require 24 sheets.

In the body of the GENERAL CODE SHEETS, continue to code a "yes" as a "1"; a "no" as a "0"; and "not known" as a blank space.

Solely binary features would yield a table labeled as shown in Fig. 2.

CODING NUMERICAL DATA

In addition to binary, or "yes-no" answers (a "1" is a positive character state, a "0" is negative), it is possible to enter numerical information associated with a feature. As shown in Fig. 3, the columns required to hold the numbers are covered by the letter "N." In the example given, the first two features (025212 and 025215) are defined as numerical questions by the "N" columns following the features. Questions which are answered by numerical values can be searched as either numerical or binary information (e.g., a delayed reaction could be coded for the day it became positive and searched for as a delayed reaction or simply as a positive attribute). Users access binary data as features 025212 and 025215 but access numeric data as features 525212 and 525215. As a convenience to the user, system programs automatically convert numerical data into binary equivalents. If the numerical value entered is "0", a "0" is entered in the binary location (e.g., 025212 and 025215). If the value is any value other than "0", a "1" is entered for the binary value. There may be occasions, however, where a numerical answer of "0" may not be equivalent to a binary "0" (i.e., the "0" is a valid measurement for the user and not a negative answer). In these cases, the user may code the binary version of the feature separately as "1" to overide the automatic conversion. The number itself is stored in a separate location denoted by a feature code number exactly 500,000 greater than the original. Thus, there is a virtual image of the first 500,000 possible features (of which approximately 10,000 have actually been assigned meanings) in the second 500,000 possible locations.

In order to record numeric results, the following convention has been adopted:

1) Record the Question Number on CODE SHEET-08/09 as usual.

2) Record the letter N (signifying numeric data) in the column(s) following the Question Number.

For example in Figure 3:

Strain 1) Lactose positive on day 4 (96 hrs.)
         Sucrose negative
         Serological titer of 1:6,400
         G - at 18 hrs.

Strain 2) Lactose positive on day 2 (48 hrs.)
         Sucrose positive on day 4 (96 hrs.)
         Serological titer of 1:12,800
         G - at 20 hrs.

Strain 3) Lactose negative
         Sucrose positive on day 1 (24 hrs.)
         Serological titer of 0
         G - at 24 hrs.

Strain 4) Lactose positive on day 1 (24 hrs.)
         Sucrose positive on day 1 (24 hrs.)
         Serological titer of 1:6,400
         G + at 24 hrs.

Note that spaces may be left between feature numbers, whereas spaces must not exist between a feature number and its numeric indicator, N. Also, zero as a numeric answer need not be aligned with the decimal point.

Serological Data:

Serological numeric data are coded as the denominator.

Thus, a 1:6,400 titer is recorded as 6400.

Directions for coding data   33

Scientific Notation:

   Scientific Notation is recorded according to "FORTRAN"
   format rules.

1) Convert the number to a base 10 exponent:

   The capital letter "E" is substituted for the number
   "10" to denote the exponent base 10. No superscript is
   used for the exponent itself.

2) Record the data in FORTRAN format (E=exponent):

   6,400 = .64E4 or 6.4E3 or 64.E2
   .0064 = .64E-2 or 6.4E-3 or 64.E-4

   In addition to filling out the CODING SHEETS as described above, it would be extremely useful for each respondent to provide a list of the methods used to obtain the answers to the various questions. This list is maintained as an ancillary text file in the computer.

RECORDING SPECIMEN AND ASSOCIATED MICROBIAL DATA

   In many research and regulatory cases, data descriptive of a specimen must be associated or coordinated with the descriptions of the strains isolated from the specimens. Some common specimen data are exemplified by "total" microbial population, numbers of special groups of organisms such as coliforms, heat resistant bacteria, halophiles, psychrophiles, etc., the presence or absence of certain organisms (e.g., *Salmonella*), incubation media, incubation conditions, geographic location of the specimen, physical and chemical condition of the specimen at the time of accession, and conditions of isolation of microbial strains from the specimen.
   All strain descriptions and data on the specimen must be linked unequivocally and simply for easy retrieval. This becomes difficult to attain where multiple conditions (such as different incubation temperatures) are used with the same medium for counts of population. Total possible combinations of medium and conditions quickly become large. The number of conditions in any one study protocol are generally less than twenty. A desirable coding system should contain informational bins only for these twenty or fewer items and should not have bins organized around combinations. Thus, the individual medium used would always have the same code number. Incubation temperature, the medium, pH, oxygen tension, or any other condition would each be encoded separately as an individual item.
   This procedure ensures that only one combination of conditions can be recorded for each medium associated with any single strain. To do otherwise would require cross-indexing of specimen descriptions with information on other strains isolated from the specimen. A method of recording the specimen data entirely compatible with the existing RKC

coding system and associated computer program design will be described here.

The first step is to assign a code number (accession number) to each specimen. (As with strain designations, the code number may contain symbols other than numbers.) It is desirable to incorporate the specimen designation as part of the strain designation field. Thus, the actual computer record describing the specimen can be coded separately as a series of entries using the same format as the strain record itself. The only difference is that the specimen designation is followed by a special character(s) instead of a strain identifier. For example, a specimen identifier might be the six characters:

ABC123

The strains could be serially assigned. The fifteenth isolate from specimen "ABC123" would be:

ABC12315

The identifier for pure specimen identifier might be:

ABC123$00

The character "$" is useful because it allows searching for specimen information while excluding strain information. It may be replaced by any legal character that is not part of a legitimate strain designation. By using such a system, all information referring to one specimen would be linked through the specimen accession number.

Where it is inappropriate to include the specimen accession designation in the identifier for a strain, the specimen designation can be entered into the "AUTHORITY OR COMMENTS" field.

It is essential to avoid using the same designations for strains and specimens. For example: Strain 1 from specimen "ABC123" might be designated: "ABC12301". If more than one specimen designation is required to manage all the specimen information, this could be confused with the first of the strain descriptions. Use of a special symbol such as "$" in the strain designation field would avoid this. If two characters are used to separate the specimen code designations as shown above, and the first is a "$", then ten numbers and twenty six letters will allow more combinations than are ever likely to be met in practice.

To summarize by example:

```
ABC123$1    SPECIMEN DESCRIPTION
ABC123$2    SPECIMEN DESCRIPTION
ABC123$3    SPECIMEN DESCRIPTION
ABC12301    SPECIMEN ABC123, STRAIN #1
ABC12302    SPECIMEN ABC123, STRAIN #2
ABC12303    SPECIMEN ABC123, STRAIN #3
ABC124$0    SPECIMEN ABC124, NO GROWTH
```

Directions for coding data 35

By using the above conventions, a search on the character string "ABC123" would find all the information on specimen "ABC123" as well as the strains isolated from the specimen. Searching on the character "$" would provide all the information on all specimens, but not the strains. Searching on "$0" would yield the specimen information where no outgrowth was detected.

A search using NOT "$" as the character string would yield strain information. The informational feature of the specimen itself is now recorded. This information is of two classes: 1) alphanumeric (i.e., words and/or abbreviations) entered in specific fields or categories or 2) numeric, entered as answers to questions on the "GENERAL CODE SHEETS" (CODE SHEETS 08/09). Thus, place names should be entered in addition to latitude and longitude or postal codes. In general, where codes exist (e.g., Standard Nomenclature of Pathology, U.S. Food and Drug Administration Commodity Codes, etc.) they should be entered along with the item they represent so that an efficient search may be performed, followed by a listing in easily understood terms. The alphanumeric information on a specimen should be repeated for each entry (specimen and strain) linked to the specimen.

The numeric data follows a different logic. Because the possible entries are so numerous, they must be encoded by assigning a number to each item or attribute and referring back to an item list (just as in the full coding system). Suggestions on mechanisms for accomplishing the reasonable assignment of item or question numbers for specimen description follow.

Some of the items to be encoded may be intentionally redundant with the alphanumeric field information because the search logic is different in the two cases. For example, latitude and longitude in an alphanumeric field are only searched on a character by character basis. If the same numbers are entered as numeric answers, they can be searched on a greater than or less than basis, thus establishing a range. With date of isolation, the logic is complicated by the fact that the year of the date is normally entered last. A range search becomes cumbersome unless the components of the date are searchable separately. A simple way of accomplishing this is to assign a separate question number to each date component. Alternatively, the date could be entered as a single number in the sequence: YYMMDD, e.g. 21 October 1984 would be encoded as 841021.

The most complicated coding problem is the recording of specimen data on total populations under multiple conditions. This is the cross-indexing problem referred to previously. Consider the case of using the same medium (e.g., TGY Agar) for total counts of psychrophiles, mesophiles, and thermophiles incubated under aerobic and anaerobic conditions at different temperatures. The simple solution of entering one number for each possible combination allows for comparisons of actual counts. However, linking counts to temperature and/or oxygen tension in a search is practically difficult. The data for temperature of incubation and oxygen tension must be coded and recorded separately if such a search is to be performed. Returning to the example, the questions to be answered might be constructed as:

098001: The total count on TGYE Agar is _____.
098002: The temperature of incubation for the total count on TGYE is _____.
098003: The TGYE Agar for total counts was incubated under aerobic conditions.
098004: Brilliant green-tetrathionate broth enrichment culture is presumptively positive for *Salmonella*.

We have just discussed some coding problems which require special coding procedures for practical, expeditious, and unambiguous data management. Microbiologists are often engaged in research or regulatory activities which are confined to special materials, environments, and conditions. In some cases, the features by which an organism is identified, or its special physiology, pathology, ecological relationship, or economic significance, may only be revealed during the course of the contemporaneous study itself. For this reason, and to allow the unrestricted and most advantageous use of the RKC Coding System, we have reserved the series of numbers from 098,000 - 099,999 for special purpose data entry in individual investigations (see Fig. 1).

The investigator may use this series at his own convenience as a temporary series in the same way as described below for the 100,000 - 199,999 series of Reference System Code numbers until permanent numbers within the RKC system can be assigned. Some examples of the 098000 - 099999 series of numbers already in use are given immediately below:

098005: Test sample is a homogenate of _____ samples.
098006: Amount of product examined is _____ grams.
098007: Direct microscopic count per gram is _____.
090008: Prudhoe crude oil is utilized.
090009: Glucose uptake in sediment samples is _____ mg/g dry weight/hr.
098010: Growth occurs on marine oil agar plus crude oil.

Although CODE SHEET 07 is designed for the entry of the geographic source of the sample and CODE SHEET 10 accommodates information on the source of the microbial strain, it often may be necessary or desirable to enlarge the scope of this type of information beyond the limited space provided by these code sheets. The use of the 098000 - 099999 series of code numbers gives the investigator wide latitude in the formulation of individually encoded descriptive statements. This also has the advantage that patterns of occurence or models of behavior can be conveniently discerned. An example of this is the formulation by E.J. Krichevsky (personal communication) of 138 descriptive statements pertinent to the pathogenicity and source of isolation of oral strains from the oral cavity. There are series of statements concerned with geographic source, animal host (rat, hamster, human), age and sex of host, human racial groups, tooth sampled, tooth surface sampled,

Directions for coding data 37

disease state associated with the sample (caries, gingivitis, periodontitis, ulcer, various tissues after treatment with specific agents, and caries in animals fed various diets).

DIRECTIONS FOR CODING REFERENCE SYSTEM DATA

The Reference System Code (RSC) is an extension of the basic RKC system for a special purpose. As described in the Introduction, the RSC is designed to record information about the origin and techniques used to develop features which are dependent upon specific reference strains. The RSC provides a mechanism of communication and coordination between central computer support personnel and users. RSC sheets are filled out by the user and examples are at the end of the Directions for Coding Data. A temporary or interim code number is assigned to unambiguously link the reference strain and technique with the feature. (Data from the *test* strains under investigation are recorded on CODE SHEET 08/09 as described above in the basic RKC system.) Interim numbers are easily converted to permanent numbers at a central facility such as the Microbial Systematics Section at the National Institutes of Health in Bethesda, Maryland. If the encoded features and techniques have been previously submitted by anyone, the current user may receive a copy of existing information (providing that the information is not privileged and it is legal to divulge it).

The interim code number field is intended to prevent re-copying data and thereby reduces transcription errors. It also reduces delay of data entry because a user of the system can assign interim numbers to his data set ranging from 097000 to 097999 while RSC sheets are edited and analyzed for assignment of permanent code numbers into the general RKC system. The RKC system uses the 100000 to 199999 series of numbers (see Fig. 1, and Sections 101-106).

The RSC is organized into sections by genus. For example, all code numbers and descriptors of *Salmonella* are in one section and features descriptive of *Streptococcus* are in another section. If the reference strain and the test strain belong to different genera, the RSC is cross-referenced. For example, a *Streptococcus* strain may be tested for sensitivity to a bacteriocin produced by a *Bacillus* strain. Organizing the RSC around genera enables the individual scientist (who is concerned with only a few genera in a given study) to locate expeditiously the extant RSC numbers and descriptors for a genus.

The following fields of information must be recorded: 1) reference strain number or designation; 2) suffix; 3) interim code number; 4) reference strain taxon; 5) producer or manufacturer of test materials; 6) determinant; 7) technique or method; 8) literature reference or comments on technique; and 9) RSC number.

The reference strain designation is recorded on all RSC sheets. We have previously defined a reference strain as the organism from which a product was obtained and used as test material. However, a reference strain may be used to produce multiple products or it may be employed in diverse kinds of tests. The reference strain designation may contain a combination of 1 to 9 numbers, characters, or symbols (i.e., AHT-B#9 is a valid strain designation). A suffix field is used to make sure that each strain designation is unique.

The reference strain taxon field is used to record the generic name and specific epithet of the reference strain. This is essential, if the names are known, to avoid confusion. If the names are not known, distinctive operational names or symbols must be assigned to all reference strains. For example, an antiserum prepared against a *Salmonella* strain might be included in a number of tests developed for the genus *Escherichia*. Although components of earlier coded data might exist in both the *Escherichia* and the *Salmonella* sections of the RSC, the reference strain taxon would be *Salmonella*, not *Escherichia*. Computer searches of the reference strain number and the reference strain taxon fields would determine if the RSC contains any information about specifically numbered or named reference strains.

The producer or manufacturer field records the name of the laboratory which produced the tested product. The producer could be a commercial source, a government agency, or a private or academic laboratory.

The determinant field identifies the tested property of the reference strain. Some examples are an antigen used in preparation of antisera or a specific protein band used as an electrophoretic standard. The determinant field is used to prepare RSC numbers, formulate the respective descriptive language for each feature for which data exists, and to determine if cross references are required for feature descriptors. In the example described earlier, a battery of tests for *Escherichia* might include antisera prepared against a *Salmonella* strain. If a code number already exists for the given *Salmonella*, the submitter of the data would be so informed. If the feature items concerning the *Salmonella* antisera are listed only in the *Salmonella* section, the *Escherichia* section of the RSC is updated by including cross references to the existing code number and its descriptor. If another submitter of data used the same *Salmonella* antisera in another typing scheme (*Arizona* typing scheme, for example), the code is again cross referenced in that section (i.e., the *Arizona* section).

The technique or method field is used to record the name of the technique or method in which the reference strain or its product are used. For example, the same reference strain may be used for serotyping and chromatography experiments. Different code numbers and descriptors are required for the properties each technique or method assesses.

The literature reference or comments on technique field is valuable as a check as to whether the RSC is sufficiently comprehensive and unambiguous. This field allows the examination of details of various methods to ensure that the submitter of data has used a specific method and not a

Directions for coding data   39

modification of that method. For example, reactions from absorbed sera would differ from reactions using unabsorbed sera. If a modification of a method is used, the submitter should comment on this in the literature reference or comments on technique field.

The final field on the RSC sheets is the RSC number. This number is assigned by central computer personnel after analysis of the RSC sheets and is the permanent number associated with a unique feature in the RSC.

Directions for coding data

CODE SHEET FOR STRAIN CHARACTERISTICS
SECTION 1 - GENERAL INFORMATION

DATE OF RESPONSE
DAY 01-31
MONTH 01-12
YEAR 00-99

DAY	MO.	YR.	EDITION OF QUESTIONNAIRE USED (From title page)
01	12	84	

NAME OF PERSON SUBMITTING INFORMATION

SURNAME (Family name): JONES

FIRST NAME AND OTHER NAMES AND/OR INITIALS (Leave one space between each name.): FRANCES ELAINE

TITLE OR RANK (e.g., Professor, Doctor, Mister, Captain, etc.): MICROBIOLOGIST

COLLECTION NUMBER IN THE WORLD DIRECTORY OF CULTURE COLLECTIONS (If known)

NIDR NIH BETHESDA MARYLAND 20205

POSTAL ADDRESS

NAME OF GENUS OR ARBITRARY NUMERICAL CODE FOR GROUP OR GROUP NAMES (See below)**

NAME OF SUBGENUS
(Insert the name only if data being submitted for strains is arranged under subgenus, otherwise leave blank.)

SPECIFIC EPITHET OF TYPE SPECIES FOR THIS GENUS

FOR HOW MANY SUBGENERA OF THE GENUS ARE YOU SUBMITTING DATA

CATEGORY OF MICROBE BEING DESCRIBED (See code*)

BT SALMONELLA

*MICROBE CODE
AL = ALGAE   BT = BACTERIA   VA = VIRUSES - ANIMAL
FN = FUNGI   BR = RICKETTSIA   VB = VIRUSES - BACTERIAL
FY = YEASTS   BM = MYCOPLASMA   VI = VIRUSES - INSECT
PZ = PROTOZOA   QQ = OTHER   VP = VIRUSES - PLANT

NOTE 1: COMPLETELY DIFFERENT SETS OF CODE SHEETS MUST BE USED FOR EACH GENUS OR GROUP.
NOTE 2: IF THE DATA BEING SUBMITTED FOR STRAINS IS ARRANGED UNDER SUBGENERA, COMPLETELY DIFFERENT SETS OF CODE SHEETS MUST BE USED FOR EACH SUBGENUS. THE GENERIC NAME MUST BE INSERTED FOR EACH SET.
NOTE 3: IF THE GENUS NAME IS NOT KNOWN AND DATA IS BEING SUBMITTED FOR ANY ARBITRARY GROUP OF ORGANISMS, USE (A) AN ARBITRARY NUMERICAL CODE, E.G., "GROUP 22" OR SIMPLY "22" OR (B) A GROUP NAME, E.G., CORYNEFORMS. SEE DIRECTIONS FOR FURTHER EXPLANATION.

FOR OFFICE USE ONLY: 01, 0201, 0202, 03

NIH-1700-6
4-78

Directions for coding data

PAGE 2

CODE SHEET FOR STRAIN CHARACTERISTICS

CODE SHEET FOR GENERAL INFORMATION (Cont.)		
001011: ARE THE ORGANISMS OF THIS GENUS PROCARYOTIC?	1	15 /
001012: HAS A TYPE SPECIES EVER BEEN ESTABLISHED FOR THIS GENUS?	2	16 /
001013: HAS A TYPE CULTURE EVER BEEN ESTABLISHED FOR THE TYPE SPECIES?	3	17 /
001014: IS THE TYPE CULTURE OF THE TYPE SPECIES (OF THE GENUS) IN LIVING FORM?	4	18
001015: IS ANY STRAIN OF THE TYPE SPECIES IN LIVING FORM?	5	19
001016: HAS A NEOTYPE STRAIN OF THE TYPE SPECIES BEEN ESTABLISHED?	6	
001017: HAS ANY STRAIN OF THE TYPE SPECIES BEEN STUDIED EXTENSIVELY IN THE LAST 20 YEARS?	7	
001018: HAVE NUCLEIC ACID HOMOLOGIES BEEN MEASURED AMONG ANY OF THE STRAINS OF SPECIES OF THIS GENUS?	8	
001019: HAVE NUCLEIC ACID HOMOLOGIES BEEN MEASURED BETWEEN ANY STRAINS OF SPECIES OF THIS GENUS AND THOSE OF OTHER GENERA?	9	20 /
001020: HAS BACTERIOPHAGE TYPING BEEN USED IN THIS GENUS?	10	
001021: HAS TRANSDUCTION BEEN DEMONSTRATED AMONG ANY STRAINS OF SPECIES OF THIS GENUS?	11	
001022: HAS TRANSDUCTION BEEN DEMONSTRATED BETWEEN ANY STRAINS OF SPECIES OF THIS GENUS AND THOSE OF OTHER GENERA?	12	
001023: HAS TRANSFORMATION BEEN DEMONSTRATED AMONG ANY STRAINS OF SPECIES OF THIS GENUS?	13	
001024: HAS TRANSFORMATION BEEN DEMONSTRATED BETWEEN ANY STRAINS OF SPECIES OF THIS GENUS AND THOSE OF OTHER GENERA?	14	

001025: HAVE STUDIES OF THE GEOGRAPHICAL DISTRIBUTION OF ANY STRAINS OF THE GENUS BEEN CARRIED OUT?	1	15 /
001026: HAVE STUDIES OF THE ECOLOGICAL DISTRIBUTION OF ANY STRAINS OF SPECIES OF THIS GENUS BEEN CARRIED OUT?	2	16 /
001027: HAVE STUDIES OF THE ECOLOGICAL INTERACTIONS WITHIN A HABITAT OF ANY STRAINS OF SPECIES OF THIS GENUS BEEN CARRIED OUT?	3	17 /
001028: HAVE STUDIES OF ECOLOGICAL DISTRIBUTION ON ANY LEVEL BEEN USED FOR TAXONOMY OF THIS GENUS?	4	18
001029: HAVE IMMUNOLOGICAL CRITERIA BEEN USED FOR TAXONOMY OF THIS GENUS?	5	19
001030: HAVE ANY STRAINS OF SPECIES OF THIS GENUS BEEN REPORTED AS PATHOGENIC? IF ANY STRAINS FOR WHICH YOU ARE SUBMITTING DATA ARE PATHOGENIC, PLEASE SUPPLY THE FOLLOWING INFORMATION ON A SEPARATE SHEET OF PAPER: (A) STRAINS BY SPECIFIC EPITHETS, (B) HOSTS (BOTH COMMON AND SCIENTIFIC NAMES), AND (C) DISEASE. IDENTIFY EACH SHEET OF PAPER WITH THE WORLD DIRECTORY OF CULTURE COLLECTIONS NUMBER IF APPLICABLE, AND ALSO THE SAME INFORMATION AS REQUIRED IN THE TOP RIGHT CORNER OF CODE SHEET-I.	6	
	7	20 /
	8	
	9	
	10	
001031: WHAT IS THE TOTAL NUMBER OF CODE SHEETS YOU ARE SUBMITTING FOR STRAINS OF THIS GENUS (OR SUBGENUS) OR ARBITRARY GROUP? (NOTE: ALL CODE SHEETS MUST BE SECURELY FASTENED TOGETHER.)	11	21 22 23 24 / 0 0 0 5
	12	
	13	71 72 73 74 75 76 77 78 / / / / / / 0 4
	14	

NIH-1700-6
4-78

GPO 929-592

Directions for coding data

STRAIN NUMBER OR DESIGNATION	MUTANT	PARENT STRAIN NUMBER OR DESIGNATION	GENUS (OPTIONAL)	SPECIFIC EPITHET	DEPOSITOR OR DONOR OF CULTURE	SURNAME: JONES
1 3 5 8 6 4 2 0 1 0 1			S L M L	ARTIS		05
2 3 5 8 6 4 2 0 5 0 1				ARTIS		05
3 1 5 6 3 8 0 1 1 0 1				NIAKHAR		05
4 1 5 6 3 8 0 1 2 0 1				ONA		05
5						05
6 K 0 0 0 A 0 0 1 7	1 1 5 6 3 8 0 1 2 0	S L M L	ONA	FDA: P. GARDNER	05
7 K 0 8 1 5 A 0 0 2 6			S L M L	MAYDAY	NIDR; DR. WALCZAK	05
8						05
9 3 0 - M		O N C T C 1 7 8	S L M L	BERTA		05
10 2 1						05
11 3 4 A B				AQUA		05
12 3 7 # 3				ARTIS		05
13						05
14						05
15						05
16						05
17						05
18						05
19						05
20						05

GENUS: SALMONELLA PAGE NUMBER: 1

CODE SHEET FOR STRAIN CHARACTERISTICS — 05
NIH-1700-1 (9-70) REV. 7-77

Directions for coding data

CODE SHEET FOR STRAIN CHARACTERISTICS-06

GENUS: SALMONELLA
SURNAME: JONES
PAGE NUMBER: 2

#	STRAIN NUMBER OR DESTINATION	COMMENTS OR AUTHORITY
1	3 5 8 6 4 2 0 1 0 1	
2	3 5 8 6 4 2 0 5 0 1	
3	1 5 6 3 8 0 1 1 0 1	
4	1 5 6 3 8 0 1 2 0 1	
5		
6	6 0 0 0 0 A 0 0 1	RESEARCH ISOLATE: CSU STUDY
7	k 0 8 1 5 A 0 0 2 6	
8		
9	3 0 - M	J. BACT 28:400-410 (1915)
10	L 2 1	
11	3 4 A B	M. L. DAVIS NIDR (1970)
12	3 4 # 3	

Directions for coding data

GENUS: SALMONELLA
SURNAME: JONES
PAGE NUMBER: 3

CODE SHEET FOR STRAIN CHARACTERISTICS—07
NIH-1700-2 REV. 7-77

Line	Strain Number or Designation	Index Bergeyana	GC Content	Culture Collection (Number if known)	Deposited Day	Deposited Month	Deposited Year	Place	Hours 0-24	Minutes 0-60	Isolated Day	Isolated Month	Isolated Year	Geographical Location of Source
1	358672010/							1			09	09	75	
2	358672050/							1			09	09	75	
3	156380/10/							1			31	06	76	
4	156380/20/							1			18	11	76	
5														
6	K000DA0017			000157				1			28	11	77	MEXICO MERIDA
7	K0815A0026							0			09	07	76	
8											09	07	76	
9	30-M													
10	L21													
11	34AB													
12	3#3													

Directions for coding data

Directions for coding data

GENUS: SALMONELLA **SURNAME:** JONES

PAGE NUMBER: 3

STRAIN NUMBER OR DESIGNATION	REMARKS	SOURCE (Plant, Animal, Organ, Soil, Type, or Other Information)
1586720101	BABY TURTLE	
2586720501	BABY TURTLE	
1563801101	COSMETIC EYE LINER DRY	
1563801201	COSMETIC EYE LINER DRY	
K000000017	DEHYDRATED NOODLE SOUP MIX	
K08154002	FRESH MILK	
30-M	WATER SAMPLE	
L21	FRIED CHICKEN	
34AB	COSMETIC EYE LINER DRY	
34#3	PIE	

CODE SHEET FOR STRAIN CHARACTERISTICS—10

Directions for coding data

Reference Strain Number or Destination	Interim Number		Genus	Reference Strain Taxon Species	Producer or Manufacturer of Reference Material
S1234	9701	A	SLML	TYPHI	DIFCO
S1234	9702	B			CDC
B1222	9703				CDC
S1234	9704	C	SFCO	AUREUS	DIFCO
ABC#12	9705				LABORATORY XYZ
A73	9706				
ABC#7	9707				
59	9708				
1	9709		STCO		MY LAB
2	9710				
3	9711				
4	9712				
#78B	9713				FDA LAB A
#37-6	9714				
C0293	9715				
4138	9716				
79D5	9717	A	ATMY	ALBUS SUBSP FUNGATUS	FRIEND'S LAB
79D5	9718	B			
79D5	9719	C			
X83	9720			BOBILI	

Directions for coding data

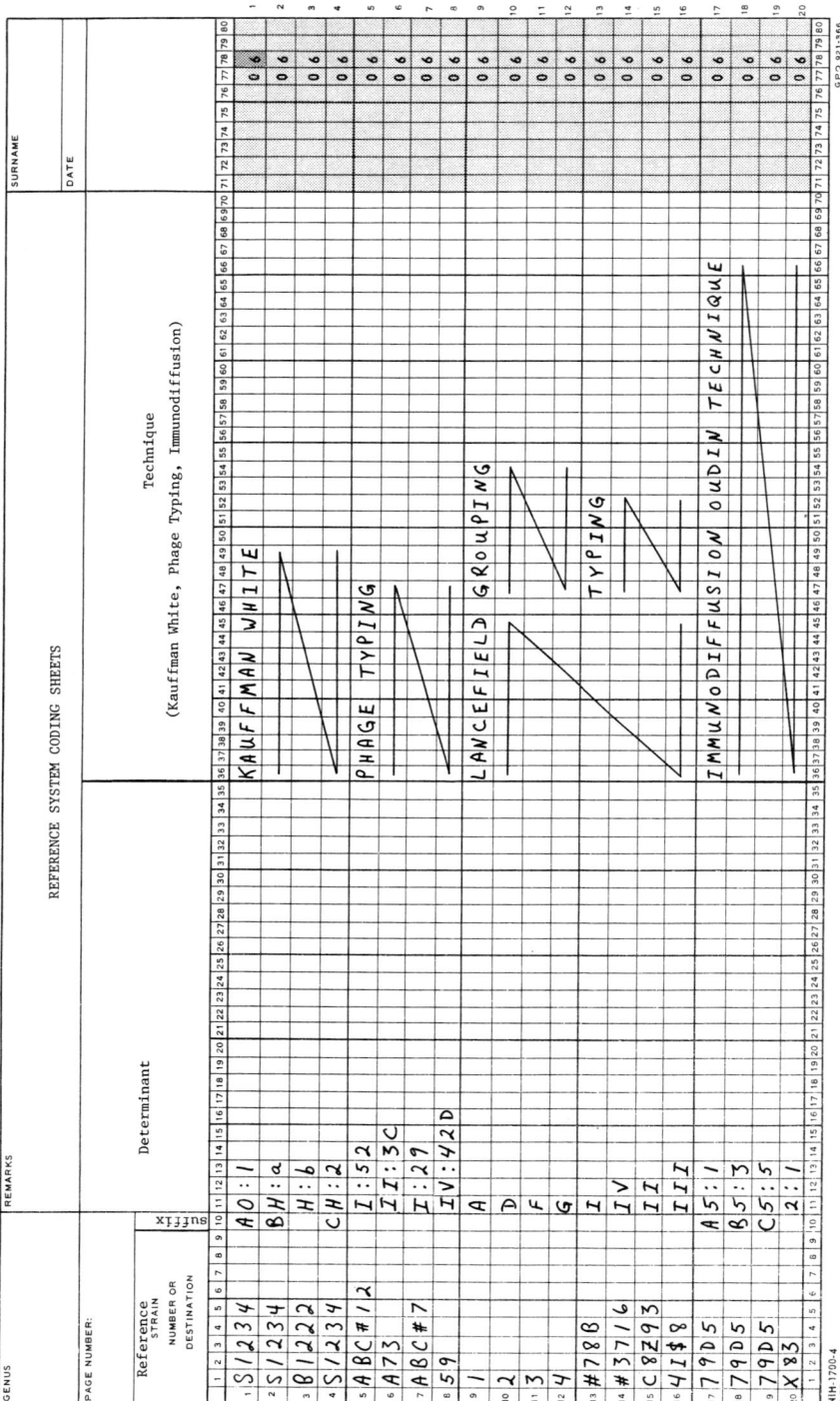

Directions for coding data

Reference STRAIN NUMBER OR DESTINATION	suffix	COMMENTS OR AUTHORITY — Literature Reference or Comments on Technique	RSC Number (office use only)
S1234	A	AEDWARDS, PR, and EWING, WH. 1955. IDENT. ENTEROBACTER.	101201
S1234	B		101701
B1223			101702
S1234	C		101655
ABC#12		COHEN, JO (ED). 1972. THE STAPHYLOCOCCI. NY: WILEY	105033
A73			105027
ABC#7			105029
59			105030
1			103701
2			103704
3			103706
4			103707
#788			
#3716			
C8293			
4188			
79D5	A		
79D5	B		
79D5	C		
X83			

SECTION 2: SPECIFIC STRAIN INFORMATION

002001: What is your laboratory strain number or designation?
002002: Is the strain a mutant of another strain which mutated in your collection? (If yes, enter "1" and if no, enter "0".)
002003: If the strain mutated in your collection, what is the parent strain designation or number?
002004: What is the specific epithet?
001997: Who was the depositor or donor of the strain in the culture collection?
002005: What is the authority for the specific epithet, i.e., Author's name, Journal, Year? If you are not a member of a formal culture collection, you probably will not be entering the Authority. Therefore, use this space for any general comments about the strain that are not covered elsewhere.
002006: Is the specific epithet listed in Index Bergeyana?
002007: What is the GC content of the strain DNA, in moles % (to the nearest whole number)?
002008: In which culture collection is the strain deposited? Use any combination of 6 letters or, if known, use the World Directory of Culture Collections Number.
001996: What day of the month was the strain deposited in a culture collection? Use 01-31 for day.
001995: What month was the strain deposited in a culture collection? Use 01 for January through 12 for December.
002009: What month and year was the strain deposited in the culture collection (for RKC1) or Year only (for RKC2)? Use the notation 01 for January through 12 for December. Use the last 2 digits of the year. The day of the month the strain was deposited may be entered in columns 20-21 of Code Sheet 07, but not on Code Sheet II.
002010: Is the geographic place of isolation known? For a negative answer simply write 0 or for a positive answer write 1.
001994: What time of day was the strain isolated? Use 00-24 for hours and 00-60 for minutes.
001993: What day was the strain isolated? Use 01-31 for day.
001992: What month was the strain isolated? Use 01 for January through 12 for December.
002011: Month and Year of the isolation (for RKC1) or Year of Isolation only (for RKC2).
001998: What was the geographic source or location where the isolation of the strain was made? Use appropriate abbreviations where available for (a) the name of the place, (b) latitude and longitude in degrees, min, s, and (c) height or depth from sea level expressed in meters.
002012: What was the specific source of isolation such as kind of water, soil, etc., species and organ and

Section 2: Specific strain information								51

tissue of plant, animal, etc. (give specific type).

[NOTE 1: Until now you have supplied information on a number of appropriate code sheets. Beginning now and for the remainder of this document give your answer to all queries (i.e., feature numbers) on either Code Sheet III or 08/09. See directions for coding.]

002013: Is the history of the strain known? Please supply what you know about the strain: (a) origin, (b) name and address of the depositor, (c) previous designations, (d) applications and, (e) names of the culture collections where the strain was previously held.

002014: Have bacteriophage(s) been demonstrated for the strain?

002015: Has transduction been demonstrated for the strain?

002016: Has transformation been demonstrated for the strain?

002017: Has the strain been used in a numerical taxonomy study?

002018: Has the strain been examined by electron microscopy?

002019: Was the electron microscopy technique of shadow casting used?

002020: Was the electron microscopy technique of negative staining used?

002021: Was the electron microscopy technique of thin sectioning used?

002022: Was the electron microscopy technique of freeze etching used?

002023: Was the scanning electron microscope used?

[NOTE 2: The various types are nomenclatural concepts. Major references defining the types are: International Code of Nomenclature of Bacteria, 1976 Revision, Published for the International Association of Microbiological Societies by the American Society of Microbiology, Washington, D. C., 1975; International Code of Botanical Nomenclature, 1979 available from the International Bureau of Plant Taxonomy and Nomenclature, Tweede Transitorium, Uithof, Utrecht, Netherlands; and International Code of Zoological Nomenclature (1964), International Trust for Zoological Nomenclature, London available from the International Trust for Zoological Nomenclature, c/o British Museum Natural History, Cromwell Road, London SW75BD, England.]

002024: The strain is the Holotype strain of the species. (It is the strain designated by the original author, when the name of the species was first published).

002025: The strain is the Monotype Strain of the Species. (The original description of the species was based on a single strain, and the author did not designate it as the holotype).

002026: The strain is the Lectotype Strain of the Species. (No holotype was designated in the original publication, and the strain is 1 of those originally used in establishing the species name. It has subsequently been designated by the original, or other, author as the type).

002027: The strain is designated as the Neotype Strain of the Species. (The strain has been proposed and accepted as the valid type strain of this species, because none of the strains on which the original description of the species was based, are extant).

002028: The strain is being considered as a lectotype or holotype strain of the species prior to aceptance, but has no official standing. (Proposed "working types" are included in this category).

002029: The strain is the Holotype Strain of the Subspecies. (It is the strain designated by the original author, when the name of the subspecies was first published).

002030: The strain is the Monotype Strain of the Subspecies because the original description of the Subspecies was based on a single strain and the author did not designate it as the holotype.

002031: The strain is designated as the Lectotype Strain of the Subspecies. (No holotype was designated in the original public strain is 1 of the those originally used in establishing the subspecies name. It has subsequently been designated by the original, or other, author as the type).

002032: The strain is the Neotype Strain of the Subspecies. (The strain has been proposed and accepted as the valid type strain of this subspecies, because none of the subspecies was based, are extant).

002033: The strain is being considered as the lectotype or holotype strain of the subspecies prior to acceptance, but has no official standing. (Proposed "working types" are included in this category.)

002034: The strain has been utilized for a specific purpose (e.g., production of an item of commerce, an antimicrobial sensitivity standard, etc.).

002035: The strain is the Syntype of the Species.
002036: The strain is the Syntype of the Subspecies.
002037: The strain has been proposed as the Neotype of the Species.

SECTION 3:   INDIVIDUAL CELL MORPHOLOGY

[NOTE 1: Feature numbers in this section apply to individual cells whether they occur free or as part of a multicellular structure or organism.]

[NOTE 2: For arrangement and physical relationships among cells, see Section 15.]

[NOTE 3: For symbiotic relationships, see Section 16.]

[NOTE 4: If cells are branched, also see Section 8.]

003001: Cells are spherical (i.e., all perpendicular axes have ratios between 6:5-1:1).
003002: Cells are cuboid or angular.
003003: Cells are reniform (bean-shaped, kidney-shaped).
003004: Cells are elliptic (ellipsoidal).
003136: Cells are elliptic (ellipsoidal), with the ratio of the long axis to the short axis between 6:1 - 3:1.
003137: Cells are elliptic (ellipsoidal), with the ratio of the long axis to the short axis between 2:1 - 3:2.
003157: Cells are citriform (lemon-shaped, ellipsoidal with a small bulge at each end).
003005: Cells are pyriform (pear-shaped).
003146: Cells are ovate (egg-shaped).
003147: Cells are ovate, with the ratio of the long axis to the short axis between 6:1 - 3:1.
003148: Cells are ovate, with the ratio of the long axis to the short axis between 2:1 - 3:2.
003149: Cells are ovate with the ratio of the long axis to the short axis between 6:5-1:1.
003006: Cells are discoid (cylinder in which the height is less than the diameter; disc-shaped).
003007: Cells are triangular.
003150: Cells are triangular, with the ratio of the long axis to the short axis between 6:1 - 3:1.
003151: Cells are triangular, with the ratio of the long axis to the short axis between 2:1 - 3:2.
003152: Cells are triangular with the ratio of the long axis to the short axis between 6:5-1:1.
003155: Cell is sagittate (arrow-headed, two equal obtuse triangles with a short side in common).
003008: Cells are rod-shaped (cylindrical).
003138: Cells are rectangular.
003139: Cells are rectangular, with the ratio of the long axis to the short axis between 6:1 - 3:1.
003140: Cells are rectangular, with the ratio of the long axis to the short axis between 2:1 - 3:2.
003141: Cells are rectangular (i.e., square) with all perpendicular axes have ratios between 6:5-1:1.
003142: Cells are rhombic.
003143: Cells are rhombic, with the ratio of the long axis to the short axis between 6:1 - 3:1.
003144: Cells are rhombic, with the ratio of the long axis to the short axis between 2:1 - 3:2.

Section 3: Individual cell morphology

003145: Cells are rhombic with the ratio of the long axis to the short axis between 6:5 - 1:1.
003025: Cells are dumbbell-shaped.
003027: Cells are ogival.
003028: Cells are apiculate.
003029: Cells are horseshoe (i.e., "U")-shaped.
003032: Cells are cordiform (heart-shaped, as the heart in a pack of cards).
003033: Cells are fusiform (spindle-shaped, biconvex lens-shaped).
003035: Cells are limaciform.
003154: Cell is vermiculate (worm-shaped, thick and almost cylindrical with irregular bends in various places).
003036: Cells are lanceolate.
003006: Cells are discoid (cylinder in which the height is less than the diameter).
003132: Cell shape is caudate.
003156: Cells are lunate (arcuate, bow-shaped, crescent-shaped).
003030: Cells curve into an almost complete circle.
003024: Transverse section of cell is crescent-shaped.
003045: Cells are ellipsoidal in transverse section.
003046: Cells are circular in transverse section.
003026: Longer axis of rod is < twice the shorter axis (cocco-bacillary).
003009: Rod axis is straight.
003010: Rod axis is irregular.
003011: Rod axis is curved in 1 plane.
003012: Rod axis is sigmoid in 1 plane.
003013: Rod axis is helical (spiral).
003014: Helical cells have axial filaments.
003015: Helical cells have crista.
003037: Cells have single spiral twist.
003038: Cells have > single spiral twist.
003031: Cells are cylindrical (uniform diameter throughout length).
003016: Rods have tapered ends.
003017: Rods have rounded ends.
003018: Rods have square ends.
003019: Rods have recurved ends (bent into a semicircular hook).
003020: Individual cell is flexuous.
003021: Cells produce tubular outgrowths (0.2-0.3 μm wide) on the end of which daughter cells are formed (also see Section 11).
003022: Internal cell contents (cytoplasm) is concentrated at 1 end of cells.
003023: Pleomorphic cells are characteristic.
003158: Cell shape from the lateral view (perpendicular to the direction of swimming) changes while swimming (as in *Euglena*).
003159: Cell shape from the apical (head on while swimming) view is quadrate.
003160: Cell shape from the apical (head on while swimming) view is circular to elliptical.

Section 3: Individual cell morphology					55

003161:	Cell shape from the apical (head on while swimming) view is fusiform.
003162:	Cell shape from the apical (head on while swimming) view is lanceolate.
003163:	Cell shape from the apical (head on while swimming) view is triangular.
003164:	Anterior (end that leads while swimming) of the cell is truncate.
003165:	Anterior (end that leads while swimming) of the cell is rounded.
003166:	Anterior (end that leads while swimming) of the cell is emarginate (notched at the margin).
003167:	Anterior (end that leads while swimming) of the cell is acute (pointed).
003168:	Anterior (end that leads while swimming) of the cell has papilla.
003169:	Anterior (end that leads while swimming) of the cell is 2 lobed.
003170:	Anterior (end that leads while swimming) of the cell is 4 lobed.
003171:	Posterior (end that trails while swimming) of the cell is rounded.
003172:	Posterior (end that trails while swimming) of the cell is lobed.
003173:	Posterior (end that trails while swimming) of the cell is acute.
003039:	Cell is constricted in middle.
003174:	Cell has a furrow (narrow trench-like depression in surface).
003175:	Cell has a gullet.
003176:	Cell has a girdle groove.
003179:	The girdle (cingulum) is displaced.
003040:	Cell is concave on 1 side.
003041:	Cell ventral surface is flat.
003042:	Cell ventral surface is concave.
003043:	Cell dorsal surface is flat.
003044:	Cell dorsal surface is convex.
003047:	Cells have an envelope.
003048:	Collar is enclosed in gelatinous material.
003049:	Cells have a central capsule.
003050:	Cells have a central capsule with no openings except for skeletal rays.
003051:	Cells have skeletal rays originating at center of cells.
003052:	Central capsule has numerous uniformaly distributed pores.
003053:	Central capsule pores restricted to 1 pole.
003054:	Central capsule has 1 opening.
003055:	Central capsule has 2 openings.
003056:	Central capsule has 3 openings.
003057:	Central capsule has 4 openings.
003058:	Central capsule has > 4 openings.
003059:	Cell has buccal cavity.
003060:	Peristome is present.
003061:	Peristome is spiral.
003062:	Peristome is triangular
003063:	Peristome is U-shaped.

003064: Peristome is located anteriorly.
003065: Peristome is located posteriorly.
003066: Peristome is located on side.
003067: Peristome has an undulating membrane.
003068: Peristome has cilia.
003069: Peristome is approximately 1/4 length of cell.
003070: Peristome is approximately 1/3 length of cell.
003071: Peristome is approximately 2/3 length of cell.
003072: Cytopharynx is open.
003073: Cytopharynx opens anteriorly.
003074: Cytopharynx opens laterally.
003076: Cytostome is present.
003077: Cytostome is triangular.
003078: Cytostome is U-shaped.
003079: Cytostome is oblong.
003080: Cytostome is rosette-like.
003081: Cytostome is slit-like.
003082: Cytostome is circular.
003083: Cytostome is ellipsoidal.
003084: Cytostome is oval.
003085: Cytostome is pyriform.
003086: Cytostome is anterior.
003087: Cytostome is posterior.
003088: Cytostome is ventral.
003089: Cytostome is terminal.
003091: Cytostome is located on the convex surface.
003092: Cytostome is located on the concave surface.
003093: The cytostome is located at the base of a trichocyst-bearing neck.
003094: Cytostome is bordered by collar.
003095: Vestibulum (vestibule) is present.
003096: Vestibulum (vestibule) is anterior.
003133: Vestibulum (vestibule) is posterior.
003097: Vestibulum (vestibule) is lateral.
003134: Cell has infundibulum.
003135: Preoral cavity is present.
003098: Cell has collarette.
003099: Cells have single collar at least 1 µm shorter than cell.
003100: Cells have single collar at least 1 µm longer than cell.
003101: Cells have 2 collars.
003102: Cells have 2 equal collars.
003103: Cell anterior end is lip-like.
003104: Cell anterior end is truncated.
003105: Cell anterior end is pointed.
003106: Cell has anterior snout (neck-like projection).
003107: Cell has sucking tube.
003108: Cell posterior end is truncated.
003109: Cell posterior end is pointed.
003110: Cell posterior end is rounded.
003111: Cells are green upon microscopic examination.
003112: Cells are blue upon microscopic examination.
003113: Cells are red upon microscopic examination.
003114: Annulus is present.
003115: Annulus is complete.
003116: Annulus descends in a left spiral.

Section 3: Individual cell morphology

003117: Sulcus is present.
003118: Sulcus is obscured.
003119: Sulcus extends to full length of body.
003120: Sulcus extends to 1/5 from posterior end.
003121: Annulus and sulcus form a triangle.
003177: The epicone equals the hypocone.
003178: The epicone is shorter than the hypocone.
003123: The epicone is longer than the hypocone.
003124: The epicone is several times longer than hypocone.
003122: The hypocone is 2 to 2.5 times longer than the epicone.
003125: The hypocone has 4 longtitudinal ridges.
003126: The epimerite is present.
003127: Epimerite has hooks.
003128: Epimerite has 1 to 4 hooks.
003129: Epimerite has 5 to 8 hooks.
003130: Epimerite has 9 to 12 hooks.
003131: Epimerite has > 12 hooks.

SECTION 4: INDIVIDUAL VEGETATIVE CELL SIZE

[NOTE 1: The ranges of cell dimensions listed below are organized into two sets. Choose the appropriate range (i.e., narrow or wide) as best suits the cells being described.

NARROW INCREMENT RANGE, LONGEST AXIS

004001: Longest axis of each cell is < 0.5 µm.
004002: Longest axis of each cell is 0.5 - 1 µm.
004003: Longest axis of each cell is 1.1 - 2.0 µm.
004004: Longest axis of each cell is 2.1 - 3.0 µm.
004005: Longest axis of each cell is 3.1 - 4.0 µm.
004006: Longest axis of each cell is 4.1 - 5.0 µm.
004007: Longest axis of each cell is 5.1 - 10 µm.

WIDE INCREMENT RANGE, LONGEST AXIS

004028: Longest axis of each cell is < 2.0 µm.
004029: Longest axis of each cell is between 2.1 and 4.0 µm.
004030: Longest axis of each cell is between 4.1 and 6.0 µm.
004031: Longest axis of each cell is between 6.1 and 10.0 µm.
004008: Longest axis of each cell is 11 - 15 µm.
004009: Longest axis of each cell is 16 - 100 µm.
004010: Longest axis of each cell is 101 - 160 µm.
004026: Longest axis of each cell is > 160 µm.

WIDE INCREMENT RANGE, SECOND LONGEST AXIS

004032: Second longest axis of each cell is < 1.5 µm.
004033: Second longest axis of each cell is between 1.6 and 3.0 µm.
004034: Second longest axis of each cell is between 3.1 and 6.0 µm.
004035: Second longest axis of each cell is > 6.0 µm.

NARROW INCREMENT RANGE, SHORTEST AXIS

004011: Shortest axis of each cell is < 0.5 µm.
004012: Shortest axis of each cell is 0.5 - 1 µm.
004013: Shortest axis of each cell is 1.1 - 2.0 µm.
004014: Shortest axis of each cell is 2.1 - 3.0 µm.
004015: Shortest axis of each cell is 3.1 - 4.0 µm.
004016: Shortest axis of each cell is 4.1 - 5.0 µm.
004017: Shortest axis of each cell is 5.1 - 10 µm.
004018: Shortest axis of each cell is 11 - 15 µm.

WIDE INCREMENT RANGE, SHORTEST AXIS

004036: Shortest axis of each cell is < 1.0 µm.
004037: Shortest axis of each cell is between 1.0 and 3.0 µm.

Section 4: Individual vegetative cell size

004038: Shortest axis of each cell is between 3.0 and 6.0 µm.
004039: Shortest axis of each cell is > 6.0 µm.
004019: Shortest axis of each cell is 16 - 100 µm.
004020: Shortest axis of each cell is 101 - 160 µm.
004027: Shortest axis of each cell is > 160 µm.

MISCELLANEOUS MEASUREMENTS

004023: Helical (spiral) cells have wave length of 1 - 2.9 µm.
004024: Helical (spiral) cells have wave length of 3 - 4.9 µm.
004025: Helical (spiral) cells have wave length of 5 - 8 µm.
004021: Cells are filterable through filters having a pore size of 0.45 µm or less.
004022: Cells are visible by light microscopy (resolution of > 0.2 µm).

SECTION 5:   INSOLUBLE INTRACELLULAR AND EXTRACELLULAR
              DEPOSITIONS

005017: Unstained cell contains refractile bodies.
005001: Cell contains calcium carbonate inclusions.
005002: Cell contains sulfur granule inclusions (see feature number 005012).
005003: Cell contains starch inclusions.
005004: Cell contains poly β-hydroxybutyric acid inclusions.
005005: Cell contains hydroxyapatite inclusions (see feature number 005016).
005006: Cell contains poly metaphosphate inclusions (volutin).
005007: Cell contains granulose inclusions.
005008: Cell contains glycogen inclusions.
005009: Cell condains lipid globules in the cell (also see Section 12 and Section 21).
005010: Cell contains discrete pigment globules.
005011: Cell contains gas vacuoles (aerosomes).
005012: Sulfur is deposited outside the cell.
005013: Iron is deposited outside the cell (also see Section 11 and Section 12).
005014: Manganese is deposited outside the cell (also see Section 11).
005015: Iodinin crystals are deposited outside the cell (also see Section 20).
005016: Hydroxyapatite is deposited outside the cell.
005018: Cell has lithosome.
005019: Cytoplasm contains granules.
005020: Cytoplasm contains barite crystals.
005021: Cytoplasm contains triurite crystals.
005022: Cell contains pigment granules.
005023: Cell contains pectin granules.

SECTION 6: ENDOSPORES AND CYSTS

006001: Endospores are produced (any refractile intracellular body capable of germination into a new vegetative cell).
006002: Endospores are produced singly per cell.
006003: Endospores are produced in pairs or larger numbers per cell.
006004: Endospore(s) is (are) round.
006005: Endospore(s) is (are) cylindrical.
006006: Endospore(s) is (are) ellipsoidal.
006007: Endospore(s) is (are) central in sporangium.
006008: Endospore(s) is (are) terminal.
006009: Endospore(s) is (are) sub-terminal.
006024: Free endospores are present.
006010: Germination of endospores is terminal.
006011: Germination of endospores is lateral.
006012: Germination of endospores is by dissolution of spore coat.
006013: Width of endospore is endospore is up to 0.9 µm.
006014: Endospores are wider than the vegetative cell (sporangium swollen).
006015: Endospores contain di-picolinic acid (pyridine-2,6-dicarboxylic acid).
006016: Endospores survive 80 C for 10 min.
006017: Endospores survive 85 C for 10 min.
006018: Endospores survive 100 C for 10 min.
006019: Endospores survive 100 C for 30 min.
006020: Parasporal crystal are formed during sporulation.
006021: Manganese cation is essential for sporulation.
006022: Antisera prepared against endospores cross react with vegetative cells.
006023: Germination of endospores is activated by sub-lethal heat.
006025: Cysts are present.
006026: Longest axis of cyst is < 0.5 µm.
006027: Longest axis of cyst is 0.5-1.0 µm.
006028: Longest axis of cyst is 1.1-1.5 µm.
006029: Longest axis of cyst is 1.6-2.0 µm.
006030: Longest axis of cyst is 2.1-2.5 µm.
006031: Longest axis of cyst is 2.6-3.0 µm.
006032: Longest axis of cyst is 3.1-3.5 µm.
006033: Longest axis of cyst is 3.6-4.0 µm.
006034: Longest axis of cyst is 4.1-4.5 µm.
006035: Longest axis of cyst is 4.6-5.0 µm.
006036: Longest axis of cyst is 5.1-7.0 µm.
006037: Longest axis of cyst is 7.1-9.0 µm.
006038: Longest axis of cyst is 9.1-11.0 µm.
006039: Longest axis of cyst is 11.1-15.0 µm.
006040: Longest axis of cyst is 15.1-20.0 µm.
006041: Longest axis of cyst is 20.1-30.0 µm.
006042: Longest axis of cyst is 30.1-40.0 µm.
006043: Longest axis of cyst is 40.1-50.0 µm.
006044: Longest axis of cyst is 50.1-75.0 µm.
006045: Longest axis of cyst is 75.1-100.0 µm.
006046: Longest axis of cyst is > 100 µm.
006047: Shortest axis of cyst is < 0.5 µm.

Section 6: Endospores and cysts

006048: Shortest axis of cyst is 0.5-1.0 µm.
006049: Shortest axis of cyst is 1.1-1.5 µm.
006050: Shortest axis of cyst is 1.6-2.0 µm.
006051: Shortest axis of cyst is 2.1-2.5 µm.
006052: Shortest axis of cyst is 2.6-3.0 µm.
006053: Shortest axis of cyst is 3.1-3.5 µm.
006054: Shortest axis of cyst is 3.6-4.0 µm.
006055: Shortest axis of cyst is 4.1-4.5 µm.
006056: Shortest axis of cyst is 4.6-5.0 µm.
006057: Shortest axis of cyst is 5.1-7.0 µm.
006058: Shortest axis of cyst is 7.1-9.0 µm.
006059: Shortest axis of cyst is 9.1-11.0 µm.
006060: Shortest axis of cyst is 11.1-15.0 µm.
006061: Shortest axis of cyst is 15.1-20.0 µm.
006062: Shortest axis of cyst is 20.1-30.0 µm.
006063: Shortest axis of cyst is 30.1-40.0 µm.
006064: Shortest axis of cyst is 40.1-50.0 µm.
006065: Shortest axis of cyst is 50.1-75.0 µm.
006066: Shortest axis of cyst is 75.1-100.0 µm.
006067: Shortest axis of cyst is > 100 µm.
006068: Cysts have a geometric surface pattern.
006069: The outline of cysts is smooth.
006070: The outline of cysts is round.
006071: The outline of cysts is oval.
006072: The outline of cysts is elongate.
006073: Cysts are uninucleate.
006074: Cysts are binucleate.
006075: Cysts are quadrinucleate.
006076: Cysts have > 4 nuclei.
006077: Cysts have a pre-formed pore.
006078: Excystment occurs by splitting of cyst wall.
006079: Excystment occurs by dissolution of cyst wall.
006080: Cysts are found in a host.
006081: Cysts are found in the liver.
006082: Cysts are found in intestinal tissues.

SECTION 7: MYXOSPORES, SPOROCYSTS, FRUITING BODIES, SPORES, SPOROZOITES, TROPHOZOITES, AND GAMETOCYTES

007001: Organisms with gliding motility produce myxospores (microcysts, resting cells) during growth.
007002: Myxospores are spherical.
007003: Myxospores are ellipsoidal.
007004: Myxospores are cylindrical.
007005: Longest axis of myxospore is 0.5 - 1.0 μm.
007006: Longest axis of myxospore is 1.1 - 1.5 μm.
007007: Longest axis of myxospore is 1.6 - 2.0 μm.
007008: Longest axis of myxospore is 2.1 - 3.0 μm.
007009: Longest axis of myxospore is 3.1 - 4.0 μm.
007010: Longest axis of myxospore is 4.1 - 5.0 μm.
007011: Longest axis of myxospore is > 5.0 μm.
007012: Shortest axis of myxospore is 0.5 - 1.0 μm.
007013: Shortest axis of myxospore is 1.1 - 1.5 μm.
007014: Shortest axis of myxospore is > 1.5 μm.
007015: Myxospores are produced from and lie at random amongst vegetative cells in the colony.
007016: Myxospores are aggregated into sessile mucoid fruiting bodies without sporocyst formation.
007017: Myxospores are aggregated into mesenteric tubular structures within a sessile fruiting body.
007018: Myxospores are aggregated into discrete sporocysts.
007076: Sporocysts are present.
007019: Sporocysts occur singly.
007020: Sporocysts are grouped into distinct fruiting bodies.
007021: Sporocysts are microscopic.
007022: Sporocysts are macroscopic.
007023: Sporocysts are angular.
007024: Sporocysts are apiculate.
007025: Sporocysts are cylindrical.
007026: Sporocysts are disc-shaped.
007027: Sporocysts are ellipsoidal.
007028: Sporocysts are spherical.
007029: Sporocysts have no regular shape.
007077: Longest axis of sporocyst is < 0.5 μm.
007030: Longest axis of sporocyst is 0.5 - 2.0 μm.
007031: Longest axis of sporocyst is 2.1 - 5.0 μm.
007032: Longest axis of sporocyst is 5.1 - 10 μm.
007033: Longest axis of sporocyst is 10.1 - 20 μm.
007034: Longest axis of sporocyst is 20.1 - 50 μm.
007035: Longest axis of sporocyst is 50.1 - 100 μm.
007036: Longest axis of sporocyst is 100.1 - 150 μm.
007078: Longest axis of sporocyst is > 150 μm.
007079: Shortest axis of sporocyst is < 0.5 μm.
007037: Shortest axis of sporocyst is 0.5 - 2.0 μm.
007038: Shortest axis of sporocyst is 2.1 - 5.0 μm.
007039: Shortest axis of sporocyst is 5.1 - 10 μm.
007040: Shortest axis of sporocyst is 10.1 - 20 μm.
007041: Shortest axis of sporocyst is 20.1 - 50 μm.
007042: Shortest axis of sporocyst is 50.1 - 100 μm.
007080: Shortest axis of sporocyst is 100.1-150 μm.
007081: Shortest axis of sporocyst is > 150 μm.

Section 7: Myxospores and fruiting bodies

007043: Sporocysts are sessile.
007082: Sporozoites are present.
007083: Sporozoites are enclosed in an oocyst.
007084: Sporocyst has 2 sporozoites.
007085: Sporocyst has 4 sporozoites.
007086: Sporocyst has 6 sporozoites.
007087: Sporocyst has 8 sporozoites.
007088: Sporocyst has > 8 sporozoites.
007089: Spores are biconical.
007090: Spores are spindle shaped.
007091: Spores are elongate.
007092: Spores are oval.
007093: Spores are truncated.
007094: Spores are thickened at the extremities.
007095: Spores are pointed at 1 end.
007096: Spores are navicular.
007097: Spores have a single posterior filament.
007098: Spores have a single polar capsule.
007099: Spores have 2 polar capsules.
007100: Spores have 3 polar capsules.
007101: Spores have 4 polar capsules.
007102: Spore capsule is bivalve.
007103: Spore capsule is trivalve.
007104: Spore capsule is composed of a single envelope.
007105: Spores have spines.
007106: Trophozoites are present.
007107: Trophozoites are spheroid.
007108: Trophozoites are cylindrical.
007109: Trophozoites are oval.
007110: Trophozoites are elongate.
007111: Trophozoites anterior end is not differentiated.
007112: Trophozoites are motile.
007113: Mature trophozoites are intracellular.
007114: In host blood cells, trophozoites are present.
007115: In host blood cells, schizonts are present.
007116: In host blood cells, ring stages are present.
007117: In host blood cells, macrogametocytes are present.
007118: In host blood cells, microgametocytes are present.
007071: Fruiting bodies are present.
007072: Fruiting bodies appear within 6 weeks.
007044: Sessile sporocysts are grouped in fruiting bodies.
007045: Sessile fruiting bodies are flat.
007046: Sessile fruiting bodies are low convex.
007047: Sessile fruiting bodies are high convex.
007048: Sessile fruiting bodies are columnar.
007049: Sessile fruiting bodies produce aerial projections of a regular form.
007050: Sessile fruiting bodies produce aerial projections of an irregular form.
007051: Sporocysts are born on stalks.
007052: Sporocysts stalks are branched.
007053: Sporocysts occur singly on a stalk.
007054: Sporocysts occur singly on a stalk branch.
007055: Sporocysts occur in whorls at the tips of stalks.
007056: Sporocysts occur in whorls at the tips of stalk branches.

Section 7: Myxospores and fruiting bodies

007057:	Sporocysts occur in compact clusters (not whorls) at or near the tips of stalks.
007058:	Sporocysts occur in compact clusters (not whorls) at or near the tips of stalk branches.
007059:	Sporocysts occur at random on stalks.
007060:	Sporocysts occur at random on stalk branches.
007061:	Fruiting bodies are butyrous.
007062:	Fruiting bodies are deliquescent.
007063:	Fruiting bodies are mucoid.
007064:	Fruiting bodies are cartilagenous.
007065:	Fruiting bodies are red.
007066:	Fruiting bodies are orange.
007067:	Fruiting bodies are yellow.
007068:	Fruiting bodies are brown.
007069:	Fruiting bodies are pink.
007070:	Fruiting bodies are some other color.
007073:	Basidiospores are present.
007074:	Zygospores are produced.
008398:	Ballistospores are produced (also see Section 8).
007075:	Only peridial hyphae are produced.

SECTION 8:   BRANCHING, HYPHAE, AND PRODUCTION OF
             ASEXUAL SPORES

[NOTE 1: The term "primary," as used here, refers to the initial growth following the germination of conidia, arthrospores, zoospores, sporangiospores, and products of hyphal fragmentation. This growth may be ALONG AND IN the substratum (substratal) or AERIAL with no growth along or in the substrate.]

[NOTE 2: The term "secondary" refers to AERIAL GROWTH originating from substratal primary growth.]

[NOTE 3: The term "conidium" is here employed for asexual spore other than the "exogenous arthrospore" and the "sporangiospore."]

[NOTE 4: The term "sporophore" refers to that portion of the growth, either of primary or secondary origin, which supports conidia or sporangia. In the absence of such supports they are "sessile."]

[NOTE 5: Mycologists do not use the terms "primary" and "secondary" hyphal growth. Also, some bacteriologists may not as well. Therefore, many groups of questions are in the sequence of 1) general questions about hyphae and related structures, 2) "primary" hyphal features, and 3) "secondary" hyphal features.]

BRANCHING HYPHAE AND SEPTATION

008001: Cells branch.
008002: Branching occurs only in pleomorphic cultures.
008003: Branch is rudimentary (limited to Y,T,V or L shapes).
008336: Mycelial growth occurs.
008662: True mycelium is produced.
008663: Pseudomycelium is produced.
008004: Primary hyphae are produced.
008005: Primary growth is in the form of branched hyphae lying on the substrate.
008006: Primary growth is in the form of branched hyphae lying in the substrate (substratal).
008007: Primary growth is in the form of aerial hyphae, the initial cell being fixed to the substrate as a holdfast.
008008: Secondary (aerial) hyphae are produced.
008009: Branching hyphae have clavate ends.
008342: Hyphae are septate.
008010: Branching hyphae are septate.
008343: Hyphae are simple-septate.
008366: Hyphae are simple-septate in advancing zone.
008344: Hyphae are nodose-septate (clamp connections present).
008345: Clamp connections are single.

Section 8: Hyphae and asexual spores

008346: Clamp connections are multiple.
008347: Hyphae have rigid right angled branches.
008011: Secondary hyphae curve back to the substratum in the form of an arch.
008350: Hyphae form short, hooked (recurved) branches which interlock to form a plectenchyma.
008351: Hyphae form thick-walled nodules which interlock to form a plectenchyma.
008352: Hyphae form contorted incrusted hyphal tips.
008353: Hyphae form cuticular cells (cells closely packed to form a pseudo-parenchyma).
008354: Hyphae form fiber hyphae.
008368: Hyphae form setal hyphae.
008355: Hyphae aggregate to form rhizomorphs.
008356: Hyphae form bulbils (knots).
008012: Septa are produced at the origins of the branches in addition to any in the parent filament.
008013: Septa are produced transversely.
008014: Septa are produced transversely and longitudinally.
008337: Hyphae differentiate.
008338: Hyphae have irregularly thickened walls (lumen diameter varies).
008339: Hyphae have scattered thick refractive areas on walls.
008340: Hyphae have projections on external surfaces of walls.
008341: Hyphae have resinous masses clinging to walls.
008348: Hyphae have swellings (outside diameter varies).
008349: Hyphae have empty (non-staining) lumen.
008412: Attached end of conidium is narrower than unattached end (i.e., truncated oval).
008406: Spiral hyphae are produced.
008407: Pectinate (1 side smooth, 1 side irregular) hyphae are produced.
008408: Favic (antler-like) hyphae are present.
008410: Proliferating organs are present.
008411: Hyphal branches are reflexive.
008413: Hyphae are straight.
008414: Hyphae are sinuous.
008415: Hyphae are contorted.
008416: Raquet hyphae are produced.
008417: Rhizoids (root like hyphae) are present.
008418: Rhizoids are found at nodes.
008419: Rhizoids are found at internodes.
008420: Stolons are present.

FRAGMENTATION AND SPORE FORMATION

008015: Hyphae fragment.
008016: Primary hyphae fragment.
008017: Secondary hyphae fragment.
008018: Fragmentation of hyphae at septa is complete.
008019: Fragmentation of hyphae at septa takes place only on terminal hyphae.
008020: Fragmentation results in exogenous arthrospores.
008021: Fragmentation produces rods.

008022: Fragmentation produces cocci.
008023: Fragmentation produces diphtheroids.
008024: Fragmentation produces motile cells.
008025: Intercalary vesicles are produced.
008026: Subterminal vesicles are produced.
008363: Chlamydospores are present.
008421: Chlamydospores are terminal.
008422: Chlamydospores are intercalary.
008423: Chlamydospores are sessile.
008424: Chlamydospores are stalked.
008425: Chlamydospores (gemmae) are spiny.
008426: Chlamydospores are pigmented.
008027: Sclerotia are produced.
008427: Sclerotia are globose (i.e., round).
008428: Sclerotia are irregular.
008429: Sclerotia are ovate.
008430: Sclerotia are cylindrical.
008431: Sclerotia are tufted.
008432: Sclerotia are pigmented.
008028: Coremia are produced.
008357: Cystidia are present.
008358: Gloecystidia are present.
008359: Setae are present.
008360: Cystidia are present on primary hyphae (vegetative hyphae).
008361: Cystidia are present in hymenium of fruiting areas.
008362: Setae are present on primary hyphae (vegetative hyphae).
008367: Setae are present in hymenium of fruiting areas.
008398: Ballistospores are produced (also see Section 7).

HYPHAL PIGMENTATION

008433: Hyphae are pigmented.
008434: Hyphae change color with age.
008435: Hyphae darken with age.
008436: Hyphae are black.
008437: Hyphae are red.
008438: Hyphae are yellow.
008439: Hyphae are green.
008440: Hyphae are gray.
008409: Hyphae are brown.
008377: Secondary hyphae are red.
008378: Secondary hyphae are yellow.
008379: Secondary hyphae are green.
008380: Secondary hyphae are blue.
008381: Secondary hyphae are violet.
008382: Secondary hyphae are white.
008441: Secondary hyphae are black.
008389: Secondary hyphae are gray.

CONIDIOPHORES

008442: Conidiophores are produced.
008390: Conidiophores are elongated.
008391: Conidiophores are spherical.

Section 8: Hyphae and asexual spores

008392: Conidiophores are branched.
008393: Conidiophores are clavate.
008443: Conidiophore branches are vesicular.
008444: Conidiophore wall is rough.
008445: Conidiophores are pigmented.
008446: Conidiophores are yellow.
008447: Conidiophores are brown.
008448: Conidiophore surface is granular.
008449: Conidiophores are thick-walled.
008450: Conidiophores arise from substrate hyphae.
008451: Conidiophores arise from aerial hyphae.
008452: Vesicle (i.e., enlarged end of conidiophore) is present.
008453: Vesicles are globose.
008454: Vesicles are clavate.

CONIDIA

008030: Conidia are produced.
008394: Microconidia are produced.
008395: Macroconidia are produced.
008396: Vilose conidia are produced.
008455: Conidia are thallic, i.e., produced by transformation of existing hyphae.
008456: Blastospores are produced.
008457: Blastospores are produced from chlamydospores.
008458: Blastospores are in acropetal chains.
008364: Arthroconidia (oidia) are present.
008459: Arthroconidia enlarge after they are initially formed.
008460: Conidia are blastic, i.e., produced as new extentions of existing hyphae.
008461: Holoblastic conidia are produced.
008397: Phialospores are produced.
008462: Conidia are produced through pores in conidiophore wall (enterotretic).
008463: Annellospores are produced.
008029: Asexual spores (conidia) are produced in pycnidia.
008464: Asexual spores (conidia) are produced in acervuli.
008465: Conidia are produced on specialized cells separate from conidiophores.
008466: There is only 1 specialized cell between conidiophore and conidium.
008467: Specialized cells taper at distal end (i.e. away from conidiophore).
008468: Specialized cells are phialides.
008469: Phialides are flask shaped.
008470: Phialides are more swollen at the base.
008471: Phialides are variable in shape.
008472: Phialides are in verticillate clusters.
008473: Phialides are lateral.
008474: Phialides are terminal.
008475: Phialides are arranged singly.
008476: Phialides are produced from chlamydospores.
008477: Phialides proliferate (i.e. occur in chains).
008478: Conidia are unicellular.

70   Section 8: Hyphae and asexual spores

008479:   Conidia of > 1 size are formed.
008480:   Conidia of > 1 type are formed.
008481:   Conidia are septate.
008482:   Conidial septa are longitudinal.
008483:   Conidial septa are transverse.
008484:   Conidial septa are longitudinal and transverse.
008485:   Conidia are unicellular.
008486:   Conidia have 1 septum.
008487:   Conidia have 2 septa.
008488:   Conidia have 3 septa.
008489:   Conidia have 4 septa.
008490:   Conidia have 5 septa.
008491:   Conidia have 6 septa.
008492:   Conidia have 7 septa.
008493:   Conidia have 8 septa.
008494:   Conidia have > 8 septa.

   CONIDIAL LOCATION AND ARRANGEMENT

008495:   Conidia are sessile.
008496:   Conidia are borne directly on sporophores
          (conidiophores).
008497:   Conidial mass (i.e. conidial head) is irregularly
          arranged.
008498:   Conidial mass is globose (i.e. spherical).
008499:   Conidial mass is columnar.
008500:   Conidial mass is loosely radiate.
008501:   Conidial mass splits with age.
008502:   Conidial mass is compact.
008365:   Conidia are produced on primary hyphae.
008031:   Conidia on primary hyphae are sessile.
008032:   Conidia on primary hyphae are on sporophores.
008033:   Conidia on primary hyphae are branched verticil-
          lately in 1 series.
008034:   Conidia on primary hyphae are branched verticil-
          lately in 2 series.
008035:   The distance between verticils of conidia on
          primary hyphae is regular.
008036:   Conidia on primary hyphae are branched
          dichotomously.
008037:   Conidia on primary hyphae are branched bifurcately.
008038:   Conidia on primary hyphae are branched
          monopodially.
008039:   Conidia on primary hyphae are branched randomly.
008040:   Conidia on primary hyphae are arranged in
          palisades.
008369:   Conidia are produced on secondary hyphae.
008041:   Conidia on secondary hyphae are sessile.
008042:   Conidia on secondary hyphae are on sporophores.
008043:   Conidia on secondary hyphae are branched verticil-
          lately in 1 series.
008044:   Conidia on secondary hyphae are branched verticil-
          lately in 2 series.
008045:   The distance between verticils of conidia on secon-
          dary hyphae is regular.
008046:   Conidia on secondary hyphae are branched
          dichotomously.

Section 8: Hyphae and asexual spores

008047: Conidia on secondary hyphae are branched bifurcately.
008048: Conidia on secondary hyphae are branched monopodially.
008049: Conidia on secondary hyphae are branched randomly.
008050: Conidia on secondary hyphae are arranged in palisades.

PROPERTIES OF CONIDIA AND CHAINS OF CONIDIA

008399: Conidia detach passively.
008503: Conidia detach actively.
008383: Conidial surface is smooth.
008384: Conidial surface is warty.
008385: Conidial surface is hairy.
008386: Conidial surface is spiny.
008504: Conidia are produced as endogenous arthrospores.
008505: Conidia occur singly.
008506: Conidia occur in chains of 2.
008507: Conidia occur in chains of 3.
008508: Conidia occur in chains of 4.
008509: Conidia occur in chains of 5.
008510: Conidia occur in chains of 6-10.
008511: Conidia occur in chains of > 10.
008512: Conidia occur in straight chains.
008513: Conidia occur in flexuous chains.
008514: Conidia occur in looped chains.
008515: Conidia occur in hooked chains.
008516: Conidia occur in short, compact, spiral chains.
008517: Conidia occur in long, extended, spiral chains.
008518: Conidia occur in open spiral chains.
008519: Conidia are produced in clusters (sympodially).
008520: Conidia are spherical.
008401: Conidia are fusiform (have tapered, pointed ends).
008402: Conidia are ellipsoidal (distal end rounded) and attached by broad septum.
008403: Conidia are pear-shaped.
008404: Conidia are club-shaped. (broader at 1 end).
008405: Conidia are tapered only at unattached end.
008521: Conidia are cylindrical (sausage shaped).
008522: Conidia are clavate.
008523: Conidia are oval.
008524: Conidia are attached to the hyphae or specialized cells by a broad septum.
008525: Conidia are radiate.
008526: Conidia are helicoid.
008527: Conidia are motile.
008528: Conidia are 0.1 - 1.0 µm long.
008529: Conidia are 1.1 - 2.0 µm long.
008530: Conidia are 2.1 - 3.0 µm long.
008531: Conidia are > 3.0 µm long.
008532: Conidia are 0.1 - 1.0 µm wide.
008533: Conidia are 1.1 - 2.0 µm wide.
008534: Conidia are 2.1 - 3.0 µm wide.
008535: Conidia are > 3.0 µm wide.
008536: Conidial masses on hyphae are dry and powdery.

Section 8: Hyphae and asexual spores

008537: Conidial masses on hyphae are imbedded in slime.
008051: Sessile conidia on primary hyphae are produced as endogenous arthrospores.
008052: Sessile conidia on primary hyphae occur singly.
008053: Sessile conidia on primary hyphae occur in chains of 2.
008054: Sessile conidia on primary hyphae occur in chains of 3.
008055: Sessile conidia on primary hyphae occur in chains of 4.
008056: Sessile conidia on primary hyphae occur in chains of 5.
008057: Sessile conidia on primary hyphae occur in chains of 6-10.
008058: Sessile conidia on primary hyphae occur in chains of > 10.
008059: Sessile conidia on primary hyphae occur in straight chains.
008060: Sessile conidia on primary hyphae occur in flexuous chains.
008061: Sessile conidia on primary hyphae occur in looped chains.
008062: Sessile conidia on primary hyphae occur in hooked chains.
008063: Sessile conidia on primary hyphae occur in short, compact, spiral chains.
008064: Sessile conidia on primary hyphae occur in long, extended, spiral chains.
008065: Sessile conidia on primary hyphae occur in open spiral chains.
008066: Sessile conidia on primary hyphae are spherical.
008067: Sessile conidia on primary hyphae are cylindrical.
008068: Sessile conidia on primary hyphae are smooth.
008069: Sessile conidia on primary hyphae are spiny.
008070: Sessile conidia on primary hyphae are hairy.
008071: Sessile conidia on primary hyphae are warty.
008072: Sessile conidia on primary hyphae are motile.
008073: Sessile conidia on primary hyphae are 0.1 - 1.0 µm long.
008074: Sessile conidia on primary hyphae are 1.1 - 2.0 µm long.
008075: Sessile conidia on primary hyphae are 2.1 - 3.0 µm long.
008076: Sessile conidia on primary hyphae are 0.1 - 1.0 µm wide.
008077: Sessile conidia on primary hyphae are 1.1 - 2.0 µm wide.
008078: Sessile conidia on primary hyphae are 2.1 - 3.0 µm wide.
008079: Sessile conidial masses on primary hyphae are dry and powdery.
008080: Sessile conidial masses on primary hyphae are embedded in slime.
008081: Conidia on sporophores on primary hyphae are produced as endogenous arthrospores.
008082: Conidia on sporophores on primary hyphae are produced by abstriction from the sporophore.

Section 8: Hyphae and asexual spores 73

008083: Conidia on sporophores on primary hyphae occur singly.
008084: Conidia on sporophores on primary hyphae occur in chains of 2.
008085: Conidia on sporophores on primary hyphae occur in chains of 3.
008086: Conidia on sporophores on primary hyphae occur in chains of 4.
008087: Conidia on sporophores on primary hyphae occur in chains of 5.
008088: Conidia on sporophores on primary hyphae occur in chains of 6-10.
008089: Conidia on sporophores on primary hyphae occur in chains of > 10.
008090: Conidia on sporophores on primary hyphae occur in straight chains.
008091: Conidia on sporophores on primary hyphae occur in flexuous chains.
008092: Conidia on sporophores on primary hyphae occur in looped chains.
008093: Conidia on sporophores on primary hyphae occur in hooked chains.
008094: Conidia on sporophores on primary hyphae occur in short, compact, spiral chains.
008095: Conidia on sporophores on primary hyphae occur in long, extended, spiral chains.
008096: Conidia on sporophores on primary hyphae occur in open spiral chains.
008097: Conidia on sporophores on primary hyphae are spherical.
008098: Conidia on sporophores on primary hyphae are cylindrical.
008099: Conidia on sporophores on primary hyphae are smooth.
008100: Conidia on sporophores on primary hyphae are spiny.
008101: Conidia on sporophores on primary hyphae are hairy.
008102: Conidia on sporophores on primary hyphae are warty.
008103: Conidia on sporophores on primary hyphae are motile.
008104: Conidia on sporophores on primary hyphae are 0.1 - 1.0 µm long.
008105: Conidia on sporophores on primary hyphae are 1.1 - 2.0 µm long.
008106: Conidia on sporophores on primary hyphae are 2.1 - 3.0 µm long.
008107: Conidia on sporophores on primary hyphae are 0.1 - 1.0 µm wide.
008108: Conidia on sporophores on primary hyphae are 1.1 - 2.0 µm wide.
008109: Conidia on sporophores on primary hyphae are 2.1 - 3.0 µm wide.
008110: Conidial masses on sporophores on primary hyphae are dry and powdery.
008111: Conidial masses on sporophores on primary hyphae are embedded in slime.
008370: Conidia on secondary hyphae occur in chains.

008371: Conidia on secondary hyphae occur in chains, within a range of 3 - 10.
008372: Conidia on secondary hyphae occur in chains, within a range of 10 - 50.
008373: Conidia on secondary hyphae occur in chains > 50.
008374: Conidia on secondary hyphae occur in spiral chains.
008375: Conidia on secondary hyphae occur in flexous chains.
008376: Conidia on secondary hyphae occur in hooked chains.
008112: Sessile Conidia on secondary hyphae are produced as endogenous arthrospores.
008113: Sessile Conidia on secondary hyphae occur singly.
008114: Sessile Conidia on secondary hyphae occur in chains of 2.
008115: Sessile Conidia on secondary hyphae occur in chains of 3.
008116: Sessile Conidia on secondary hyphae occur in chains of 4.
008117: Sessile Conidia on secondary hyphae occur in chains of 5.
008118: Sessile Conidia on secondary hyphae occur in chains of 6-10.
008119: Sessile Conidia on secondary hyphae occur in chains of > 10.
008120: Sessile Conidia on secondary hyphae occur in straight chains.
008121: Sessile Conidia on secondary hyphae occur in flexuous chains.
008122: Sessile Conidia on secondary hyphae occur in looped chains.
008123: Sessile Conidia on secondary hyphae occur in hooked chains.
008124: Sessile Conidia on secondary hyphae occur in short, compact, spiral chains.
008125: Sessile Conidia on secondary hyphae occur in long, extended, spiral chains.
008126: Sessile Conidia on secondary hyphae occur in open spiral chains.
008127: Sessile Conidia on secondary hyphae are spherical.
008128: Sessile Conidia on secondary hyphae are cylindrical.
008129: Sessile Conidia on secondary hyphae are smooth.
008130: Sessile Conidia on secondary hyphae are spiny.
008131: Sessile Conidia on secondary hyphae are hairy.
008132: Sessile Conidia on secondary hyphae are warty.
008133: Sessile Conidia on secondary hyphae are motile.
008134: Sessile Conidia on secondary hyphae are 0.1 - 1.0 µm long.
008135: Sessile Conidia on secondary hyphae are 1.1 - 2.0 µm long.
008136: Sessile Conidia on secondary hyphae are 2.1 - 3.0 µm long.
008137: Sessile Conidia on secondary hyphae are 0.1 - 1.0 µm wide.
008138: Sessile Conidia on secondary hyphae are 1.1 - 2.0 µm wide.

Section 8: Hyphae and asexual spores 75

008139: Sessile Conidia on secondary hyphae are 2.1 - 3.0 µm wide.
008140: Sessile Conidial masses on secondary hyphae are dry and powdery.
008141: Sessile Conidial masses on secondary hyphae are imbedded in slime.
008142: Conidia on sporophores on secondary hyphae are produced as endogenous arthrospores.
008143: Conidia on sporophores on secondary hyphae are produced by abstriction from the sporophore.
008144: Conidia on sporophores on secondary hyphae occur singly.
008145: Conidia on sporophores on secondary hyphae occur in chains of 2.
008146: Conidia on sporophores on secondary hyphae occur in chains of 3.
008147: Conidia on sporophores on secondary hyphae occur in chains of 4.
008148: Conidia on sporophores on secondary hyphae occur in chains of 5.
008149: Conidia on sporophores on secondary hyphae occur in chains of 6-10.
008150: Conidia on sporophores on secondary hyphae occur in chains of > 10.
008151: Conidia on sporophores on secondary hyphae occur in straight chains.
008152: Conidia on sporophores on secondary hyphae occur in flexuous chains.
008153: Conidia on sporophores on secondary hyphae occur in looped chains.
008154: Conidia on sporophores on secondary hyphae occur in hooked chains.
008155: Conidia on sporophores on secondary hyphae occur in short, compact, spiral chains.
008156: Conidia on sporophores on secondary hyphae occur in long, extended, spiral chains.
008157: Conidia on sporophores on secondary hyphae occur in open spiral chains.
008158: Conidia on sporophores on secondary hyphae are spherical.
008159: Conidia on sporophores on secondary hyphae are cylindrical.
008160: Conidia on sporophores on secondary hyphae are smooth.
008161: Conidia on sporophores on secondary hyphae are spiny.
008162: Conidia on sporophores on secondary hyphae are hairy.
008163: Conidia on sporophores on secondary hyphae are warty.
008164: Conidia on sporophores on secondary hyphae are motile.
008165: Conidia on sporophores on secondary hyphae are 0.1-1.0 µm long.
008166: Conidia on sporophores on secondary hyphae are 1.1-2.0 µm long.

76 Section 8: Hyphae and asexual spores

008167: Conidia on sporophores on secondary hyphae are 2.1-3.0 μm long.
008168: Conidia on sporophores on secondary hyphae are 0.1-1.0 μm wide.
008169: Conidia on sporophores on secondary hyphae are 1.1-2.0 μm wide.
008170: Conidia on sporophores on secondary hyphae are 2.1-3.0 μm wide.
008171: Conidial masses on sporophores on secondary hyphae are dry and powdery.
008172: Conidial masses on sporophores on secondary hyphae are embedded in slime.

SPORANGIA AND SPORANGIOSPORES

008173: Asexual spores are produced in sporangia (spore vesicles).
008664: During reproduction autospores are formed.
008666: Cell forms statospores during reproduction.

SPORANGIAL LOCATION AND ARRANGEMENT

008538: Sporangia are sessile.
008539: Sporangia are on sporophores.
008540: Sporangia are branched verticillately in 1 series.
008541: Sporangia are branched verticillately in 2 series.
008542: The distance between verticils of sporangia is regular.
008543: Sporangia are branched dichotomously.
008544: Sporangia are branched bifurcately.
008545: Sporangia are branched monopodially.
008546: Sporangia are branched randomly.
008547: Sporangia are arranged in palisades.
008387: Sporangia are produced on primary hyphae.
008174: Sporangia on primary hyphae are sessile.
008175: Sporangia on primary hyphae are on sporophores.
008176: Sporangia on primary hyphae are branched verticillately in 1 series.
008177: Sporangia on primary hyphae are branched verticillately in 2 series.
008178: The distance between verticils of sporangia on primary hyphae is regular.
008179: Sporangia on primary hyphae are branched dichotomously.
008180: Sporangia on primary hyphae are branched bifurcately.
008181: Sporangia on primary hyphae are branched monopodially.
008182: Sporangia on primary hyphae are branched randomly.
008183: Sporangia on primary hyphae are arranged in palisades.
008388: Sporangia are produced on secondary hyphae.
008184: Sporangia on secondary hyphae are sessile.
008185: Sporangia on secondary hyphae are on sporophores.
008186: Sporangia on secondary hyphae are branched verticillately in 1 series.

Section 8: Hyphae and asexual spores

008187: Sporangia on secondary hyphae are branched verticillately in 2 series.
008188: The distance between verticils of sporangia on secondary hyphae is regular.
008189: Sporangia on secondary hyphae are branched dichotomously.
008190: Sporangia on secondary hyphae are branched bifurcately.
008191: Sporangia on secondary hyphae are branched monopodially.
008192: Sporangia on secondary hyphae are branched randomly.
008193: Sporangia on secondary hyphae are arranged in palisades.

PROPERTIES OF SPORANGIA

008194: The walls of sporangia are loose and clearly separated from the spores.
008195: The walls of sporangia form a tight sheath around the spores.
008196: Sporangiospores are released from a pore in the sporangial wall.
008197: Sporangiospores are released through a rupture in the sporangial wall.
008198: The sporangial wall disintegrates prior to the release of spores.
008199: The sporangial wall disintegrates concurrently with the release of spores.
008200: The sporangial wall remains after the release of spores.
008201: Sporangia separate from the sporophore at the point of connection.
008202: Sporangia separate from the sporophore at a point below their connection.
008548: Collarette is formed after release of sporangiospores.
008549: Sporangia are apiculate.
008550: Sporangia are bell-shaped.
008551: Sporangia are clavate.
008552: Sporangia are cylindrical.
008553: Sporangia are digital.
008554: Sporangia are fusiform.
008555: Sporangia are irregular.
008556: Sporangia are oval.
008557: Sporangia are pod-like.
008558: Sporangia are spherical.
008559: Sporangia are urceolate.
008560: Sporangia are vermiforme.
008561: Columella is formed.
008562: Apophysis is present.
008563: Sporangial walls are smooth.
008564: Sporangial walls are spiny (echinulate).
008565: Sporangial walls are warty (tubercular).
008566: Sporangia occur singly.
008567: Sporangia occur in clusters.

Section 8: Hyphae and asexual spores

008568: Sporangia occur in rows.
008569: Sporangia are produced acropetally.
008570: Sporangia are produced laterally.
008571: Sporangia are 1.0 - 2.0 µm long.
008572: Sporangia are 2.1 - 5.0 µm long.
008573: Sporangia are 5.1 - 10.0 µm long.
008574: Sporangia are 11.0 - 15.0 µm long.
008575: Sporangia are 16.0 - 20.0 µm long.
008576: Sporangia are 21.0 - 30.0 µm long.
008577: Sporangia are 31.0 - 50.0 µm long.
008578: Sporangia are 1.0 - 2.0 µm wide.
008579: Sporangia are 2.1 - 3.0 µm wide.
008580: Sporangia are 3.1 - 4.0 µm wide.
008581: Sporangia are 4.1 - 5.0 µm wide.
008582: Sporangia are 5.1 - 10.0 µm wide.
008583: Sporangia are 11.0 - 15.0 µm wide.
008584: Sporangia are 16.0 - 20.0 µm wide.
008585: Sporangia are 21.0 - 30.0 µm wide.
008203: Sporangia on primary hyphae are apiculate.
008204: Sporangia on primary hyphae are bell-shaped.
008205: Sporangia on primary hyphae are clavate.
008206: Sporangia on primary hyphae are cylindrical.
008207: Sporangia on primary hyphae are digital.
008208: Sporangia on primary hyphae are fusiform.
008209: Sporangia on primary hyphae are irregular.
008210: Sporangia on primary hyphae are oval.
008211: Sporangia on primary hyphae are pod-like.
008212: Sporangia on primary hyphae are spherical.
008213: Sporangia on primary hyphae are urceolate.
008214: Sporangia on primary hyphae are vermiforme.
008215: Sporangial walls on primary hyphae are smooth.
008216: Sporangial walls on primary hyphae are spiny (echinulate).
008217: Sporangial walls on primary hyphae are warty (tubercular).
008218: Sporangia on primary hyphae occur singly.
008219: Sporangia on primary hyphae occur in clusters.
008220: Sporangia on primary hyphae occur in rows.
008221: Sporangia on primary hyphae are produced acropetally.
008222: Sporangia on primary hyphae are produced laterally.
008223: Sporangia on primary hyphae are 1.0 - 2.0 µm long.
008224: Sporangia on primary hyphae are 2.1 - 5.0 µm long.
008225: Sporangia on primary hyphae are 5.1 - 10.0 µm long.
008226: Sporangia on primary hyphae are 11.0 - 15.0 µm long.
008227: Sporangia on primary hyphae are 16.0 - 20.0 µm long.
008228: Sporangia on primary hyphae are 21.0 - 30.0 µm long.
008229: Sporangia on primary hyphae are 31.0 - 50.0 µm long.
008230: Sporangia on primary hyphae are 1.0 - 2.0 µm wide.

Section 8: Hyphae and asexual spores   79

008231: Sporangia on primary hyphae are 2.1 - 3.0 µm
         wide.
008232: Sporangia on primary hyphae are 3.1 - 4.0 µm
         wide.
008233: Sporangia on primary hyphae are 4.1 - 5.0 µm
         wide.
008234: Sporangia on primary hyphae are 6.0 - 10.0 µm
         wide.
008235: Sporangia on primary hyphae are 11.0 - 15.0 µm
         wide.
008236: Sporangia on primary hyphae are 16.0 - 20.0 µm
         wide.
008237: Sporangia on primary hyphae are 21.0 - 30.0 µm
         wide.
008238: Sporangia on secondary hyphae are apiculate.
008239: Sporangia on secondary hyphae are bell-shaped.
008240: Sporangia on secondary hyphae are clavate.
008241: Sporangia on secondary hyphae are cylindrical.
008242: Sporangia on secondary hyphae are digital.
008243: Sporangia on secondary hyphae are fusiform.
008244: Sporangia on secondary hyphae are irregular.
008245: Sporangia on secondary hyphae are oval.
008246: Sporangia on secondary hyphae are pod-like.
008247: Sporangia on secondary hyphae are spherical.
008248: Sporangia on secondary hyphae are urceolate.
008249: Sporangia on secondary hyphae are vermiforme.
008250: Sporangial walls on secondary hyphae are smooth.
008251: Sporangial walls on secondary hyphae are spiny
         (echinulate).
008252: Sporangial walls on secondary hyphae are warty
         (tubercular).
008253: Sporangia on secondary hyphae occur singly.
008254: Sporangia on secondary hyphae occur in clusters.
008255: Sporangia on secondary hyphae occur in rows.
008256: Sporangia on secondary hyphae are produced acrope-
         tally.
008257: Sporangia on secondary hyphae are produced
         laterally.
008258: Sporangia on secondary hyphae are 1.0 - 2.0 µm
         long.
008259: Sporangia on secondary hyphae are 2.1 - 5.0 µm
         long.
008260: Sporangia on secondary hyphae are 5.1 - 10.0 µm
         long.
008261: Sporangia on secondary hyphae are 11.0 - 15.0 µm
         long.
008262: Sporangia on secondary hyphae are 16.0 - 20.0 µm
         long.
008263: Sporangia on secondary hyphae are 21.0 - 30.0 µm
         long.
008264: Sporangia on secondary hyphae are 31.0 - 50.0 µm
         long.
008265: Sporangia on secondary hyphae are 1.0 - 2.0 µm
         wide.
008266: Sporangia on secondary hyphae are 2.1 - 3.0 µm
         wide.

008267: Sporangia on secondary hyphae are 3.1 - 4.0 μm wide.
008268: Sporangia on secondary hyphae are 4.1 - 5.0 μm wide.
008269: Sporangia on secondary hyphae are 6.0 - 10.0 μm wide.
008270: Sporangia on secondary hyphae are 11.0 - 15.0 μm wide.
008271: Sporangia on secondary hyphae are 16.0 - 20.0 μm wide.
008272: Sporangia on secondary hyphae are 21.0 - 30.0 μm wide.

SPORANGIOSPORES

008273: Sporangiospores are motile.
008274: Sporangiospores have single polar flagella.
008275: Sporangiospores have tufts of polar flagella.
008276: Sporangiospores have lateral flagella.
008586: Sporangiospores occur singly.
008587: Sporangiospores occur in chains of 2.
008588: Sporangiospores occur in chains of 3.
008589: Sporangiospores occur in chains of 4.
008590: Sporangiospores occur in chains of 5.
008591: Sporangiospores occur in chains of > 5.
008592: Sporangiospores occur in masses.
008593: Chains of sporangiospores are straight.
008594: Chains of sporangiospores are coiled.
008595: Sporangiospores are spherical.
008596: Sporangiospores are cylindrical.
008597: Sporangiospores are 0.1 - 1.0 μm long.
008598: Sporangiospores are 1.1 - 2.0 μm long.
008599: Sporangiospores are 2.1 - 3.0 μm long.
008600: Sporangiospores are 3.1 - 4.0 μm long.
008601: Sporangiospores are 4.1 - 5.0 μm long.
008602: Sporangiospores are 0.1 - 1.0 μm wide.
008603: Sporangiospores are 1.1 - 2.0 μm wide.
008604: Sporangiospores are 2.1 - 3.0 μm wide.
008277: Sporangiospores on primary hyphae occur singly.
008278: Sporangiospores on primary hyphae occur in chains of 2.
008279: Sporangiospores on primary hyphae occur in chains of 3.
008280: Sporangiospores on primary hyphae occur in chains of 4.
008281: Sporangiospores on primary hyphae occur in chains of 5.
008282: Sporangiospores on primary hyphae occur in chains of > 5.
008283: Sporangiospores on primary hyphae occur in masses.
008284: Chains of sporangiospores on primary hyphae are straight.
008285: Chains of sporangiospores on primary hyphae are coiled.
008286: Sporangiospores on primary hyphae are spherical.
008287: Sporangiospores on primary hyphae are cylindrical.

Section 8: Hyphae and asexual spores   81

008288: Sporangiospores on primary hyphae are 0.1-1.0 µm long.
008289: Sporangiospores on primary hyphae are 1.1-2.0 µm long.
008290: Sporangiospores on primary hyphae are 2.1-3.0 µm long.
008291: Sporangiospores on primary hyphae are 3.1-4.0 µm long.
008292: Sporangiospores on primary hyphae are 4.1-5.0 µm long.
008293: Sporangiospores on primary hyphae are 0.1-1.0 µm wide.
008294: Sporangiospores on primary hyphae are 1.1-2.0 µm wide.
008295: Sporangiospores on primary hyphae are 2.1-3.0 µm wide.
008296: Sporangiospores on secondary hyphae occur singly.
008297: Sporangiospores on secondary hyphae occur in chains of 2.
008298: Sporangiospores on secondary hyphae occur in chains of 3.
008299: Sporangiospores on secondary hyphae occur in chains of 4.
008300: Sporangiospores on secondary hyphae occur in chains of 5.
008301: Sporangiospores on secondary hyphae occur in chains of > 5.
008302: Sporangiospores on secondary hyphae occur in masses.
008303: Chains of sporangiospores on secondary hyphae are straight.
008304: Chains of sporangiospores on secondary hyphae are coiled.
008305: Sporangiospores on secondary hyphae are spherical.
008306: Sporangiospores on secondary hyphae are cylindrical.
008307: Sporangiospores on secondary hyphae are 0.1 - 1.0 µm long.
008308: Sporangiospores on secondary hyphae are 1.1 - 2.0 µm long.
008309: Sporangiospores on secondary hyphae are 2.1 - 3.0 µm long.
008310: Sporangiospores on secondary hyphae are 3.1 - 4.0 µm long.
008311: Sporangiospores on secondary hyphae are 4.1 - 5.0 µm long.
008312: Sporangiospores on secondary hyphae are 0.1 - 1.0 µm wide.
008313: Sporangiospores on secondary hyphae are 1.1 - 2.0 µm wide.
008314: Sporangiospores on secondary hyphae are 2.1 - 3.0 µm wide.
008605: Sporangioles are present.
008606: Sporangioles are spiny.

ZOOSPORES

008665: Cell produces zoospores for sexual reproduction.
008315: Cells in a multilocular structure (resulting from transverse and longitudinal septation) produce zoospores.
008607: Zoospores are spherical.
008608: Zoospores are cylindrical.
008609: Zoospores are 0.1 - 1.0 µm long.
008610: Zoospores are 1.1 - 2.0 µm long.
008611: Zoospores are 2.1 - 3.0 µm long.
008612: Zoospores are 3.1 - 4.0 µm long.
008613: Zoospores are 4.1 - 5.0 µm long.
008614: Zoospores are 0.1 - 1.0 µm wide.
008615: Zoospores are 1.1 - 2.0 µm wide.
008616: Zoospores are 2.1 - 3.0 µm wide.
008316: Zoospores on primary hyphae are spherical.
008317: Zoospores on primary hyphae are cylindrical.
008318: Zoospores on primary hyphae are 0.1 - 1.0 µm long.
008319: Zoospores on primary hyphae are 1.1 - 2.0 µm long.
008320: Zoospores on primary hyphae are 2.1 - 3.0 µm long.
008321: Zoospores on primary hyphae are 3.1 - 4.0 µm long.
008322: Zoospores on primary hyphae are 4.1 - 5.0 µm long.
008323: Zoospores on primary hyphae are 0.1 - 1.0 µm wide.
008324: Zoospores on primary hyphae are 1.1 - 2.0 µm wide.
008325: Zoospores on primary hyphae are 2.1 - 3.0 µm wide.
008326: Zoospores on secondary hyphae are spherical.
008327: Zoospores on secondary hyphae are cylindrical.
008328: Zoospores on secondary hyphae are 0.1 - 1.0 µm long.
008329: Zoospores on secondary hyphae are 1.1 - 2.0 µm long.
008330: Zoospores on secondary hyphae are 2.1 - 3.0 µm long.
008331: Zoospores on secondary hyphae are 3.1 - 4.0 µm long.
008332: Zoospores on secondary hyphae are 4.1 - 5.0 µm long.
008333: Zoospores on secondary hyphae are 0.1 - 1.0 µm wide.
008334: Zoospores on secondary hyphae are 1.1 - 2.0 µm wide.
008335: Zoospores on secondary hyphae are 2.1 - 3.0 µm wide.

Section 8: Hyphae and asexual spores 83

SEXUAL REPRODUCTION

008617: Sexual reproduction occurs.
008618: Organism is homothallic (male and female on same mycelium).
008619: Organism is heterothallic (male and female on separate mycelium exhibit interfertility).
014009: Interfertility is bi-polar.
014010: Interfertility is tetra-polar.
008620: Gametangia are formed.
008621: Gametangia are morphologically similar to vegetative hyphae (i.e. no sexual differentiation).
008622: Male gametangia are produced.
008623: Male gametes are produced.
008624: Conidia act as male gametes.
008625: Female gametangia are produced.
008626: Female gametes are produced.
008627: Zygospores are produced.
008628: Zygospores are thick-walled.
008629: Zygospores are roughened.
008630: Zygospores are spiny.
008631: Only 1 supporting cell is present.
008632: Two supporting cells are present.
008633: Hyphae are developed on supporting cells.
008634: Asci are produced.
008635: Ascus is cylindrical.
008636: Ascus is round.
008637: Ascus is in locule.
008638: Ascospores are oval.
008639: Ascospores are round.
008640: Four ascospores are produced.
008641: Eight ascospores are produced.
008642: Paraphyses are produced.
008643: Ascus is phototrophic.
008644: Ascus tip is phototrophic.
008645: Paraphyses are phototrophic.
008646: Fruit body is phototrophic.
008647: Asci develop in perithecium.
008648: Asci develop in apothecium.
008649: Asci develop in cleistothecium.
008650: Cleistothecial wall is entire.
008651: Cleistothecial wall is composed of peridial hyphae.
008652: Peridial hyphae are dumbbell-shaped.
008653: Peridial hyphae are irregular on 1 side only.
008654: Fruit body is in stroma.
008655: Teleospores are formed.
008656: Probasidium are formed.
008657: Probasidium septate.
008658: Sporidia are formed.
008659: Sporidia are lateral.
008660: Sporidia are terminal.
008661: Sporidia bud.

SECTION 9: STALKS

009027: Stalks are present.
009001: Cells are borne on stalks.
009002: Stalks are attached to 1 pole of the cell.
009003: Stalks are attached in a sub-polar position on the cell.
009004: Stalks are attached to the side of the cell (other than sub-polar).
009005: The stalk is an extension of the cell wall.
009006: Stalks develop at the flagellated end of a cell.
009007: The stalk is a secretion of the cell.
009008: One end of the stalk is attached to a substratum.
009009: Holdfasts are produced.
009010: The holdfast is located at the tip of the stalk.
009011: The holdfast is produced on the cell at a point from the origin of the stalk.
009012: Stalks (or pelicle) are (is) very short, almost reduced to a simple holdfast.
009013: Stalks are elongated.
009028: Stalk is twice as long as body.
009029: Stalk is stiff.
009030: Stalk contains spasmoneme (contractile fiber).
009031: Stalk is filamentous.
009032: Pedicle is flexible.
009014: Stalks are branched.
009015: Stalks are dichotomously branched.
009033: Stalk branches contract singly.
009034: Stalk branches contract together.
009016: Stalks are dumbbell-shaped in cross section.
009017: Stalks are circular in cross section.
009018: Stalks are ribbon-like.
009019: Stalks are horn-shaped.
009020: Stalks are lobate.
009021: Stalks are slender.
009022: Stalks are twisted.
009023: Stalks contain ferric hydroxide.
009024: Stalks are soluble in HCl.
009025: Cells are permanently attached to the stalks and in this state are non-motile.
009026: Motile daughter cells are liberated from the mother stalk cell.

SECTION 10: SHEATHS

[NOTE 1: The term "sheath" as used here does not apply to the tightly fitting sporangial walls encountered in the Actinomycetales (for this see Section 8).]

010001: Sheaths are present.
010002: Several filaments, trichomes, or chains of cells are enclosed in a common sheath.
010003: There is one filament, trichome, or chain of cells to a sheath.
010004: Sheathed filaments are attached by a holdfast.
010005: Sheath is uniform in width throughout.
010006: Sheaths taper from wide base to narrow tip.
010007: Sheaths taper from narrow base to wide tip.
010008: Sheaths split longitudinally releasing the filaments.
010009: Electron microscopy reveals sheaths have uniform structure.
010010: Electron microscopy reveals sheaths have alveolar structure.

SECTION 11: CAPSULES

011001: Capsule is present.
011002: Capsule is predominantly polysaccharide.
011003: Capsule polysaccharide is antigenic.
011004: Capsule is predominantly polypeptide.
011005: Each cell has an individual capsule (capsules do not fuse).
011006: Cells are separated from each other within a common capsule (zoogloeae).
011007: Zoogloeae are encysted.
011008: Cells are in pairs in a common capsule.
011009: Cells are in chains in a common capsule.
011010: Cells are in compact masses surrounded by a common capsule.
011011: Secondary capsules are formed.
011012: Attachment to a substrate is by means of capsular material.
011013: Thinly capsulated cells have marginal thickening ("torus") of the capsule.
011014: "Torus" appears to completely surround the cell.
011015: "Torus" is open at 1 end like a horseshoe.
011016: "Torus" is impregnated with iron compounds (also see Section 5).
011017: Iron compounds are incorporated in cell capsules (also see Section 5).
011018: Iron compounds are deposited on cell capsule surface (also see Section 5).
011019: Manganese compounds are incorporated in cell capsules (also see Section 5).
011020: Manganese compounds are deposited on cell capsule surface (also see Section 5).

SECTION 12: STAIN REACTIONS

012001: Cells stain by Gram methods.
012021: Cells are Gram positive.
012022: Cells are Gram negative.
012023: Cells are Gram variable.
012002: Cells are Gram positive in early exponential growth.
012003: Cells are Gram negative in early exponential growth.
012004: Cells are Gram variable in early exponential growth.
012005: Cells lose Gram positive character after cessation of active growth (cells in stationary phase).
012006: Cells appear barred when stained by Gram methods.
012027: Individual cells stain unevenly when stained by Gram methods.
012007: Cells stain with Giemsa.
012008: Morphology is best determined by dissolving iron or manganese compounds from cells in dilute mineral acid followed by staining with Schiff's reagent (also see Sections 5 and 11).
012024: Micro-colonies bind neutral red.
012009: Cells are acid fast by Ziehl-Neelsen method.
012010: Cells are alcohol fast by modified Ziehl-Neelsen method.
012026: Ziehl-Neelsen acid fast stain reveals alternate bands of dark and light staining in the cell.
012011: Methylene blue stains reveal inclusion granules (also see Section 5).
012012: Methylene blue stains reveal bipolar bodies (also see Section 3).
012013: Methylene blue stains reveal barred structures.
012025: Individual cells stained with safranin exhibit different intensities.
012014: Sudan black B reveals intracellular lipids (fat bodies) (also see Sections 5 and 21).
012015: Iodine reveals intracellular polysaccharides (glycogen, "granulose", etc.) (also see Section 5).
012018: After growth on sucrose, iodine reveals intracellular polysaccharides (glycogen, "granulose", etc).
012019: After growth on D-Glucose, iodine reveals intracellular polysaccharides (glycogen, "granulose", etc).
012020: After growth on raffinose, iodine reveals intracellular polysaccharides (glycogen, "granulose", etc).
012016: Cellulose stain reveals cellulose compounds.
012017: Prussian blue test reveals iron compounds (also see Section 5 and 11).

SECTION 13: MOTILITY, FLAGELLATION, AND EXTERNAL ORGANELLES

GENERAL MODE OF LOCOMOTION

013001: Cells are motile.
013041: Cell motility is predominantly unidirectional.
013042: Cells oscillate (twitch) in place.
013043: Cells rotate in position.
013044: Cells tumble as they move.
013045: Cell motility is random (darting) in direction.
013046: Cell motility follows an undulating path.
013047: Locomotion occurs by markedly eruptive cytoplasmic flow.
013048: Cells move and/or feed by protoplasmic flow, with or without discrete pseudopodia.
013002: Cells are motile by flagella.
013049: Flagellum is directed backwards when swimming.
013050: Flagellum moves in circular motion.
013051: Only anterior end of flagellum moves.
013052: Flagella beat in unison.
013003: Cells are motile by spirochaetal motility (rotatory flexion).
013004: Cells demonstrate creeping or gliding motility on a solid surface.
013005: Groups of cells creep or glide as an aggregate unit (also see Sections 16).
013040: Cells glide on concave or ventral side on solid surface.
013006: Conidia are motile.
013007: Cells are motile within a sheath.
013008: Whole colonies of flagellated cells are motile (also see Section 16).

CHARACTERISTICS OF FLAGELLA

013038: Cells are flagellated but not motile.
013039: Flagellum is in a sheath.
013009: Cells have flagella.
013010: Flagella are polar.
013011: A single flagellum is located at 1 pole.
013012: A single flagellum is located at each pole.
013013: Several flagella are located at 1 pole.
013014: Several flagella are located at both poles.
013015: Flagella are sub-polar.
013016: A single flagellum is located near the pole.
013017: Several flagella are located near the pole.
013018: Flagellum (or flagella) is (are) inserted frankly laterally (from the middle of the cell).
013019: Cell has a single lateral flagellum.
013020: Cell has several lateral flagella.
013021: Flagella are attached to the concave curve of the cell.
013022: Cells are peritrichous.
013023: Two or more flagella of distinctly different appearance are in different locations on the cell.

Section 13: Motility and external organelles

013053: Cells have only anterior (end that leads while swimming) flagella.
013350: Insertion of flagellum is in anterior (end that leads while swimming) depression.
013351: Insertion of flagellum is anterior (end that leads while swimming) and ventral.
013352: Insertion of flagellum or flagella is posterior (end that trails while swimming).
013054: Flagella are equal in length.
013055: Flagellum is at least 1 μm longer than cell.
013056: Flagellum is at least 1 μm shorter than cell.
013057: Flagellum length is within 1 μm of cell length.
013058: Flagellum length is within 1 μm of being twice the cell length.
013343: The longer, only, or girdle flagellum is shorter than the body length.
013344: The longer, only, or girdle flagellum is equal to the body length.
013345: The longer, only, or girdle flagellum is greater than one but less than two times the body length.
013346: The longer, only, or girdle flagellum is greater than two times the body length.
013347: In transmission electron microscopy, a cross section of the longest flagellum is banded (appears as successive ribbonlike or cordlike structures).
013348: In transmission electron microscopy, the longest flagellum has mastigonemes (fine hairlike appendages).
013349: The longer girdle flagellum has scales.
013059: Cell has a total of 1 flagellum.
013060: Cell has a total of 2 flagella.
013061: Cell has a total of 3 flagella.
013062: Cell has a total of 4 flagella.
013063: Cell has a total of 5 flagella.
013064: Cell has a total of 6 flagella.
013065: Cell has a total of 7 flagella.
013066: Cell has a total of 8 flagella.
013067: Cell has a total of 9 flagella.
013068: Cell has a total of 10 flagella.
013069: Cell has a total of > 10 flagella.
013070: Cell has 2 flagella, 1 anterior and 1 trailing.
013071: Cell has 3 flagella, 1 anterior and 2 trailing.
013072: Cell has 1 flagellum undulating in cytostome.
013024: Average wavelength of flagella is < 1.0 μm.
013025: Average wavelength of flagella is 1.00 - 1.49 μm.
013026: Average wavelength of flagella is 1.50 - 1.99 μm.
013027: Average wavelength of flagella is 2.00 - 2.49 μm.
013028: Average wavelength of flagella is 2.50 - 2.99 μm.
013029: Average wavelength of flagella is 3.00 - 3.49 μm.
013030: Average wavelength of flagella is 3.50 - 4.00 μm.
013031: Average wavelength of flagella is > 4.0 μm.
013032: Average amplitude of flagella is < 0.30 μm.
013033: Average amplitude of flagella is 0.30 - 0.39 μm.
013034: Average amplitude of flagella is 0.40 - 0.49 μm.
013035: Average amplitude of flagella is 0.50 - 0.59 μm.
013036: Average amplitude of flagella is 0.60 - 0.70 μm.

013037: Average amplitude of flagella is > 0.70 μm.
013353: The shorter flagellum or flagella is thinner than the longer one.
013354: The longer flagellum is thinner than the shorter one.

PROTOPLASMIC EXTRUSIONS - PROTOPLASMIC PSEUDOPODIA AND AMEBOID PROTOPLASMIC FLOW

013073: Cells have pseudopodia.
013074: Pseudopodia are short, blunt.
013075: Pseudopodia are broad.
013076: Pseudopodia are conical.
013077: Pseudopodia move in wave-like pattern.
013078: Pseudopodia are formed slowly.
013079: Cells have lobopodia.
013080: Cells have large lobopodia.
013081: Lobopodia are anastomose.
013082: Cells have filopodia (sharply pointed pseudopodia with no microtubular axonemes).
013083: Filopodia are branched.
013084: Filopodia are straight.
013085: Filopodia are anastomose.
013086: Filopodia arise from broad hyaline lobe or zone.
013087: Filopodia arise directly from main mass of granuloplasm.
013088: Filopodia radiate from all sides.
013089: Filopodia are formed from a limited area.
013090: Cells have rhizopodia (reticulopodia).
013091: Rhizopodia radiate from a central mass of cytoplasm.
013092: Cells have axopodia (sharply pointed, radiating pseudopodia with microtubular axonemes).
013093: Cells have 16-20 axopodia.
013094: Cells are exclusively amoeboid (monophasic).

CILIA AND CIRRI

013095: Cells have cilia.
013355: Other appendages include pseudocilia.
013096: Cell has polygonally sculptured surface between somatic ciliary rows (argyrome system, silverline system).
013097: Cell has sutures (systeme secant).
013098: Cilia are present only in early developmental stages.
013099: Cilia are located on anterior 1/3 to 1/2 of cell.
013100: Ciliation is uniform over cell surface.
013101: Ciliation is confined to ventral surface.
013102: Ciliation is confined to dorsal surface.
013103: Ciliary zones encircle body.
013104: Cell has locomotor fringe (trochal band).
013105: Ciliation is confined to anterior half.
013106: Cilia are in oblique rows.
013107: Cilia are in spiral rows.
013108: Cilia are in longitudinal rows.

Section 13: Motility and external organelles

013109: One to 10 ciliary rows (kineties, meridians) are present.
013110: Eleven to 20 ciliary rows (kineties, meridians) are present.
013111: Twenty-one to 30 ciliary rows (kineties, meridians) are present.
013112: Thirty-one to 40 ciliary rows (kineties, meridians) are present.
013113: Forty-one to 50 ciliary rows (kineties, meridians) are present.
013114: More than 50 ciliary rows (kineties, meridians) are present.
013115: Cell has postoral meridian (kineties).
013308: Cell has 1 postoral meridian (kineties).
013309: Cell has 2 postoral meridians (kineties).
013310: Cell has > 2 postoral meridians (kineties).
013116: Cell has director meridian (kineties).
013117: Cell has germinal row.
013118: Somatic cell ciliature is compound.
013119: Somatic cell has simple ciliature.
013120: Somatic cell ciliature is scopulary.
013121: Cell has atrial ciliature.
013122: Cell has somatic ciliature.
013311: Somatic cilia within a kinety are single.
013312: Somatic cilia within a kinety are paired.
013313: Somatic cilia within a kinety are multiple and clustered as cirri.
013314: Somatic cilia within a kinety are multiple but linear.
013123: Cell has vestibular ciliature.
013124: Cell ciliature is circumoral.
013125: Cell ciliature is cirral.
013126: Cell has marginal cilia.
013127: Cell ciliature is coronal.
013128: Cell has pectinelle cilia.
013129: Cell has oral ciliature.
013130: Cell has postoral suture.
013131: Cell has perioral ciliature.
013132: Cell ciliature is buccal.
013133: Cell has prebuccal ciliature.
013134: Cell has perizonal cilature.
013135: Cell ciliature is peristomial.
013136: Cell has saltatorial cilia.
013137: Longest cilia are at posterior end.
013138: Longest cilia are at anterior end.
013139: Caudal cilia are present.
013140: Thigmotactic cilia are present.
013141: Thigmotactic cilia are anterior.
013142: Thigmotactic cilia are posterior.
013143: Cell has bristles.
013144: Cell has sensory bristles.
013145: Cell has clavate cilia.
013146: Cirri are present (cilia fused into 1 structure).
013147: Cirri are frontal.
013148: Cirri are ventral.
013149: Cirri are anal.
013150: Cirri are caudal.

013151: Cirri are marginal.
013152: Cell "walks" on cirri.

MEMBRANELLAE AND EXTERNAL OR PARORAL
MEMBRANE-LIKE STRUCTURES

013153: Membranella (oral polykinetid, 2 to several rows of cilia that function as a single unit) are present.
013154: Cell has long cilia that border and encircle the oral area.
013155: Cell has extensor membrane.
013156: Cell has hetermembranelle.
013157: Cell has cirromembranelle.
013158: Cell has paramembranelle.
013159: Cytostomal membranellae is present.
013160: Cell has alpha membranoid.
013161: Cell has beta membranoid.
013162: Cell has gamma membranoid.
013163: Cell has zeta membranoid.
013164: Cell has adoral zone of membranelles (AZM).
013165: Cell has paroral membrane (oral dikinetid).

EXTERNAL ARCHITECTURE:

-GENERAL

013166: On the right of the adoral zone is a narrow non-ciliated zone.
013167: The adoral zone is parallel to the cell axis.
013168: The adoral zone winds counter-clockwise to the cytostome.
013169: Undulating membrane is present.
013170: Partial undulating membrane is present.
013171: Cell has anarchic field.
013172: Cell has ogival field.
013173: Cell has apical funnel.
013174: Cell has brosse.
013175: Cell has capitulum.
013176: Cell has circumoral connective.
013177: Oral dikinetid is stichodyad.
013178: Oral monokinetid is stichomonad.
013179: Oral dikinetid is diplostichomonad.
013181: Cell has kinetosomes (basal granule, blepharoplast).
013182: Dyad kinetosomes (basal granule, blepharoplast) are present.
013183: Erratic kinetosomes (basal granule, blepharoplast) are present.
013184: Cell has barren kinetosomes (basal granule, blepharoplast).
013185: Cell has supernumerary kinetosomes (basal granule, blepharoplast).
013186: Cell has kinety.
013187: Cell has biolar kinety.
013188: Cell has diplokinety.
013189: Cell has polykinety.

Section 13: Motility and external organelles 93

013190: Cell has peroral kinety.
013315: Right oral dikinetid consists of 3 clearly different segments of which the middle segment is stichodyadic.
013316: Right oral dikinetid consists of a single segment.
013317: Right oral dikinetid consists of 2 segments with only the posterior s segment somatically differentiated.
013318: Cell has oral monokinetids.
013319: Oral monokinetids are preoral.
013320: Oral monokinetids are paroral.
013321: Oral monokinetids are post oral.
013322: Oral monokinetids are adoral.
013323: Ciliation is present on oral monokinetids.
013324: Oral monokinetids are to the right of the oral area.
013325: Oral monokinetids are to the left of the oral area.
013326: Cell has oral dikinetids.
013327: Oral dikinetids are preoral.
013328: Oral dikinetids are paroral.
013329: Oral dikinetids are post oral.
013330: Oral dikinetids are adoral.
013331: Ciliation is present on oral dikinetids.
013332: Ciliation is present on only 1 of the oral dikinetids.
013333: Ciliation is present on both oral dikinetids.
013334: Cell has oral polykinetids.
013335: Oral polykinetids are preoral.
013336: Oral polykinetids are paroral.
013337: Oral polykinetids are post oral.
013338: Oral polykinetids are adoral.
013339: Ciliation is present on oral polykinetids.
013340: Ciliation is present on all oral polykinetids.
013341: Ciliation is partially present on oral polykinetids.
013342: Oral polykinetids are in complete rows.
013191: Cell has ophryokinety.
013192: Cell has haplokinety.
013193: Cell has circumoral kineties.
013194: Cell has kinetofragments.
013195: Cell has perilemma.
013196: Cell has epistomial disc.
013197: Cell has reorganization band.
013198: Cell has intermeridional connectives.
013199: Cell has flange.
013200: Cell has frange.
013201: Cell has denticle.
013202: Cell has operculum.
013203: Cell has papilla.
013204: Cell has paves.
013205: Cell has peniculus.
013206: Cell has podite.
013207: Cell has rostrum.
013208: Cell has scopuloid.
013209: Cell has scopula.
013210: Cell has scutica.
013211: Cell has scutico-vestige.

Section 13: Motility and external organelles

013212: Cell has syncilium.
013213: Cell has velum.
013214: Cell has an atrium.
013215: Cell has vestibulum.
013216: Cell has rhabdos.
013217: Cell has rosette.
013218: Cell has ambyihymenium.
013219: Cell has synhymenium.
013220: Cell surface has spines.
013360: Cell has spines on scales.
013221: Cell has radial spines.
013222: Cell has spine-like projections on the posterior end.
013356: Cell surface has scales.
013357: Cell surface has mineral scales.
013358: Cell surface has organic scales.
013359: Scales are elaborated in a consistent manner.
013223: Cell has infundibulum.
013224: Cell has quadrulus.
013225: Cell has rhabdos.
013226: Cell has striated bands.
013227: Cell is polystomic.

-TENTACLES

013228: Tentacles are present.
013229: Tentacles are arranged radially.
013230: Tentacles are in bundles (fascicles).
013231: Tentacles occur singly.
013232: Tentacles occur in bundles of 1-3.
013233: Tentacles occur in bundles of 4-6.
013234: Tentacles are located anteriorly.
013235: Cells have 2 tentacles.
013236: Cells have 3 tentacles.
013237: Cells have 4 tentacles.
013238: Cells have 5 tentacles.
013239: Cells have 6 tentacles.
013240: Cells have 7 tentacles.
013241: Cells have 8 tentacles.
013242: Cells have 9 tentacles.
013243: Cells have 10 tentacles.
013244: Cells have > 10 tentacles.
013246: Cell has tentaculoids.

-SHELLS AND LORICA

013247: Cells have shell (lorica, test).
013248: Cell fills posterior half of shell.
013249: Shell is larger than cell.
013250: Shell is filled by cell.
013251: Shell has 2 arms.
013252: Shell has > 2 arms.
013253: Shell is open at both ends.
013254: Shell has single opening.
013255: Shell has aperture with tooth present.
013256: Shell has a central chamber.
013257: Shell is single-chambered.

Section 13: Motility and external organelles 95

```
013258:   Shell is two-chambered.
013259:   Second chamber of shell is tubular.
013260:   Second chamber of shell branches.
013261:   Second chamber of shell is partially divided.
013262:   Chambers have chamberlets.
013263:   All shell chambers are in a rectilinear series.
013264:   Shell chambers are loosely attached.
013265:   Shell has a stalk.
013266:   Shell has flagellum passing thru apical pore.
013267:   Shell is red.
013268:   Shell is brown.
013269:   Shell is yellow.
013270:   Shell is calcareous.
013361:   Cell surface has coccoliths.
013271:   Shell is made of cellulose.
013272:   Shell is chitin-like.
013273:   Shell is siliceous (contains silicon).
013274:   Shell is composed of similar sized sand grains.
013275:   Shell is composed of sand grains of mixed size.
013276:   Shell has glassy lustre.
013277:   Shell is flattened.
013278:   Shell is coiled.
013279:   Shell is trochoid.
013280:   Shell is discoid.
013281:   Shell is globular.
013282:   Shell is made of plates.
013283:   Shell is spheroid.
013284:   Shell is smooth.
013285:   Shell is sculptured.
013286:   Shell is oval.
013287:   Shell is cup-like.
013288:   Shell is cylindrical.
013289:   Shell is tubular.
013290:   Shell is campanulate.
013291:   Shell is fusiform.
013292:   Shell has pores.
013293:   Shell is thin.

          -PELLICLE, MUCOUS ENVELOPE, ADHESIVE EXTERNAL
           ORGANELLES, ATTACHMENTS, ETC.

013294:   Mucous envelope is present.
013367:   Extracellular glycoproteins are produced.
013295:   Pellicle is present.
013296:   Pellicle is present and rigid.
013297:   Pellicle is present and flexible.
013298:   Pellicle is striated longitudinally.
013299:   Pellicle is striated spirally.
013300:   Pellicle is striated obliquely.
013301:   Pellicle is granulated.
013302:   Cell has pellicular alveolus.
013303:   Cell has pellicular crest.
013304:   Cell has pellicular pores.
013305:   Hooks are present.
013306:   Cell has adhesive disc.
013307:   Cell has adhesive organelle.
013368:   Pili (fimbriae) are present.
```

013369: Pili have an adhesive property.
013370: Cells adhere to solid surfaces.
014048: Pilus (or pili) has sex or mating function.
013362: Other appendages include haptonema.
013363: Haptonema is shorter than the body length.
013364: Haptonema is equal to the body length.
013365: Haptonema is greater than one but less than two times the body length.
013366: Haptonema is greater than two times the body length.

SECTION 14: MODE OF CELL DIVISION

[NOTE 1: For conidia and sporangia spore formation see Section 7. For arthrospore formation or mycelial fragmentation see Section 8.]

014046:	Cell is prokaryotic.
014047:	Cell is mesokaryotic.
014048:	Cell is eukaryotic.
014001:	Individual cells divide transversely (along the shortest axis) by binary fission.
014045:	Cells reproduce by desmoschisis (vegetative cell division).
014002:	Individual cells divide longitudinally (along the longest axis) by binary fission.
014013:	Exoerythrocytic schizogony occurs.
014014:	Erythrocytic schizogony occurs.
014015:	Schizogony occurs in epitheial cells.
014016:	Gametes develop within a gametocyst.
014017:	Sporogonic cycle occurs.
014018:	Conjugation occurs.
014045:	Pilus (or pili) has sex or mating function.
008617:	Sexual reproduction occurs.
014019:	Cell undergoes autogamy.
014020:	Zygotes are motile.
014021:	Syngamy occurs.
014003:	Cells reproduce by budding directly from mother cell.
014004:	Daughter cells bud on tubular outgrowth from mother cells (also see Section 7).
014012:	One or more daughter cells can form at 1 pole or both poles of the cell.
014008:	Monopolar budding occurs (buds occur only at 1 site on the mother cell).
014011:	Bipolar budding occurs (buds occur only at opposite ends of the mother cell).
014022:	Cell undergoes evaginative budding (evaginogemmy).
014023:	Cell undergoes exogenous budding (exogemmy).
014044:	Reproduction is by vermiform exogenous budding.
014024:	Cell undergoes homothetogenic fission.
014025:	Cell undergoes polygemmic fission.
014026:	Cell undergoes symmetrogenic fission.
014027:	Cell is a karyonide.
014028:	Cell is monogemmic.
014029:	Cell undergoes monotomic division.
014030:	Cell undergoes buccokinetal stomatogenesis.
014031:	Cell undergoes parakinetal stomatogenesis.
014032:	Cell undergoes telokinetal stomatogenesis.
014033:	Cell undergoes apokinetal stomatogensis.
014034:	Cell undergoes strobilations.
014035:	Cell undergoes palintomy.
014036:	Cell is polyenergid.
014037:	Cell has mating type.
014038:	Cell belongs to a syngen.
014039:	Cellular fission results in proter.
014040:	Cell produces tomite.
014041:	Cell produces swarmer.

Section 14: Mode of cell division

014042: Cell produces telotroch.
014043: Cell has phoront stage.
014005: Microfilaments give rise to elementary bodies (round bodies).
008618: Organism is homothallic (male and female on same mycelium).
008619: Organism is heterothallic (male and female on separate mycelium exhibit interfertility).
014009: Interfertility is bi-polar.
014010: Interfertility is tetra-polar.

SECTION 15: ARRANGEMENT

[NOTE 1: For symbiotic arrangements see Section 16.]

015001: Cells occur singly.
015002: Cells occur in pairs.
015003: Cells are arranged in an angular fashion after division (snapping).
015004: Cells occur in chains.
015005: Cells are arranged in irregular aggregates.
015006: Cells are arranged in two-dimensional tetrads.
015007: Cells are arranged in cubical packets (three-dimensional).
015008: Cells are arranged in palisade fashion.
015009: Cells occur in flat sheets.
015010: Cells occur in rosettes.
015011: Cells are connected in a regular open network.
015012: Cells form half-circles.
015013: Curved rods join ends to form a complete ring.
015025: Cells radiate from a common center (i.e., asteroidal arrangement).
015024: Growth tends to be in serpentine, cord-like masses in which the cells show a parallel orientation.
015014: Organisms are multicellular; character may be observed without staining.
015015: Organisms are multicellular; character is revealed only after staining.
015016: Aggregates of cells break up into small clusters.
015017: Organisms are filamentous (i.e., > 10 µm, with little or no indentation). [For branched filaments also see Section 8.]
015023: In filaments the long axis of the individual cell is perpendicular to the long axis of the filament.
015018: Filaments are twisted into bundles.
015019: Filaments lie loosely, longitudinally together in slightly spirally twisted rolls.
015020: Filaments are firmly attached to a substrate.
015021: Filaments are swollen at the tip.
015022: Cells are arranged in cross-hatched configuration (Lieskeella).
015026: Discrete colonies form in liquid medium.
015027: Two to 10 cells are in colony.
015028: Eleven to 20 cells are in colony.
015029: Twenty-one to 30 cells are in colony.
015030: Thirty-one to 40 cells are in colony.
015031: Forty-one to 50 cells are in colony.
015032: Fifty-one to 60 cells are in colony.
015033: Sixty-one to 70 cells are in colony.
015034: Seventy-one to 80 cells are in colony.
015035: Eighty-one to 90 cells are in colony.
015036: Ninety-one to 100 cells are in colony.
015037: More than 100 cells are in colony.
015038: Arrangement of cells forms a colony at end of a stalk.
015039: Arrangement of cells forms a colony at end of branched stalk.

015040: Arrangement of cells forms a colony in gelatinous matrix.
015041: Arrangement of cells forms a colony united laterally.
015042: Arrangement of cells forms a spherical colony.
015043: Arrangement of cells forms an ellipsoidal colony.
015044: Arrangement of cells forms a wheel-like colony.
015045: Arrangement of cells forms a catenoid colony.
015046: Arrangement of cells forms an arboroid colony (dendrite).

SECTION 16: CULTURAL CONDITIONS, INHIBITORS, NUTRITION, GROWTH, AND LIFE CYCLES

SOLID MEDIA

016405: Isolated agar colonies are < 1 mm diameter within 48 hrs.
016406: Isolated agar colonies are 1-2 mm diameter within 48 hrs.
016407: Isolated agar colonies are 2-6 mm diameter within 48 hrs.
016408: Isolated agar colonies are > 6 mm diameter within 48 hrs.
016001: Isolated agar colonies are < 1 mm diameter within a week.
016002: Isolated agar colonies are 1-2 mm diameter within a week.
016003: Isolated agar colonies are 2-6 mm diameter within a week.
016004: Isolated agar colonies are > 6 mm diameter within a week.
016357: Isolated agar colonies are < 1 mm diameter within 10 d.
016358: Isolated agar colonies are 1-2 mm diameter within 10 d.
016359: Isolated agar colonies are 2-6 mm diameter within 10 d.
016360: Isolated agar colonies are > 6 mm diameter within 10 d.
016277: Isolated agar colonies are consistently variable in size upon serial transfers.
016292: Growth from inoculum covers plate within 1 week.
016293: Growth from inoculum covers plate within 2 weeks.
016294: Growth from inoculum covers plate within 3 weeks.
016295: Growth from inoculum covers plate within 4 weeks.
016296: Growth from inoculum covers plate within 5 weeks.
016297: Growth from inoculum covers plate within 6 weeks.
016298: Growth from inoculum covers plate within 7 or more weeks.
016211: Agar macro-colonies are opalescent or iridescent.
016005: Agar macro-colonies are translucent.
016006: Agar macro-colonies are transparent.
016007: Agar macro-colonies are opaque.
016008: Agar macro-colony margin is entire.
016009: Agar macro-colony margin is erose.
016010: Agar macro-colony margin is filamentous (rhizoid).
016011: Agar macro-colony margin is irregular.
016012: Agar macro-colony margin is undulate or lobate.
016361: Agar macro-colony margin is lobate.
016362: Agar macro-colony margin is undulate.
016189: Agar macro-colony is convex.
016013: Agar macro-colony is low convex.
016014: Agar macro-colony is high convex.
016015: Agar macro-colony is convoluted.
016016: Agar macro-colony is flat (membranous).
016018: Agar macro-colony is umbonate.
016439: Agar macro-colony exhibits concentric lines.

Section 16: Cultural conditions

016017: Agar macro-colony is raised but not convex.
016188: Edges of macro-colony are raised higher than center (crater-shape).
016231: Agar micro-colonies are opalescent or iridescent.
016232: Agar micro-colonies are translucent.
016233: Agar micro-colonies are transparent.
016234: Agar micro-colonies are opaque.
016235: Agar micro-colony margin is entire.
016236: Agar micro-colony margin is erose.
016237: Agar micro-colony margin is filamentous (rhizoid).
016238: Agar micro-colony margin is irregular.
016239: Agar micro-colony margin is undulate or lobate.
016441: Agar micro-colony margin is lobate.
016442: Agar micro-colony margin is undulate.
016240: Agar micro-colony is convex.
016241: Agar micro-colony is low convex.
016242: Agar micro-colony is high convex.
016243: Agar micro-colony is convoluted.
016244: Agar micro-colony is flat (membranous).
016247: Agar micro-colony is umbonate.
016440: Agar micro-colony exhibits concentric lines.
016245: Agar micro-colony is raised but not convex.
016246: Edges of micro-colony are raised higher than center (crater-shape).
016019: Colony swarming is exhibited on agar (dispersion of individual members of a population due to active motility).
016363: Colony spreading is exhibited on agar (growth extends several mm or more beyond the point of inoculation).
016213: Growth of colonies result in pitting of agar.
016020: Entire colony moves as a unit (also see Section 13).
016021: Colonial dissociation occurs regularly.
016349: Colony dissociation occurs with a frequency of >1%.
016350: Colonial morphology varies with cultural and nutritional conditions.
016022: Colony consistency is butyrous (soft, buttery).
016023: Colony consistency is viscid (mucoid).
016024: Colony consistency is elastic.
016025: Colony consistency is cartilaginous (rubbery).
016026: Colony consistency is brittle (breaks up into granules).
016027: Colony surface is glistening.
016028: Colony surface is dull (matte).
016029: Colony surface is powdery, dry.
016030: Colony surface is smooth.
016031: Colony surface is rough.
016032: Colorless crystals are formed within colony during growth.
016170: Colony forms an easily dispersible sediment in an exudate.
016033: Mycoplasma, or L-form, colony is formed (also see Section 3).

Section 16: Cultural conditions 103

LIQUID MEDIA

016206: Maximum turbidity in liquid cultures is slight.
016207: Maximum turbidity in liquid cultures is moderate.
016208: Maximum turbidity in liquid cultures is heavy.
[NOTE 1: Opacity tubes are described in: World Health Technical Report Series, 1953, vol. 68, p. 7. The tubes are available from Burroughs Wellcome & Co., London, England).]
016034: Maximum turbidity in liquid cultures approximates Brown's Opacity Tube No. 1.
016035: Maximum turbidity in liquid cultures approximates Brown's Opacity Tube No. 2.
016036: Maximum turbidity in liquid cultures approximates Brown's Opacity Tube No. 3.
016037: Maximum turbidity in liquid cultures approximates Brown's Opacity Tube No. 4.
016038: Maximum turbidity in liquid cultures approximates Brown's Opacity Tube No. 5.
016039: Maximum turbidity in liquid cultures approximates Brown's Opacity Tube No. 6.
016040: Maximum turbidity in liquid cultures approximates Brown's Opacity Tube No. 7.
016203: Maximum turbidity in liquid cultures approximates Brown's Opacity Tube No. 8.
016204: Maximum turbidity in liquid cultures approximates Brown's Opacity Tube No. 9.
016205: Maximum turbidity in liquid cultures approximates Brown's Opacity Tube No. 10.
016190: Turbidity of liquid culture is evenly dispersed.
016041: Cells form an easily dispersible sediment in liquid culture.
016042: Colonies form in bladder-like masses resembling puff-balls in liquids.
016043: Floccular growth occurs in liquid culture.
016351: Granular growth occurs in liquid culture.
016044: Ring growth on the wall of the tube occurs in liquid culture.
016045: Culture grows on walls of container without clouding the medium.
016046: Pellicle is formed in liquid culture.
016365: Growth occurs only as surface mat in liquid culture.
016047: Liquid culture becomes viscous slime.
016166: Growth in liquid culture containing 5% or more sucrose results in gelling of the liquid.
016167: Growth in liquid culture containing 5% or more sucrose results in liquid becoming viscous slime.
016168: Growth in liquid culture containing 5% or more sucrose results in cells adhering to the walls of the container without clouding of the medium.
016169: Growth in liquid culture containing 5% or more sucrose does NOT result in any viscosity change or adherance of cells.

Section 16: Cultural conditions

GROWTH KINETICS

016048: Fastest mean generation time is < 1 hr.
016049: Fastest mean generation time is 1-4 hrs.
016050: Fastest mean generation time is 4-12 hrs.
016051: Fastest mean generation time is > 12 hrs.
016052: Termination of exponential growth phase, under optimal conditions, occurs within 48 hrs.
016451: Fastest mean generation time during autotrophic growth is > 48 hours.
016452: Fastest mean generation time during autotrophic growth is > 24 and < 48 hours.
016453: Fastest mean generation time during autotrophic growth is > 12 and < 24 hours.
016454: Fastest mean generation time during autotrophic growth is > 8 and < 12 hours.
016455: Fastest mean generation time during autotrophic growth is > 4.8 and < 8 hours.
016456: Fastest mean generation time during autotrophic growth is < 4.8 hours.

pH LIMITS OF GROWTH

016194: Growth takes place at an initial pH of 10.0.
016165: Growth takes place at an initial pH of 9.6.
016053: Growth takes place at an initial pH of 9.0.
016352: Growth takes place at an initial pH of 8.5.
016187: Growth takes place at an initial pH of 8.0.
016377: Growth takes place at an initial pH of 7.5.
016054: Growth takes place at an initial pH of 7.0.
016376: Growth takes place at an initial pH of 6.5.
016055: Growth takes place at an initial pH of 6.0.
016288: Growth takes place at an initial pH of 5.7.
016193: Growth takes place at an initial pH of 5.5.
016056: Growth takes place at an initial pH of 5.0.
016375: Growth takes place at an initial pH of 4.8.
016172: Growth takes place at an initial pH of 4.5.
016057: Growth takes place at an initial pH of 4.0.
016058: Growth takes place at an initial pH of 3.5.
016457: Growth takes place at an initial pH of 3.0.
016458: Maximal growth rate occurs at an initial pH of 3.0.
016459: Maximal growth rate occurs at an initial pH of 4.0.
016460: Maximal growth rate occurs at an initial pH of 5.0.
016461: Maximal growth rate occurs at an initial pH of 6.0.
016462: Maximal growth rate occurs at an initial pH of 7.0.
016463: Maximal growth rate occurs at an initial pH of 8.0.

RELATIONSHIP TO OXYGEN AND CARBON DIOXIDE

016209: Amino Acid nutritional requirements vary with oxygen tension.
016059: Growth occurs from loop inoculum spread on surface of solid media incubated in air.
016450: Growth occurs on the surface of solid medium (agar).
016217: Growth takes place only in the presence of oxygen or air (obligate aerobe).

Section 16: Cultural conditions

016218: Growth is prevented by presence of oxygen or air (obligate anaerobe).
016219: Growth takes place either aerobically or anaerobically (facultative).
016220: Growth is dependent on reduced oxygen tension (micro-aerophilic).
016060: In 1.5-2.0% previously solidified agar, inoculated by stab, growth is confined to the surface or a depth from the surface of approximately no greater than 1 mm (i.e., an obligate aerobe).
016061: In 1.5-2.0% previously solidified agar, inoculated by stab, growth takes place at all depths in tube. (facultative)
016062: In 1.5-2.0% previously solidified agar, inoculated by stab, growth begins BELOW THE SURFACE when incubated in air.
016063: In 1.5-2.0% previously solidified agar, inoculated by seeding or by stab, incubated in air, growth is largely confined to a linear dimension of approximately 5 cm from the bottom of the tube in a 16 x 150 mm tube filled with medium to a depth of 9-10 cm (i.e., obligate anaerobe).
016064: Organisms will grow optimally from relatively small inocula (1000 cells or less/ml) only if strict anaerobic operating conditions are observed.
016065: Increased partial carbon dioxide pressure decreases the lag period.
016066: Increased partial carbon dioxide pressure decreases the mean generation time.
016067: Increased partial carbon dioxide pressure increases total cell yield.
016068: Carbon dioxide is fixed (also see Section 34).
016210: Amino Acid nutritional requirements vary with carbon dioxide tension.

LIGHT REQUIREMENTS FOR PHOTOSYNTHESIS

016069: Organisms are photosynthetic.
016070: Regardless of medium or any other experimental condition, growth occurs only in the light.
016071: Growth occurs anaerobically in the light.
016072: Growth occurs anaerobically in the dark.
016073: Growth occurs aerobically in the light.
016074: Growth occurs aerobically in the dark.
016464: Glucose is a carbon source for growth in the dark.
016465: Acetate is a carbon source for growth in the dark.
016466: Glycine is a carbon source for growth in the dark.
016467: Glycerol is a carbon source for growth in the dark.
016468: Growth occurs at a light intensity of 1,000 lux.
016469: Growth occurs at a light intensity of 3,000 lux.
016470: Growth occurs at a light intensity of 10,000 lux.
016471: Growth occurs at a light intensity of 20,000 lux.
016472: Maximal growth rate occurs at a light intensity of 1000 lux.
016473: Maximal growth rate occurs at a light intensity of 3000 lux.

016474: Maximal growth rate occurs at a light intensity of 10,000 lux.
016475: Maximal growth rate occurs at a light intensity of 20,000 lux.

TOLERANCE TO INHIBITORY SUBSTANCES
(also see Sections 18,19/40 and 23)

016289: Organism grows in media containing 0.02% sodium azide.
016086: Organism grows in media containing 0.04% sodium azide.
016286: Organism grows in media containing 0.05% sodium azide.
016378: Organism grows in media containing 0.0005% basic fuchsin.
016344: Organism grows in media containing 0.001% basic fuchsin.
016345: Organism grows in media containing 0.002% basic fuchsin.
016400: Organism grows in media containing 0.003% basic fuchsin.
016077: Organism grows in media containing 0.004% basic fuchsin.
016171: Organism grows in media containing 0.1% benzyl-viologen.
016370: Organism grows in media containing 0.125% bile salts.
016186: Organism grows in media containing 1.25% bile salts.
016164: Organism grows in media containing 2.5% bile salts.
016214: Organism grows in media containing 5% bile salts.
016075: Organism grows in media containing 10% bile salts.
016215: Organism grows in media containing 5% bile salts and 0.1% sodium desoxycholate.
016278: Cells are soluble in 2%-10% sodium desoxycholate.
016283: Organism grows in media containing 0.01% desoxycholate.
016404: Organism grows in media containing 0.1% sodium desoxycholate.
016331: Organism grows in media containing 0.5% sodium desoxycholate.
016332: Organism grows in media containing 1.0% sodium desoxycholate.
016314: Organism grows in media containing 0.01% Biebrich scarlet.
016078: Organism grows in media containing 0.0005% brilliant green.
016381: Organism grows in media containing 0.001% brilliant green.
016195: Organism grows in media containing 0.00125% brilliant green.
016398: Organism grows in media containing 0.0009% cetrimide.
016399: Organism grows in media containing 0.01% cetrimide.
016356: Organism grows in media containing 0.03% cetrimide.
016355: Organism grows in media containing 0.05% cetrimide.

Section 16: Cultural conditions

016284: Organism grows in media containing 0.1% cetrimide.
016185: Organism grows in media containing 0.00004% crystal violet.
016449: Organism grows in media containing 0.0001% crystal violet.
016397: Organism grows in media containing 0.0004% crystal violet.
016079: Organism grows in media containing 0.001% crystal violet.
016382: Organism grows in media containing 0.005% crystal violet.
016083: Organism grows in media containing 0.0075% potassium cyanide.
016371: Organism grows in media containing 0.01% potassium cyanide.
016415: Organism grows in media containing 0.075% potassium cyanide.
016364: Organism grows in media containing 0.1% potassium cyanide.
016394: Organism grows in media containing 0.75% potassium cyanide.
016313: Organism grows in media containing 0.01% eosin yellow.
016317: Organism grows in media containing 1 µg/ml ethambutol.
016318: Organism grows in media containing 2 µg/ml ethambutol.
016319: Organism grows in media containing 5 µg/ml ethambutol.
016373: Organism grows in media containing 15% ethanol.
016320: Organism grows in media containing 10 µg/ml ethionamide.
016321: Organism grows in media containing 20 µg/ml ethionamide.
016322: Organism grows in media containing 40 µg/ml ethionamide.
016260: Organism grows in media containing 62.5 µg/ml hydroxylamine hydrochloride.
016261: Organism grows in media containing 125.0 µg/ml hydroxylamine hydrochloride.
016262: Organism grows in media containing 250.0 µg/ml hydroxylamine hydrochloride.
016263: Organism grows in media containing 500.0 µg/ml hydroxylamine hydrochloride.
016191: Organism grows in media containing 0.001 M iodoacetate.
016192: Acid is produced in media containing 0.001 M iodoacetate.
016267: Organism grows in media containing 0.2 µg/ml isoniazid.
016250: Organism grows in media containing 1 µg/ml isoniazid.
016268: Organism grows in media containing 5.0 µg/ml isoniazid.
016251: Organism grows in media containing 10 µg/ml isoniazid.

Section 16: Cultural conditions

016252: Organism grows in media containing 100 µg/ml isoniazid.
016076: Organism grows in media containing 0.005% sodium lauryl sulfate.
016087: Organism grows in media containing 0.25% lithium ion.
016290: Organism grows in media containing 10 µg/ml of lysozyme.
016173: Organism grows in media containing 0.3% sodium malonate.
016414: Organism grows in media containing 0.000002% malachite green.
016396: Organism grows in media containing 0.00002% malachite green.
016395: Organism grows in media containing 0.0001% malachite green.
016310: Organism grows in media containing 0.01% malachite green.
016311: Organism grows in media containing 0.01% methyl violet.
016368: Organism grows in media containing 0.01% methylene blue.
016080: Organism grows in media containing 0.1% methylene blue.
016340: Organism grows in media containing 0.5% nicotinamide.
016325: Organism grows in media containing 0.01 M nitrite.
016273: Organism grows in media containing 0.0125 M nitrite.
016274: Organism grows in media containing 0.025 M nitrite.
016326: Organism grows in media containing 0.05 M nitrite.
016327: Organism grows in media containing 0.1 M nitrite.
016270: Organism grows in media containing 0.005% oleate.
016089: Organism grows in media containing 0.025% oleate.
016271: Organism grows in media containing 0.05% oleate.
016272: Organism grows in media containing 0.1% oleate.
016323: Organism grows in media containing 0.5 µg/ml paraminosalicylate (PAS).
016253: Organism grows in media containing 1 µg/ml paraminosalicylate (PAS).
016269: Organism grows in media containing 2 µg/ml paraminosalicylate (PAS).
016254: Organism grows in media containing 10 µg/ml paraminosalicylate (PAS).
016324: Organism grows in media containing 100 µg/ml paraminosalicylate (PAS).
016255: Organism grows in media containing 0.2% paraminosalicylate (PAS).
016335: Organism grows in media containing 3000 µg/ml paraminobenzoic acid.
016264: Organism grows in media containing 500 µg/ml paranitrobenzoic acid.
016393: Organism grows in media containing 0.1% phenol.
016258: Organism grows in media containing 0.1% picric acid.

Section 16: Cultural conditions

016259: Organism grows in media containing 0.2% picric acid.
016088: Organism grows in media containing 0.75% potassium ion.
016265: Organism grows in media containing 5% propylene glycol.
016081: Organism grows in media containing 0.001% pyronin.
016312: Organism grows in media containing 0.01% pyronin B.
016329: Organism grows in media containing 0.03% pyronin B.
016330: Organism grows in media containing 0.05% pyronin B.
016256: Organism grows in media containing 0.5 mg/ml sodium salicylate.
016257: Organism grows in media containing 1.0 mg/ml sodium salicylate.
016328: Organism grows in media containing 2.0 mg/ml sodium salicylate.
016084: Organism grows in media containing 0.4% sodium selenite.
016383: Organism grows in media containing 0.1% Teepol (BDH Chem. Ltd; Poole, England).
016384: Organism grows in media containing 0.2% Teepol (BDH Chem. Ltd; Poole, England).
016385: Organism grows in media containing 0.3% Teepol (BDH Chem. Ltd; Poole, England).
016386: Organism grows in media containing 0.4% Teepol (BDH Chem. Ltd; Poole, England).
016387: Organism grows in media containing 0.5% Teepol (BDH Chem. Ltd; Poole, England).
016388: Organism grows in media containing 0.7% Teepol (BDH Chem. Ltd; Poole, England).
016380: Organism grows in media containing 4% sodium taurocholate.
016085: Organism grows in media containing 0.02% potassium tellurite.
016285: Organism grows in media containing 0.3% potassium tellurite.
016281: Organism grows in media containing 0.01% Tergitol 7.
016336: Organism grows in media containing 1 µg/ml thiacetazone (TB-1,TSC).
016337: Organism grows in media containing 2 µg/ml thiacetazone (TB-1,TSC).
016338: Organism grows in media containing 5 µg/ml thiacetazone (TB-1,TSC).
016276: Organism grows in media containing 10 µg/ml thiacetazone (TB-1,TSC).
016339: Organism grows in media containing 100 µg/ml thiacetazone (TB-1,TSC).
016341: Organism grows in media containing 0.001% thionin.
016342: Organism grows in media containing 0.002% thionin.
016082: Organism grows in media containing 0.0033% thionin.
016343: Organism grows in media containing 0.004% thionin.
016315: Organism grows in media containing 1 µg/ml thiophene-2-carboxylic acid hydrazide (TCH,T2H).
016248: Organism grows in media containing 10 µg/ml thiophene-2-carboxylic acid hydrazide (TCH,T2H).

110   Section 16: Cultural conditions

016316: Organism grows in media containing 25 µg/ml
        thiophene-2-carboxylic acid hydrazide (TCH,T2H).
016334: Organism grows in media containing 0.03% toluidine
        blue.
016282: Organism grows in media containing 0.001%
        triphenyltetrazolium chloride (TTC).
016287: Organism grows in media containing 0.01% triphenyl-
        tetrazolium chloride (TTC).
016266: Organism grows in media containing 400 µg/ml
        (0.04%) triphenyltetrazolium chloride(TTC).
016275: Organism grows in media containing 2.5% Tween 80.
016090: Glucose, autoclaved in the medium, inhibits growth.
016091: Peptone or amino acids added to a mineral salts
        medium, inhibits growth.
016163: Growth is inhibited by Optocin (ethyl hydrocuprein
        hydrochloride), disc 5 µg.
016196: Growth is inhibited by mercuric chloride, (disc)
        1.0 µg.
016197: Growth is inhibited by mercuric chloride, (disc)
        10.0 µg.
016348: Growth is inhibited by mercuric chloride, (disc)
        20.0 µg.

    NUTRITIONAL CHARACTERISTICS - GROWTH IN COMPLEX MEDIA

016092: The organisms have been isolated and grown in pure
        (axenic) culture.
016443: The organisms are grown as a monoxenic culture.
016444: The organisms are grown as a polyxenic culture.
016445: The organisms are grown as a xenic culture.
016093: Cells have not been cultivated free of living host
        cells.
016094: Organisms are intracellular parasites of man or
        other vertebrates.
016095: Ascitic fluid, serum, or other body fluid is re-
        quired for growth.
016389: Rumen fluid is required for growth.
016096: Bovine serum albumin plus fatty acid or fatty acid
        ester stimulate growth.
016379: Growth is stimulated by the presence of 0.1% Tween
        80.
016212: At least 1 vitamin (growth factor) is required for
        growth.
016097: Mixtures of vitamins (growth factors), purines or
        pyrimidines, amino acids and fatty acids will
        support growth through 5 or more serial transfers.
016098: Mixtures of vitamins (growth factors), purines or
        pyrimidines, amino acids will support growth
        through 5 or more serial transfers.
016099: Mixtures of vitamins (growth factors), and amino
        acids will support growth through 5 or more serial
        transfers.
016100: Mixtures of amino acids can serve as the source of
        C and N for growth through 5 or more serial trans-
        fers.
016101: A single amino acid can serve as the sole source of

Section 16: Cultural conditions   111

          C and N for growth through 5 or more serial trans-
          fers (also see Section 29).
016367:   Casein can serve as sole source of carbon and
          nitrogen.
016291:   A mixture of meat extract and peptones (e.g.,
          nutrient broth) supports growth through 5 or more
          serial transfers.
016249:   Urea can be used as the sole source of nitrogen.
016346:   Urea can be used as the sole source of carbon.
016347:   Urea can be used as the sole source of carbon and
          nitrogen.
016308:   Growth on autoclaved agar containing malt extract
          results in an intensified brown coloration of the
          media.
016309:   Growth on autoclaved agar containing malt extract
          results in bleaching (decolorization) of the media.
016102:   Optimal growth occurs in sugar-free peptones.
016103:   Added fermentable carbohydrate is required for
          optimal growth.
016104:   Filter-sterilized glucose, when added aseptically
          to sugar-free media, promotes optimal growth;
          without added glucose continued growth is rela-
          tively poor or absent.
016105:   Glucose must be autoclaved in the medium to support
          optimal growth in an otherwise sugar-free medium.
016106:   P-Aminobenzoic acid is required for growth.
016107:   Biotin is required for growth.
016354:   Choline is required for growth.
016108:   Folic acid is required for growth.
016109:   Lipoic acid is required for growth.
016353:   Menadione is required for growth.
016372:   D-Mevalonic acid is required for growth.
016110:   Niacin (nicotinic acid) is required for growth.
016391:   Orotic acid is required for growth.
016111:   Pantothenic acid is required for growth.
016280:   Pantothenic acid can be used as the sole source of
          carbon.
016112:   Pyridoxal or pyridoxamine is required for growth.
016113:   Riboflavin is required for growth.
016114:   Thiamine is required for growth.
016115:   Trigonelline is utilized.
016390:   Tween 80 is required for growth.
016116:   Vitamin B12 (cyanocobalamin) is required for
          growth.
016117:   Hemin (X factor) is required for growth.
016118:   Nicotinamide-Adenine di- (or tri-phosphate)
          nucleotide (V factor) is required for growth.
016119:   Adenine is required for growth.
016120:   Cytosine is required for growth.
016121:   Guanine is required for growth.
016122:   Hypoxanthine is required for growth.
016123:   Thymine is required for growth.
016124:   Uracil is required for growth.
016392:   Uric acid is required for growth.
016125:   Xanthine is required for growth.
016126:   Sterols required for growth.
016369:   Gelling agent (eg., agar) is required for growth.

016127: Acetyl cholesterone is utilized.
016128: Cholesterol (δ-4-cholesterone) is utilized.
016129: Desoxycortisone is utilized.
016130: Digitonin is utilized.
016131: Testosterone is utilized.
016279: Testosterone can be used as the sole source of carbon.

NUTRITIONAL CHARACTERISTICS - GROWTH ON MINERAL SALTS MEDIA

016476: Growth is supported by a chemically defined medium.
016132: Synthetic mineral salts media will support growth in the absence of vitamins (growth factors) and organic nitrogen compounds, pro-vided other sources of carbon, nitrogen and energy are supplied.
016133: Synthetic mineral salts media will support growth in the absence of organic nitrogen compounds, provided other sources of carbon, nitrogen, energy and vitamins (growth factors) are supplied.
016134: Natural sea water is required for growth.
016135: Addition of complex organic compounds (such as in animal and plant tissue extracts) to a mineral salts medium inhibits growth (renders it toxic).
016136: Molecular nitrogen can be used as the sole source of nitrogen.
016137: Ammonium salts can serve as the sole source of nitrogen for growth.
016138: Nitrate can serve as the sole source of nitrogen for growth.
016477: Sodium nitrate as nitrogen source gives a faster growth rate than ammonium chloride.
016139: Nitrite can serve as the sole source of nitrogen for growth.
016140: Molecular hydrogen can serve as the sole source of energy.
016141: An inorganic sulfur compound can serve as the sole source of energy (also see Section 24).
016142: Ammonium ion can serve as the sole source of energy.
016143: Nitrite can serve as the sole source of energy.
016144: Carbon monoxide can be used as the sole source of carbon and energy.
016145: Carbon dioxide can be used as the sole source of carbon.
016146: Methane can be used as the sole source of carbon and energy.
016147: Methanol can be used as the sole source of carbon and energy.

NUTRITIONAL CHARACTERISTICS - INGESTION OF BIOLOGICAL MATERIAL

016416: Cell is omnivorous.
016417: Cell is carnivorous.
016418: Cell is cannibalistic.
016419: Cell is microphagic.

Section 16: Cultural conditions

016420: Cell is phagotropic.
016421: Cell is macrophagic.
016422: Cell is histophagic.
016423: Cell is xylophagous.
016424: Cell is algivorous.

BIOASSOCIATIONS - SYMBIOSIS OR PARASITISM

016148: Bacteria are parasites on other bacteria (*Bdellovibrio*).
016149: Organisms grow in symbiosis.
016150: Organisms grow in symbiosis with bacteria.
016151: Cells grow in symbiosis with short ovoid bacteria in a barrel-shaped arrangement.
016152: Cells grow in symbiosis with large cylindrical bacterial cells.
016153: Cells grow in symbiosis with protozoa.
016154: Cells grow in symbiosis with leguminous plants.
016155: Cells which grow in symbiosis with leguminous plants produce root nodules.
016156: Cells which grow in symbiosis with leguminous plants produce leaf nodules.
016157: Cells grow in symbiosis with non-leguminous plants.
016158: Cells which grow in symbiosis with non-leguminous plants produce root nodules.
016159: Cells which grow in symbiosis with non-leguminous plants produce leaf nodules.
016160: Organisms grow in symbiosis with invertebrate animals.
016425: Cell is mutualistic.
016426: Cells are parasitic.
016427: Parasite is directly transmitted from 1 host to another.

BIOASSOCIATIONS - VECTORS

016161: Insects are vector hosts for the organism.
016162: Arthropods are vector hosts for the organism.
016446: Host of cell is a mollusc.
016447: Host of cell is a sea urchin.
016448: Host of cell is a ciliate.

LIFE CYCLES

016428: Life history is monophasic.
016429: Life history is diphasic.
016430: Life history is polyphasic.
016431: Life history includes amoeboid and flagellate phases.
016478: The cell is benthically attached at some stage in it's life cycle.
016479: The cell is pseudofilamentous (cells loosely held together in uniseriate row without sharing crosswalls) at some stage in it's life cycle.
016480: Cells are planktonic during their entire life.

HABITAT

016432: Cell habitat is littoral.
016433: Cell habitat is cavernicolous.
016434: Cell habitat is neritic.
016435: Cell habitat is sanguicolous.
016436: Cell habitat is inquiline.
016437: Cell habitat is interstitial.
016438: Cell is psammophilic.

DIFFERENTIAL MEDIA

016174: Growth on Triple Sugar Iron Agar results in alkaline (red) slant.
016175: Growth on Triple Sugar Iron Agar results in acid (yellow) slant.
016176: Growth on Triple Sugar Iron Agar results in no change of slant.
016177: Growth on Triple Sugar Iron Agar results in alkaline (red) butt.
016178: Growth on Triple Sugar Iron Agar results in acid (yellow) butt.
016179: Growth on Triple Sugar Iron Agar results in no change in butt.
016180: Growth on Triple Sugar Iron Agar results in gas production.
016181: Growth occurs on MacConkey Agar.
016182: Colonies on MacConkey Agar are brick red in color.
016333: Colonies on MacConkey Agar are yellow or yellowish-pink color.
016183: Growth occurs on Levine Eosine Methylene Blue Agar.
016184: Colonies on Levine Eosine Methylene Blue Agar have a sheen.
016401: Growth occurs on bismuth sulfite agar (Wilson-Blair medium).
016402: Colonies on bismuth sulfite agar (Wilson-Blair medium) are brown to black in color.
016403: Colonies on bismuth sulfite agar (Wilson-Blair medium) have a metallic sheen.
016374: Growth occurs on Mueller Hinton Medium.
016198: Skim milk supports growth.
016410: Coagulation occurs in Iron Milk Medium.
016411: Digestion of milk proteins occurs in Iron Milk Medium.
016412: Gas production occurs in Iron Milk Medium.
016413: Blackening occurs in Iron Milk Medium.
016202: Coagulated egg white is digested.
016366: Coagulated egg yolk is liquified.
016409: Lipids in egg yolk are hydrolyzed.
016199: Growth occurs on Salmonella Shigella Agar.
016216: Colonies on Salmonella-Shigella Agar exhibit blackening.
016221: Growth occurs on Christensen citrate medium.
016222: Growth occurs on Endo agar.
016223: Colonies on Endo agar are red.
016200: Growth occurs on Loefler's Medium.

Section 16: Cultural conditions

016201: Serum digestion occurs on Loefler's Medium.
016224: Growth in Lysine Iron Agar results in alkaline (purple) slant.
016225: Growth in Lysine Iron Agar results in no change of slant.
016226: Growth in Lysine Iron Agar results in acid (yellow) slant.
016227: Growth in Lysine Iron Agar results in alkaline (purple) butt.
016228: Growth in Lysine Iron Agar results in no change of butt.
016229: Growth in Lysine Iron Agar results in acid (yellow) butt.
016230: Growth in Lysine Iron Agar results in gas production.
016299: Growth on Kligler's Iron Agar results in alkaline (red) slant.
016300: Growth on Kligler's Iron Agar results in acid (yellow) slant.
016301: Growth on Kligler's Iron Agar results in no change of slant.
016302: Growth on Kligler's Iron Agar results in alkaline (red) butt.
016303: Growth on Kligler's Iron Agar results in acid (yellow) butt.
016304: Growth on Kligler's Iron Agar results in no change in butt.
016305: Growth on Kligler's Iron Agar results in gas production.
016307: Growth occurs in brilliant green lactose bile broth.
016306: Gas is produced in brilliant green lactose bile broth (confirmatory coliform test).

SECTION 17:   VEGETATIVE CELL TEMPERATURE RELATIONS

017001: The optimal temperature range for growth is 0-10 C.
017002: The optimal temperature range for growth is 11-20 C.
017003: The optimal temperature range for growth is 21-30 C.
017004: The optimal temperature range for growth is 31-40 C.
017005: The optimal temperature range for growth is 41-50 C.
017006: The optimal temperature range for growth is 51-60 C.
017007: The optimal temperature range for growth is 61-70 C.
017008: The optimal temperature range for growth is 71-80 C.
017009: The optimal temperature range for growth is 81-90 C.
017010: The optimal temperature range for growth is 91-100 C.
017081: Maximum cell yield obtained occurs at 5 C.
017082: Maximum cell yield obtained occurs at 10 C.
017083: Maximum cell yield obtained occurs at 15 C.
017084: Maximum cell yield obtained occurs at 20 C.
017085: Maximum cell yield obtained occurs at 25 C.
017086: Maximum cell yield obtained occurs at 30 C.
017087: Maximum cell yield obtained occurs at 35 C.
017011: Growth occurs at 0 C.
017057: Growth occurs at 1 C.
017058: Growth occurs at 2 C.
017038: Growth occurs at 3 C.
017049: Growth occurs at 4 C.
017032: Growth occurs at 5 C.
017059: Growth occurs at 6 C.
017039: Growth occurs at 7 C.
017040: Growth occurs at 8 C.
017060: Growth occurs at 9 C
017012: Growth occurs at 10 C.
017061: Growth occurs at 11 C.
017062: Growth occurs at 12 C.
017063: Growth occurs at 13 C.
017064: Growth occurs at 14 C.
017013: Growth occurs at 15 C.
017065: Growth occurs at 16 C.
017066: Growth occurs at 17 C.
017067: Growth occurs at 18 C.
017068: Growth occurs at 19 C.
017037: Growth occurs at 20 C.
017069: Growth occurs at 21 C.
017036: Growth occurs at 22 C.
017047: Growth occurs at 23 C.
017070: Growth occurs at 24 C.
017014: Growth occurs at 25 C.
017071: Growth occurs at 26 C.
017072: Growth occurs at 27 C.
017048: Growth occurs at 28 C.

Section 17: Vegetative cell temperature relations

```
017073:  Growth occurs at 29 C.
017033:  Growth occurs at 30 C.
017050:  Growth occurs at 31 C.
017041:  Growth occurs at 32 C.
017052:  Growth occurs at 33 C.
017042:  Growth occurs at 34 C.
017034:  Growth occurs at 35 C.
017074:  Growth occurs at 36 C.
017015:  Growth occurs at 37 C.
017075:  Growth occurs at 38 C.
017051:  Growth occurs at 39 C.
017043:  Growth occurs at 40 C.
017016:  Growth occurs at 41 C.
017035:  Growth occurs at 42 C.
017045:  Growth occurs at 43 C.
017044:  Growth occurs at 44 C.
017053:  Growth occurs at 44.5 C.
017017:  Growth occurs at 45 C.
017076:  Growth occurs at 46 C.
017077:  Growth occurs at 47 C.
017078:  Growth occurs at 48 C.
017079:  Growth occurs at 49 C.
017018:  Growth occurs at 50 C.
017080:  Growth occurs at 51 C
017046:  Growth occurs at 52 C.
017019:  Growth occurs at 55 C.
017020:  Growth occurs at 60 C.
017021:  Growth occurs at 65 C.
017022:  Growth occurs at 70 C.
017023:  Growth occurs at 80 C.
017024:  Growth occurs at 90 C.
017025:  Growth occurs at 100 C.
017055:  Organisms survive 55 C for 30 min.
017026:  Organisms survive 60 C for 30 min.
017054:  Organisms survive 60 C for 4 hrs.
017027:  Organisms survive 70 C for 2.5 min.
017028:  Organisms survive 72 C for 15 min.
017029:  Organisms survive 80 C for 10 min.
017030:  Organisms survive 100 C for 10 min.
017031:  Organisms survive 100 C for 30 min.
```

SECTION 18: SODIUM CHLORIDE & OTHER OSMOTIC AGENTS - TOLERANCE AND REQUIREMENTS

SODIUM CHLORIDE

018001: No growth occurs in media containing < 50 mM sodium ion.
018002: No growth occurs in the presence of sodium ion.
018028: Added NaCl is required for growth.
018026: Growth occurs in the presence of 0.1% NaCl.
018035: Growth occurs in the presence of 0.2% NaCl (2 parts/thousand).
018003: Growth occurs in the presence of 0.5% NaCl.
018024: Growth occurs in the presence of 1% NaCl (10 parts/thousasnd).
018025: Growth occurs in the presence of 1.5% NaCl (15 parts/thousand).
018019: Growth occurs in the presence of 2% NaCl (20 parts/thousand).
018004: Growth occurs in the presence of 3% NaCl (30 parts/thousand).
018005: Growth occurs in the presence of 4% NaCl (40 parts/thousand).
018006: Growth occurs in the presence of 5% NaCl (50 parts/thousand).
018021: Growth occurs in the presence of 6% NaCl (60 parts/thousand).
018020: Growth occurs in the presence of 6.5% NaCl (65 parts/thousand).
018007: Growth occurs in the presence of 7% NaCl (70 parts/thousand).
018022: Growth occurs in the presence of 7.5% NaCl (75 parts/thousand).
018023: Growth occurs in the presence of 8% NaCl (80 parts/thousand).
018008: Growth occurs in the presence of 10% NaCl (100 parts/thousand).
018009: Growth occurs in the presence of 15% NaCl (150 parts/thousand).
018010: Growth occurs in the presence of 20% NaCl (200 parts/thousand).
018011: Growth occurs in the presence of 25% NaCl (250 parts/thousand).
018027: Growth occurs in the presence of 32% NaCl (320 parts/thousand).
018012: 0.5%-5% NaCl is essential for growth.
018013: 6-10% NaCl is essential for growth.
018014: 11-15% NaCl is essential for growth.
018015: 16-20% NaCl is essential for growth.
018016: 21-25% NaCl is essential for growth.
018017: More than 25% NaCl is essential for growth.
018036: Maximal growth rate in the presence of 0.1% NaCl (1 part/thousand).
018037: Maximal growth rate in the presence of 0.2% NaCl (2 parts/thousand).

Section 18: Osmotic agents

018038: Maximal growth rate in the presence of 1% NaCl (10 parts/thousand).
018039: Maximal growth rate in the presence of 1.5% NaCl (15 parts/thousand).
018040: Maximal growth rate in the presence of 2% NaCl (20 parts/thousand).
018041: Maximal growth rate in the presence of 3% NaCl (30 parts/thousand).
018018: Synthetic sea salt substitutes for sea water requirement.

OTHER OSMOTIC AGENTS

018029: Growth occurs in the presence of 35% glucose.
018030: Growth occurs in the presence of 40% glucose.
018031: Growth occurs in the presence of 45% glucose.
018032: Growth occurs in the presence of 50% glucose.
018033: Growth occurs in the presence of 55% glucose.
018034: Growth occurs in the presence of 60% glucose. (also see Section 16).

SECTION 19/40: ANTIBIOTIC SENSITIVITY

[NOTE 1: Section 19/40 is *not to be used* for antibiotic DISC sensitivity *relative to clinical applications*. For coding antibiotic DISC sensitivities relative to clinical applications (e.g., Kirby-Bauer Methods) use Section 35.]

[NOTE 2: Two systems of dilutions in liquid media for testing of sensitivity to antibiotics are presented in tabular form, followed by disc sensitivity and special liquid concentrations not fitting in either of the two dilution series.]

[NOTE 3: Concentrations are in µg/ml except for Penicillin G, Bacitracin, and Polymixin B as stated in the antibiotic listing where units denotes units/ml.]

[NOTE 4: The items in the Table have the following forms:

Organism is susceptible to ampicillin concentration in medium greater than 1600 µg/ml.

Organism is susceptible to ampicillin concentration in medium of 0.025 µg/ml.]

CONCENTRATIONS

ANTIBIOTIC	>1600	0.025	0.05	0.1	0.2
Ampicillin	019002	019003	019004	019005	019006
Carbenicillin	019299	019300	019301	019302	019303
Cephaloglycine	019318	019319	019320	019321	019322
Cephaloridine	019337	019338	019339	019340	019341
Cephalothin	019025	019026	019027	019028	019029
Chloromycetin (Chloramphenicol)	019045	019046	019047	019048	019049
Clindamycin	019356	019357	019358	019359	019360
Colistin (Coly-mycin)	019066	019067	019068	019069	019070
Erythromycin (Ilotycin)	019087	019088	019089	019090	019091
Gentamicin	019375	019376	019377	019378	019379
Kanamycin	019107	019108	019109	019110	019111
Lincomycin	019393	019394	019395	019396	019397
Nalidixic Acid	019130	019131	019132	019133	019134
Neomycin (Mycifradin)	019150	019151	019152	019153	019154
Nitrofurantoin (Furadantin/Macrodantin)	019170	019171	019172	019173	019174
Oxacillin	019190	019191	019192	019193	019194
Penicillin G units	019212	019213	019214	019215	019216
Streptomycin or Dihydrostreptomycin	019236	019237	019238	019239	019240
Sulfisoxazole (Gantrisin)	019256	019257	019258	019259	019260

Section 19/40: Antibiotic sensitivity   121

Tetracycline
   (Achromycin)   019276   019277   019278   019279   019280
Vancomycin
   (Vancocyn)   019411   019412   019413   019414   019415

                                 CONCENTRATIONS

ANTIBIOTIC   0.4   0.8   1.6   3

Ampicillin   019007   019008   019009   019010
Carbenicillin   019304   019305   019306   019307
Cephaloglycine   019323   019324   019325   019326
Cephaloridine   019342   019343   019344   019345
Cephalothin   019030   019031   019032   019033
Chloromycetin
   (Chloramphenicol)   019050   019051   019052   019053
Clindamycin   019361   019362   019363   019364
Colistin
   (Coly-mycin)   019071   019072   019073   019074
Erythromycin
   (Ilotycin)   019092   019093   019094   019095
Gentamicin   019380   019381   019382   019383
Kanamycin   019112   019113   019114   019115
Lincomycin   019398   019399   019400   019401
Nalidixic Acid   019135   019136   019137   019138
Neomycin
   (Mycifradin)   019155   019156   019157   019158
Nitrofurantoin (Furadan-
   tin/Macrodantin)   019175   019176   019177   019178
Oxacillin   019195   019196   019197   019198
Penicillin G units   019217   019218   019219   019220
Streptomycin or Dihydro-
   streptomycin   019241   019242   019243   019244
Sulfisoxazole
   (Gantrisin)   019261   019262   019263   019264
Tetracycline
   (Achromycin)   019281   019282   019283   019284
Vancomycin
   (Vancocyn)   019416   019417   019418   019419

                                 CONCENTRATIONS

ANTIBIOTIC   6   12   24   50   100

Ampicillin   019011   019012   019013   019014   019015
Carbenicillin   019308   019309   019310   019311   019312
Cephaloglycine   019327   019328   019329   019330   019331
Cephaloridine   019346   019347   019348   019349   019350
Cephalothin   019034   019035   019036   019037   019038
Chloromycetin
   (Chloramphenicol)   019054   019055   019056   019057   019058
Clindamycin   019365   019366   019367   019368   019369
Colistin
   (Coly-mycin)   019075   019076   019077   019078   019079
Erythromycin
   (Ilotycin)   019096   019097   019098   019099   019100

Section 19/40: Antibiotic sensitivity

Gentamicin	019384	019385	019386	019387	019388
Kanamycin	019116	019117	019118	019119	019120
Lincomycin	019402	019403	019404	019405	019406
Nalidixic Acid	019139	019140	019141	019142	019143
Neomycin (Mycifradin)	019159	019160	019161	019162	019163
Nitrofurantoin (Furadantin/Macrodantin)	019179	019180	019181	019182	019183
Oxacillin	019199	019200	019201	019202	019203
Penicillin G units	019221	019222	019223	019224	019225
Streptomycin or Dihydrostreptomycin	019245	019246	019247	019248	019249
Sulfisoxazole (Gantrisin)	019265	019266	019267	019268	019269
Tetracycline (Achromycin)	019285	019286	019287	019288	019289
Vancomycin (Vancocyn)	019420	019421	019422	019423	019424

CONCENTRATIONS

ANTIBIOTIC	200	400	800	1600
Ampicillin	019016	019017	019018	019019
Carbenicillin	019313	019314	019315	019316
Cephaloglycine	019332	019333	019334	019335
Cephaloridine	019351	019352	019353	019354
Cephalothin	019039	019040	019041	019042
Chloromycetin (Chloramphenicol)	019059	019060	019061	019062
Clindamycin	019370	019371	019372	019373
Colistin (Coly-mycin)	019080	019081	019082	019083
Erythromycin (Ilotycin)	019101	019102	019103	019104
Gentamicin	019389	019390	019391	019392
Kanamycin	019121	019122	019123	019124
Lincomycin	019407	019408	019409	019410
Nalidixic Acid	019144	019145	019146	019147
Neomycin (Mycifradin)	019164	019165	019166	019167
Nitrofurantoin (Furadantin/Macrodantin)	019184	019185	019186	019187
Oxacillin	019204	019205	019206	019207
Penicillin G units	019226	019227	019228	019229
Streptomycin or Dihydrostreptomycin	019250	019251	019252	019253
Sulfisoxazole (Gantrisin)	019270	019271	019272	019273
Tetracycline (Achromycin)	019290	019291	019292	019293
Vancomycin (Vancocyn)	019425	019426	019427	019428

Section 19/40: Antibiotic sensitivity 123

CONCENTRATIONS

ANTIBIOTIC	0.0312	0.0625	0.125	0.25	0.5
Amoxicillin	019512	019513	019514	019515	019516
Ampicillin	019529	019530	019531	019532	019533
Bacitracin units	019546	019547	019548	019549	019550
Capreomycin	019563	019564	019565	019566	019567
Carbenicillin	019580	019581	019582	019583	019584
Cefoxitin	019597	019598	019599	019600	019601
Celesticetin	019614	019615	019616	019617	019618
Cephalexin	019648	019649	019650	019651	019652
Cephaloglycine	019631	019632	019633	019634	019635
Cephaloridine	019665	019666	019667	019668	019669
Cephalothin	019682	019683	019684	019685	019686
Cephapirin	040350	040351	040352	040353	040354
Chloromycetin (Chloramphenicol)	019699	019700	019701	019702	019703
Chlortetracycline (Aureomycin)	019716	019717	019718	019719	019720
Clindamycin	019733	019734	019735	019736	019737
Cloxacillin	019750	019751	019752	019753	019754
Colistin (Coly-mycin)	019767	019768	019769	019770	019771
Dicloxacillin	019784	019785	019786	019787	019788
Erythromycin (Ilotycin)	019801	019802	019803	019804	019805
Fusidic Acid	019818	019819	019820	019821	019822
Gentamicin	019835	019836	019837	019838	019839
Kanamycin	019852	019853	019854	019855	019856
Lincomycin	019869	019870	019871	019872	019873
Methicillin	019886	019887	019888	019889	019890
Metronidazole	019903	019904	019905	019906	019907
Minocycline	019920	019921	019922	019923	019924
Nafcillin	019937	019938	019939	019940	019941
Nalidixic Acid	019954	019955	019956	019957	019958
Neomycin (Mycifradin)	019971	019972	019973	019974	019975
Nitrofurantoin (Furadan- tin/Macrodantin)	040333	040334	040335	040336	040337
Novobiocin (Albamycin)	040006	040007	040008	040009	040010
Oleandomycin	040023	040024	040025	040026	040027
Oxacillin	040040	040041	040042	040043	040044
Oxytetracycline (Tetramycin, Terramycin)	040057	040058	040059	040060	040061
Paromomycin	040074	040075	040076	040077	040078
Penicillin G units	040091	040092	040093	040094	040095
Phenethicillin	040108	040109	040110	040111	040112
Phenoxymethyl Penicillin	040125	040126	040127	040128	040129
Polymyxin B units (Aerosporin)	040142	040143	040144	040145	040146
Rifampin (Rifampicin)	040159	040160	040161	040162	040163
Spectinomycin	040176	040177	040178	040179	040180

Section 19/40: Antibiotic sensitivity

Streptomycin or Dihydro-streptomycin	040193	040194	040195	040196	040197
Subtilin	040210	040211	040212	040213	040214
Sulfamethoxazole/Trimethoprim	040227	040228	040229	040230	040231
Sulfisoxazole (Gantrisin)	040244	040245	040246	040247	040248
Tetracycline (Achromycin)	040261	040262	040263	040264	040265
Thiostrepton	040278	040279	040280	040281	040282
Tobramycin	040295	040296	040297	040298	040299
Vancomycin (Vancocyn)	040312	040313	040314	040315	040316

CONCENTRATIONS

ANTIBIOTIC	1	2	4	8
Amoxicillin	019517	019518	019519	019520
Ampicillin	019534	019535	019536	019537
Bacitracin units	019551	019552	019553	019554
Capreomycin	019568	019569	019570	019571
Carbenicillin	019585	019586	019587	019588
Cefoxitin	019602	019603	019604	019605
Celesticetin	019619	019620	019621	019622
Cephalexin	019653	019654	019655	019656
Cephaloglycine	019636	019637	019638	019639
Cephaloridine	019670	019671	019672	019673
Cephalothin	019687	019688	019689	019690
Cephapirin	040355	040356	040357	040358
Chloromycetin (Chloramphenicol)	019704	019705	019706	019707
Chlortetracycline (Aureomycin)	019721	019722	019723	019724
Clindamycin	019738	019739	019740	019741
Cloxacillin	019755	019756	019757	019758
Colistin (Coly-mycin)	019772	019773	019774	019775
Dicloxacillin	019789	019790	019791	019792
Erythromycin (Ilotycin)	019806	019807	019808	019809
Fusidic Acid	019823	019824	019825	019826
Gentamicin	019840	019841	019842	019843
Kanamycin	019857	019858	019859	019860
Lincomycin	019874	019875	019876	019877
Methicillin	019891	019892	019893	019894
Metronidazole	019908	019909	019910	019911
Minocycline	019925	019926	019927	019928
Nafcillin	019942	019943	019944	019945
Nalidixic Acid	019959	019960	019961	019962
Neomycin (Mycifradin)	019976	019977	019978	019979
Nitrofurantoin (Furadantin/Macrodantin)	040338	040339	040340	040341
Novobiocin (Albamycin)	040011	040012	040013	040014
Oleandomycin	040028	040029	040030	040031

Section 19/40: Antibiotic sensitivity

Oxacillin	040045	040046	040047	040048
Oxytetracycline (Tetramycin, Terramycin)	040062	040063	040064	040065
Paromomycin	040079	040080	040081	040082
Penicillin G units	040096	040097	040098	040099
Phenethicillin	040113	040114	040115	040116
Phenoxymethyl Penicillin	040130	040131	040132	040133
Polymyxin B units (Aerosporin)	040147	040148	040149	040150
Rifampin (Rifampicin)	040164	040165	040166	040167
Spectinomycin	040181	040182	040183	040184
Streptomycin or Dihydro-streptomycin	040198	019510	040200	040201
Subtilin	040215	040216	040217	040218
Sulfamethoxazole/ Trimethoprim	040232	040233	040234	040235
Sulfisoxazole (Gantrisin)	040249	040250	040251	040252
Tetracycline (Achromycin)	040266	040267	040268	040269
Thiostrepton	040283	040284	040285	040286
Tobramycin	040300	040301	040302	040303
Vancomycin (Vancocyn)	040317	040318	040319	040320

CONCENTRATIONS

ANTIBIOTIC	16	32	64	128	256
Amoxicillin	019521	019522	019523	019524	019525
Ampicillin	019538	019539	019540	019541	019542
Bacitracin units	019555	019556	019557	019558	019559
Capreomycin	019572	019573	019574	019575	019576
Carbenicillin	019589	019590	019591	019592	019593
Cefoxitin	019606	019607	019608	019609	019610
Celesticetin	019623	019624	019625	019626	019627
Cephalexin	019657	019658	019659	019660	019661
Cephaloglycine	019640	019641	019642	019643	019644
Cephaloridine	019674	019675	019676	019677	019678
Cephalothin	019691	019692	019693	019694	019695
Cephapirin	040359	040360	040361	040362	040363
Chloromycetin (Chloramphenicol)	019708	019709	019710	019711	019712
Chlortetracycline (Aureomycin)	019725	019726	019727	019728	019729
Clindamycin	019742	019743	019744	019745	019746
Cloxacillin	019759	019760	019761	019762	019763
Colistin (Coly-mycin)	019776	019777	019778	019779	019780
Dicloxacillin	019793	019794	019795	019796	019797
Erythromycin (Ilotycin)	019810	019811	019812	019813	019814
Fusidic Acid	019827	019828	019829	019830	019831
Gentamicin	019844	019845	019846	019847	019848
Kanamycin	019861	019862	019863	019864	019865

Section 19/40: Antibiotic sensitivity

Lincomycin	019878	019879	019880	019881	019882
Methicillin	019895	019896	019897	019898	019899
Metronidazole	019912	019913	019914	019915	019916
Minocycline	019929	019930	019931	019932	019933
Nafcillin	019946	019947	019948	019949	019950
Nalidixic Acid	019963	019964	019965	019966	019967
Neomycin (Mycifradin)	019980	019981	019982	019983	019984
Nitrofurantoin (Furadantin/Macrodantin)	040342	040343	040344	040345	040346
Novobiocin	040015	040016	040017	040018	040019
(Albamycin)	040015	040016	040017	040018	040019
Oleandomycin	040032	040033	040034	040035	040036
Oxacillin	040049	040050	040051	040052	040053
Oxytetracycline (Tetramycin, Terramycin)	040066	040067	040068	040069	040070
Paromomycin	040083	040084	040085	040086	040087
Penicillin G units	040100	040101	040102	040103	040104
Phenethicillin	040117	040118	040119	040120	040121
Phenoxymethyl Penicillin	040134	040135	040136	040137	040138
Polymyxin B units (Aerosporin)	040151	040152	040153	040154	040155
Rifampin (Rifampicin)	040168	040169	040170	040171	040172
Spectinomycin	040185	040186	040187	040188	040189
Streptomycin or Dihydrostreptomycin	040202	040203	040204	040205	040206
Subtilin	040219	040220	040221	040222	040223
Sulfamethoxazole/Trimethoprim	040236	040237	040238	040239	040240
Sulfisoxazole (Gantrisin)	040253	040254	040255	040256	040257
Tetracycline (Achromycin)	040270	040271	040272	040273	040274
Thiostrepton	040287	040288	040289	040290	040291
Tobramycin	040304	040305	040306	040307	040308
Vancomycin (Vancocyn)	040321	040322	040323	040324	040325

CONCENTRATIONS

ANTIBIOTIC	512	1024	2048
Amoxicillin	019526	019527	019528
Ampicillin	019543	019544	019545
Bacitracin units	019560	019561	019562
Capreomycin	019577	019578	019579
Carbenicillin	019594	019595	019596
Cefoxitin	019611	019612	019613
Celesticetin	019628	019629	019630
Cephalexin	019662	019663	019664
Cephaloglycine	019645	019646	019647
Cephaloridine	019679	019680	019681
Cephalothin	019696	019697	019698
Cephapirin	040364	040365	040366
Chloromycetin			

Section 19/40: Antibiotic sensitivity

```
  (Chloramphenicol)    019713   019714   019715
Chlortetracycline
  (Aureomycin)         019730   019731   019732
Clindamycin            019747   019748   019749
Cloxacillin            019764   019765   019766
Colistin
  (Coly-mycin)         019781   019782   019783
Dicloxacillin          019798   019799   019800
Erythromycin
  (Ilotycin)           019815   019816   019817
Fusidic Acid           019832   019833   019834
Gentamicin             019849   019850   019851
Kanamycin              019866   019867   019868
Lincomycin             019883   019884   019885
Methicillin            019900   019901   019902
Metronidazole          019917   019918   019919
Minocycline            019934   019935   019936
Nafcillin              019951   019952   019953
Nalidixic Acid         019968   019969   019970
Neomycin
  (Mycifradin)         019985   019986   019987
Nitrofurantoin (Furadan-
  tin/Macrodantin)     040347   040348   040349
Novobiocin             040020   040021   040022
  (Albamycin)          040020   040021   040022
Oleandomycin           040037   040038   040039
Oxacillin              040054   040055   040056
Oxytetracycline (Tetramycin,
  Terramycin)          040071   040072   040073
Paromomycin            040088   040089   040090
Penicillin G units     040105   040106   040107
Phenethicillin         040122   040123   040124
Phenoxymethyl
  Penicillin           040139   040140   040141
Polymyxin B  units
  (Aerosporin)         040156   040157   040158
Rifampin (Rifampicin)
                       040173   040174   040175
Spectinomycin          040190   040191   040192
Streptomycin or Dihydro-
  streptomycin         040207   040208   040209
Subtilin               040224   040225   040226
Sulfamethoxazole/
  Trimethoprim         040241   040242   040243
Sulfisoxazole
  (Gantrisin)          040258   040259   040260
Tetracycline
  (Achromycin)         040275   040276   040277
Thiostrepton           040292   040293   040294
Tobramycin             040309   040310   040311
Vancomycin
  (Vancocyn)           040326   040327   040328
```

AMPICILLIN

019001: Organism is susceptible to ampicillin concentration (disc) 2 µg.
019429: Organism exhibits intermediate resistance to ampicillin concentration (disc) 2 µg.
019430: Organism is susceptible to ampicillin concentration (disc) 10 µg.
019431: Organism exhibits intermediate resistance to ampicillin concentration (disc) 10 µg.
040199: Organism is susceptible to ampicillin concentration (disc) 30 µg.
040004: Organism exhibits intermediate resistance to ampicillin concentration (disc) 30 µg.

BACITRACIN

019988: Organism is susceptible to bacitracin concentration (disc) 0.02 unit.
019991: Organism exhibits intermediate resistance to bacitracin concentration (disc) 0.02 unit.
019504: Organism is susceptible to bacitracin concentration (disc) 0.04 unit.
019505: Organism exhibits intermediate resistance to bacitracin concentration (disc) .04 unit.
019020: Organism is susceptible to bacitracin concentration (disc) 0.4 unit.
019432: Organism exhibits intermediate resistance to bacitracin concentration (disc) 0.4 unit.
019021: Organism is susceptible to bacitracin concentration (disc) 2 units.
019433: Organism exhibits intermediate resistance to bacitracin concentration (disc) 2 units.
019022: Organism is susceptible to bacitracin concentration (disc) 5 units.
019434: Organism exhibits intermediate resistance to bacitracin concentration (disc) 5 units.
019023: Organism is susceptible to bacitracin concentration (disc) 10 units.
019435: Organism exhibits intermediate resistance to bacitracin concentration (disc) 10 units.

CAPREOMYCIN

040369: Organism is susceptible to capreomycin concentration 10 µg/ml.
040370: Organism is susceptible to capreomycin concentration 20 µg/ml.

CARBENICILLIN

019298: Organism is susceptible to carbenicillin concentration (disc) 50 µg.
019436: Organism exhibits intermediate resistance to carbenicillin concentration (disc) 50 µg.
040381: Organism is susceptible to carbenicillin concentration (disc) 100 µg.

Section 19/40: Antibiotic sensitivity

040382: Organism exhibits intermediate resistance to carbenicillin concentration (disc) 100 µg.

CEPHALOGLYCINE

019317: Organism is susceptible to cephaloglycine concentration (disc) 30 µg.
019437: Organism exhibits intermediate resistance to cephaloglycine concentration (disc) 30 µg.

CEFOXITHIN

040379: Organism is susceptible to cefoxithin concentration (disc) 30 µg.
040380: Organism exhibits intermediate resistance to cefoxithin concentration (disc) 30 µg.

CEPHALORIDINE

019336: Organism is susceptible to cephaloridine concentration (disc) 30 µg.
019438: Organism exhibits intermediate resistance to cephaloridine concentration (disc) 30 µg.

CEPHALOTHIN

019024: Organism is susceptible to cephalothin concentration (disc) 10 µg.
019439: Organism exhibits intermediate resistance to cephalothin concentration (disc) 10 µg.
019440: Organism is susceptible to cephalothin concentration (disc) 30 µg.
019441: Organism exhibits intermediate resistance to cephalothin concentration (disc) 30 µg.

CHLOROMYCETIN (CHLORAMPHENICOL)

019496: Organism is susceptible to chloromycetin (chloramphenicol) concentration (disc) 2.5 µg.
019497: Organism exhibits intermediate resistance to chloromycetin (chloramphenicol) concentration (disc) 2.5 µg.
019043: Organism is susceptible to chloromycetin (chloramphenicol) concentration (disc) 5 µg.
019443: Organism exhibits intermediate resistance to chloromycetin (chloramphenicol) concentration (disc) 5 µg.
019498: Organism is susceptible to chloromycetin (chloramphenicol) concentration (disc) 10 µg.
019499: Organism exhibits intermediate resistance to chloromycetin (chloramphenicol) concentration (disc) 10 µg.
019044: Organism is susceptible to chloromycetin (chloramphenicol) concentration (disc) 30 µg.
019444: Organism exhibits intermediate resistance to chloromycetin (chloramphenicol) concentration (disc) 30 µg.

CHLORTETRACYCLINE (AUREOMYCIN)

019484: Organism is susceptible to chlortetracycline (aureomycin) concentration (disc) 5 µg.
019485: Organism exhibits intermediate resistance to chlortetracycline (aureomycin) concentration (disc) 5 µg.
019063: Organism is susceptible to chlortetracycline (aureomycin) concentration (disc) 30 µg.
019445: Organism exhibits intermediate resistance to chlortetracycline (aureomycin) concentration (disc) 30 µg.

CLINDAMYCIN

019355: Organism is susceptible to clindamycin concentration (disc) 2 µg.
019442: Organism exhibits intermediate resistance to clindamycin concentration (disc) 2 µg.

COLISTIN (COLY-MYCIN)

019064: Organism is susceptible to colistin concentration (disc) 2 µg.
019446: Organism exhibits intermediate resistance to colistin concentration (disc) 2 µg.
019065: Organism is susceptible to colistin concentration (disc) 10 µg.
019447: Organism exhibits intermediate resistance to colistin concentration (disc) 10 µg.

2,4-DIAMINO-6,7-DIISOPROPYLPTERIDINE
(0/129 VIBRIOSTAT)

019084: Organism is susceptible to 2,4-diamino-6,7-diisopropylpteridine (0/129 vibriostat) crystals on agar.
040329: Organism is susceptible to 2,4-diamino-6,7-diisopropylpteridine (0/129 vibriostat) concentration (disc) 20 µg.
040330: Organism is susceptible to 2,4-diamino-6,7-diisopropylpteridine (0/129 vibriostat) concentration (disc) 40 µg.

CYCLOHEXIMIDE (ACTIDIONE)

040389: Organism is susceptible to Cycloheximide (Actidione) concentration in medium of 0.01 µg/ml.
040390: Organism exhibits intermediate resistance to Cycloheximide (Actidione) concentration in medium of 0.01 µg/ml.
040391: Organism is susceptible to Cycloheximide (Actidione) concentration in medium of 0.05 µg/ml.
040392: Organism exhibits intermediate resistance to Cycloheximide (Actidione) concentration in medium of 0.05 µg/ml.

Section 19/40: Antibiotic sensitivity   131

040393: Organism is susceptible to Cycloheximide (Actidione) concentration in medium of 0.1 µg/ml.
040394: Organism exhibits intermediate resistance to Cycloheximide (Actidione) concentration in medium of 0.1 µg/ml.
040395: Organism is susceptible to Cycloheximide (Actidione) concentration in medium of 1 µg/ml.
040396: Organism exhibits intermediate resistance to Cycloheximide (Actidione) concentration in medium of 1 µg/ml.

ERYTHROMYCIN (ILOTYCIN)

019085: Organism is susceptible to erythromycin (ilotycin) concentration (disc) 2 µg.
019448: Organism exhibits intermediate resistance to erythromycin (ilotycin) concentration (disc) 2 µg.
019086: Organism is susceptible to erythromycin (ilotycin) concentration (disc) 15 µg.
019449: Organism exhibits intermediate resistance to erythromycin (ilotycin) concentration (disc) 15 µg.
019506: Organism is susceptible to erythromycin (ilotycin) concentration (disc) 60 µg.
019507: Organism exhibits intermediate resistance to erythromycin (ilotycin) concentration (disc) 60 µg.

GENTAMICIN

019374: Organism is susceptible to gentamicin concentration (disc) 10 µg.
019450: Organism exhibits intermediate resistance to gentamicin concentration (disc) 10 µg.

KANAMYCIN

019105: Organism is susceptible to kanamycin concentration (disc) 5 µg.
019451: Organism exhibits intermediate resistance to kanamycin concentration (disc) 5 µg.
019106: Organism is susceptible to kanamycin concentration (disc) 30 µg.
019452: Organism exhibits intermediate resistance to kanamycin concentration (disc) 30 µg.
019508: Organism is susceptible to kanamycin concentration (disc) 1000 µg.
019509: Organism exhibits intermediate resistance to kanamycin concentration (disc) 1000 µg.

LINCOMYCIN

019125: Organism is susceptible to lincomycin concentration (disc) 2 µg.
019453: Organism exhibits intermediate resistance to lincomycin concentration (disc) 2 µg.

MANDELAMINE (METHENEAMINE MANDELATE)

019126: Organism is susceptible to mandelamine (metheneamine mandelate) concentration (disc) 3 mg.
019454: Organism exhibits intermediate resistance to mandelamine (metheneamine mandelate) concentration
(disc) 3 mg.

METHICILLIN

019127: Organism is susceptible to methicillin concentration (disc) 5 µg.
019455: Organism exhibits intermediate resistance to
methicillin concentration (disc) 5 µg.

NAFCILLIN

019492: Organism is susceptible to nafcillin concentration
(disc) 1 µg.
019493: Organism exhibits intermediate resistance to nafcillin concentration (disc) 1 µg.

NALIDIXIC ACID

019128: Organism is susceptible to nalidixic acid concentration (disc) 5 µg.
019456: Organism exhibits intermediate resistance to
nalidixic acid concentration (disc) 5 µg.
019129: Organism is susceptible to nalidixic acid concentration (disc) 30 µg.
019457: Organism exhibits intermediate resistance to
nalidixic acid concentration (disc) 30 µg.

NEOMYCIN (MYCIFRADIN)

019148: Organism is susceptible to neomycin (mycifradin)
concentration (disc) 5 µg.
019458: Organism exhibits intermediate resistance to neomycin (mycifradin) concentration (disc) 5 µg.
019149: Organism is susceptible to neomycin (mycifradin)
concentration (disc) 30 µg.
019459: Organism exhibits intermediate resistance to neomycin (mycifradin) concentration (disc) 30 µg.

NITROFURANTOIN (FURADANTIN/MACRODANTIN)

040385: Organism is susceptible to nitrofurantoin concentration (disc) 30 µg.
040386: Organism exhibits intermediate resistance to
nitrofurantoin concentration (disc) 30 µg.
019168: Organism is susceptible to nitrofurantoin concentration (disc) 100 µg.
019460: Organism exhibits intermediate resistance to
nitrofurantoin concentration (disc) 100 µg.
019169: Organism is susceptible to nitrofurantoin concentration (disc) 300 µg.
019461: Organism exhibits intermediate resistance to

nitrofurantoin concentration (disc) 300 µg.

NOVOBIOCIN (ALBAMYCIN)

019486: Organism is susceptible to novobiocin (albamycin) concentration (disc) 5 µg.
019487: Organism exhibits intermediate resistance to novobiocin (albamycin) concentration (disc) 5 µg.
019992: Organism is susceptible to novobiocin (albamycin) concentration (disc) 10 µg.
019993: Organism exhibits intermediate resistance to novobiocin (albamycin) concentration (disc) 10 µg.
019188: Organism is susceptible to novobiocin (albamycin) concentration (disc) 30 µg.
019462: Organism exhibits intermediate resistance to novobiocin (albamycin) concentration (disc) 30 µg.

OLEANDOMYCIN

019494: Organism is susceptible to oleandomycin concentration (disc) 15 µg.
019495: Organism exhibits intermediate resistance to oleandomycin concentration (disc) 15 µg.

OXACILLIN

019189: Organism is susceptible to oxacillin concentration (disc) 1 µg.
019463: Organism exhibits intermediate resistance to oxacillin concentration (disc) 1 µg.
019464: Organism is susceptible to oxacillin concentration (disc) 2 µg.
019465: Organism exhibits intermediate resistance to oxacillin concentration (disc) 2 µg.

OXYTETRACYLINE (TETRAMYCIN, TERRAMYCIN)

019502: Organism is susceptible to oxytetracycline (tetramycin, terramycin) concentration (disc) 2.5 µg.
019503: Organism exhibits intermediate resistance to oxytetracycline (tetramycin, terramycin) concentration (disc) 2.5 µg.
040331: Organism is susceptible to oxytetracycline (tetramycin, terramycin) concentration (disc) 5 µg.
040332: Organism exhibits intermediate resistance to oxytetracycline (tetramycin, terramycin) concentration (disc) 5 µg.
019208: Organism is susceptible to oxytetracycline (tetramycin, terramycin) concentration (disc) 30 µg.
019466: Organism exhibits intermediate resistance to oxytetracycline (tetramycin, terramycin) concentration (disc) 30 µg.

PENICILLIN G

019209: Organism is susceptible to penicillin G concentration (disc) 1 unit.

019467: Organism exhibits intermediate resistance to penicillin G concentration (disc) 1 unit.
019210: Organism is susceptible to penicillin G concentration (disc) 2 units.
019468: Organism exhibits intermediate resistance to penicillin G concentration (disc) 2 units.
019994: Organism is susceptible to penicillin G concentration (disc) 2.5 units.
019995: Organism exhibits intermediate resistance to penicillin G concentration (disc) 2.5 units.
040377: Organism is susceptible to penicillin G concentration (disc) 5 units.
040378: Organism exhibits intermediate resistance to penicillin G concentration (disc) 5 units.
019211: Organism is susceptible to penicillin G concentration (disc) 10 units.
019469: Organism exhibits intermediate resistance to penicillin G concentration (disc) 10 units.

POLYMYXIN B (AEROSPORIN)

040375: Organism is susceptible to polymyxin B (aerosporin) concentration (disc) 5 units.
040376: Organism exhibits intermediate resistance to polymyxin B (aerosporin) concentration (disc) 5 units.
019500: Organism is susceptible to polymyxin B (aerosporin) concentration (disc) 30 units.
019501: Organism exhibits intermediate resistance to polymyxin B (aerosporin) concentration (disc) 30 units.
019230: Organism is susceptible to polymyxin B (aerosporin) concentration (disc) 50 units.
019470: Organism exhibits intermediate resistance to polymyxin B (aerosporin) concentration (disc) 50 units.
019231: Organism is susceptible to polymyxin B (aerosporin) concentration (disc) 300 units.
019471: Organism exhibits intermediate resistance to polymyxin B (aerosporin) concentration (disc) 300 units.

PUROMYCIN

019232: Organism is susceptible to puromycin concentration (disc) 200 µg.
019472: Organism exhibits intermediate resistance to puromycin concentration (disc) 200 µg.

RIFAMPIN (RIFAMPICIN)

019989: Organism is susceptible to rifampin (rifampicin) concentration (disc) 15 µg.
019990: Organism exhibits intermediate resistance to rifampin (rifampicin) concentration (disc) 15 µg.
040005: Organism is susceptible to rifampin (rifampicin) concentration (NOT disc) 20 µg/ml.
040367: Organism is susceptible to rifampin (rifampicin) concentration (NOT disc) 40 µg/ml.
040368: Organism is susceptible to rifampin (rifampicin) concentration (NOT disc) 80 µg/ml.

Section 19/40: Antibiotic sensitivity 135

STREPTOMYCIN OR DIHYDROSTREPTOMYCIN

019233: Organism is susceptible to streptomycin concentration (disc) 2.0 µg.
019473: Organism exhibits intermediate resistance to streptomycin concentration (disc) 2.0 µg.
019234: Organism is susceptible to streptomycin concentration (disc) 2.5 µg.
019474: Organism exhibits intermediate resistance to streptomycin concentration (disc) 2.5 µg.
040003: Organism is susceptible to streptomycin concentration (NOT disc) 5 µg/ml.
019511: Organism is susceptible to streptomycin concentration (NOT disc) 10 µg/ml.
019235: Organism is susceptible to streptomycin concentration (disc) 10 µg.
019475: Organism exhibits intermediate resistance to streptomycin concentration (disc) 10 µg.
019996: Organism is susceptible to streptomycin concentration (disc) 30 µg.
019997: Organism exhibits intermediate resistance to streptomycin concentration (disc) 30 µg.
040371: Organism is susceptible to streptomycin concentration (disc) 50 µg.
040372: Organism exhibits intermediate resistance to streptomycin concentration (disc) 50 µg.

SULFADIAZINE

019488: Organism is susceptible to sulfadiazine concentration (disc) 1 mg.
019489: Organism exhibits intermediate resistance to sulfadiazine concentration (disc) 1 mg.

SULFISOXAZOLE (GANTRISIN)

019490: Organism is susceptible to sulfisoxazole (gantrisin) concentration (disc) 5 µg.
019491: Organism exhibits intermediate resistance to sulfisoxazole (gantrisin) concentration (disc) 5 µg.
019254: Organism is susceptible to sulfisoxazole (gantrisin) concentration (disc) 0.1 mg.
019476: Organism exhibits intermediate resistance to sulfisoxazole (gantrisin) concentration (disc) 0.1 mg.
019255: Organism is susceptible to sulfisoxazole (gantrisin) concentration (disc) 0.25 mg.
019477: Organism exhibits intermediate resistance to sulfisoxazole (gantrisin) concentration (disc) 0.25 mg.

TETRACYCLINE (ACHROMYCIN)

019998: Organism is susceptible to tetracycline (achromycin) concentration (disc) 2.5 µg.
019999: Organism exhibits intermediate resistance to tetracycline (achromycin) concentration (disc) 2.5 µg.

019274: Organism is susceptible to tetracycline (achromycin) concentration (disc) 5 µg.
019478: Organism exhibits intermediate resistance to tetracycline (achromycin) concentration (disc) 5 µg.
040001: Organism is susceptible to tetracycline (achromycin) concentration (disc) 10 µg.
040002: Organism exhibits intermediate resistance to tetracycline (achromycin) concentration (disc) 10 µg.
019275: Organism is susceptible to tetracycline (achromycin) concentration (disc) 30 µg.
019479: Organism exhibits intermediate resistance to tetracycline (achromycin) concentration (disc) 30 µg.

TOBRAMYCIN

040383: Organism is susceptible to tobramycin concentration (disc) 10 µg.
040384: Organism exhibits intermediate resistance to tobramycin concentration (disc) 10 µg.

TRIPLE SULFA
(SULFADIAZINE/SULFAMETHAZINE/SULFAMERAZINE)

040373: Organism is susceptible to triple sulfa (sulfadiazine/sulfamethazine/sulfamerazine) concentration (disc) 0.1 mg.
040374: Organism exhibits intermediate resistance to triple sulfa (sulfadiazine/sulfamethazine/sulfamerazine) concentration (disc) 0.1 mg.
040387: Organism is susceptible to triple sulfa (sulfadiazine/sulfamethazine/sulfamerazine) concentration (disc) 0.25 mg.
040388: Organism exhibits intermediate resistance to triple sulfa (sulfadiazine/sulfamethazine/sulfamerazine) concentration (disc) 0.25 mg.
019294: Organism is susceptible to triple sulfa (sulfadiazine/sulfamethazine/sulfamerazine) concentration (disc) 1 mg.
019480: Organism exhibits intermediate resistance to triple sulfa (sulfadiazine/sulfamethazine/sulfamerazine) concentration (disc) 1 mg.

VANCOCYN (VANCOMYCIN)

019295: Organism is susceptible to vancocyn (vancomycin) concentration (disc) 5.0 µg.
019481: Organism exhibits intermediate resistance to vancocyn (vancomycin) concentration (disc) 5.0 µg.
019296: Organism is susceptible to vancocyn (vancomycin) concentration (disc) 7.5 µg.
019482: Organism exhibits intermediate resistance to vancocyn (vancomycin) concentration (disc) 7.5 µg.
019297: Organism is susceptible to vancocyn (vancomycin) concentration (disc) 30.0 µg.
019483: Organism exhibits intermediate resistance to vancocyn (vancomycin) concentration (disc) 30.0 µg.

SECTION 20: PIGMENTS AND ODORS

[NOTE 1: For definition of colors use a standard reference such as Munsel, Color-harmony or Ridgeway standards.]

[NOTE 2: For colors of fruiting bodies, also see Section 7.]

PIGMENTS

020001: Colonies are pure (paper) white on solid medium.
020002: Colonies are gray on solid medium.
020003: Colonies have a characteristic greenish hue as revealed by oblique light.
020059: Fluorescent pigment is observable with Wood's light (long ultraviolet).
020060: Fluorescent pigment is observable with short wavelength ultraviolet light (*ca* 260 nm).
020004: Colonies fluoresce with Wood's light (long ultraviolet).
020058: Colonies fluoresce with short wavelength ultraviolet light (*ca* 260 nm).
020005: Extracellular fluorescent pigment is observed with Wood's light (long ultraviolet).
020006: Extracellular fluorescent pigment is observed with short wavelength ultraviolet light (*ca* 260 nm).
020007: Colonies are luminescent in the dark.
020008: Pigments are produced only in the light (photochromogenicity).
020061: Pigment production is stimulated by light.
020087: Pigment production is independent of light (scotochromogenicity).
020009: Underside of colony is pigmented.
020066: Underside of colony is blue.
020067: Underside of colony is yellow (golden).
020068: Underside of colony is green.
020069: Underside of colony is red.
020070: Underside of colony is orange.
020071: Underside of colony is violet (purple).
020072: Underside of colony is brown.
020073: Underside of colony is black.
020074: Pigment on underside of colony is pH sensitive.
020010: Pigment on underside of colony changes color when the pH is lowered below pH 3.0.
020011: Pigment on underside of colony changes color when the pH is raised above pH 12.0.
020012: Cells contain green photosynthetic pigments.
020013: Cells contain bacteriochlorophyll 770 nm.
020014: Cells contain bacteriochlorophyll 870 nm.
020015: Cells contain Chlorobium chlorophyll 660.
020016: Cells contain Chlorobium chlorophyll 650.
020017: Cells contain chlorophyll a.
020084: Cells contain chlorophyll b.
020085: Cells contain chlorophyll c.
020105: Cells contain chlorophyll d.

Section 20: Pigments and odors

020018: Cells contain or produce pigments other than chlorophylls.
020088: Cells contain phycocyanin.
020089: Cells contain phycoerythrin.
020090: Cells contain α-carotene.
020091: Cells contain β-carotene.
020092: Cells contain echinenone.
020093: Cells contain lutein.
020094: Cells contain zeanthin.
020095: Cells contain antheraxanthin.
020096: Cells contain neoxanthin.
020097: Cells contain fucoxanthin.
020098: Cells contain diatoxanthin.
020099: Cells contain diadinoxanthin.
020100: Cells contain peridinin.
020101: Cells contain alloxanthin.
020102: Cells contain myxoxanthophyll.
020103: Cells contain oxcillaxanthin.
020104: Cells contain violaxanthin.
003111: Cells are green upon microscopic examination.
003112: Cells are blue upon microscopic examination.
003113: Cells are red upon microscopic examination.
020019: Diffusible (water-soluble) pigments are produced.
020020: Diffusible blue pigments are produced.
020021: Diffusible yellow pigments are produced.
020022: Diffusible green pigments are produced.
020023: Diffusible red pigments are produced.
020024: Diffusible orange pigments are produced.
020025: Diffusible violet (purple) pigments are produced.
020026: Diffusible brown pigments are produced.
020027: Diffusible black pigments are produced.
020028: Pyocyanin is produced.
020029: Indochrome is produced.
020030: Fluorescein is produced.
020031: Melanin is produced.
020032: Iodinin is produced (also see Section 5).
020033: Chlorcraphin is produced.
020034: Oxychlororaphin is produced.
020075: Diffusible pigment is pH sensitive.
020035: Diffusible pigment changes color when pH is lowered below 3.0.
020036: Diffusible pigment changes color when pH is raised above 12.0.
020037: Non-diffusible pigments are produced.
020076: Non-diffusible pigment is pH sensitive.
020077: Non-diffusible pigment changes color when pH is lowered below 3.0.
020078: Non-diffusible pigment changes color when pH is raised above 12.0.
020080: Non-diffusible pigment occurs only in the center of the colony.
020081: Non-diffusible pigment occurs in concentric rings within the colony.
020056: Spores on surface of colony are pigmented (non-diffusible).
020038: Non-diffusible red pigments are produced.
020039: Non-diffusible brown pigments are produced.

Section 20: Pigments and odors   139

020057: Non-diffusible black pigments are produced.
020040: Non-diffusible green pigments are produced.
020041: Non-diffusible violet (purple) pigments are
        produced.
020042: Non-diffusible blue pigments are produced.
020043: Non-diffusible golden (yellow) pigments are
        produced.
020044: Non-diffusible orange pigments are produced.
020045: Cells contain carotenoid pigments.
020086: Carotenoid pigments are contained in a plastid.
020046: Cells contain spirilloxanthin (series A,B,C).
020047: Cells contain xanthine.
020048: Cells contain okenone.
020049: Cells contain chlorobactene (A,B).
020050: Cells contain tetrahydroxyspirilloxanthine.
020051: Indigoidine is produced.
020052: Violacein is produced.
020053: Prodigiosin is produced.
020054: Heme pigments are produced.
020055: Heme pigments are coupled to a protein.
020083: Black pigments are produced on laked blood agar.

   ODORS

020062: Detectable odor is present.
020063: Detected odor is fragrant (eg., fruity, win-
        tergreen).
020064: Detected odor is musty.
020065: Detected odor is antiseptic (eg., iodoform).
020079: Detected odor is putrefactive.
020082: Detected odor is reminiscent of ammonia (ammonia-
        cal).

SECTION 21: CELL CONTENTS

CELL BOUND LIPIDS, FATTY ACIDS, AND STEROLS

021204: Lipid is > 15% of cell dry weight.
021205: Sterol is < 0.1% of cell dry weight.
021206: Fatty acid is > 7% of cell dry weight.
021207: The most abundant type of cell associated fatty acids are unsaturated.
021208: The most abundant type of cell associated fatty acids are monounsaturated.
021209: The most abundant type of cell associated fatty acids are diunsaturated.
021210: The most abundant type of cell associated fatty acids are triunsaturated.
021211: The most abundant fatty acid has 14 carbons.
021212: The most abundant fatty acid has 16 carbons.
021213: The most abundant fatty acid has 18 carbons.
021214: The most abundant fatty acid has 20 carbons.
021215: The most abundant fatty acid has 22 carbons.
021216: The cell fatty acid present in largest quantity has 16 carbons and no double bonds (saturated,16:0).
021217: The cell fatty acid present in largest quantity has 16 carbons and 1 double bond (16:1).
021218: The cell fatty acid present in largest quantity has 18 carbons and 1 double bond (18:1).
021219: The cell fatty acid present in largest quantity has 18 carbons and 2 double bonds (18:2).
021220: The cell fatty acid present in largest quantity has 18 carbons and 3 double bonds (18:3).
021221: The cell fatty acid present in largest quantity has 18 carbons and 4 double bonds (18:4).
021222: The cell fatty acid present in largest quantity has 20 carbons and 1 double bond (20:1).
021223: The cell fatty acid present in largest quantity has 20 carbons and 5 double bonds (20:5).
021224: The cell fatty acid present in largest quantity has 22 carbons and 6 double bonds (22:6).
021225: A saturated 12 carbon chain (12:0) is >5% of the cell fatty acids.
021226: A saturated 14 carbon chain (14:0) is >5% of the cell fatty acids.
021227: A monounsaturated 16 carbon chain (16:1) is >5% of the cell fatty acids.
021228: A diunsaturated 16 carbon chain (16:2) is >5% of the cell fatty acids.
021229: A triunsaturated 16 carbon chain (16:3) is >5% of the cell fatty acids.
021230: A 16 carbon chain with 4 double bonds (16:4) is >5% of the cell fatty acids.
021231: A monounsaturated 18 carbon chain (18:1) is >5% of the cell fatty acids.
021232: A diunsaturated 18 carbon chain (18:2) is >5% of the cell fatty acids.
021233: A triunsaturated 18 carbon chain (18:3) is >5% of the cell fatty acids.

Section 21: Cell contents 141

021234: A 18 carbon chain with 4 double bonds (18:4) is >5% of the cell fatty acids.
021235: A monounsaturated 20 carbon chain (20:1) is >5% of the cell fatty acids.
021236: A 20 carbon chain with five double bonds (20:5) is >5% of the cell fatty acids.
021237: A 22 carbon chain with 6 double bonds (22:6) is >5% of the cell fatty acids.
021238: Cells contain sterols.
021239: C-27 sterols are the most abundant of the cell sterols.
021240: C-28 sterols are the most abundant of the cell sterols.
021241: C-29 sterols are the most abundant of the cell sterols.
021242: Sterol produced in largest amount has a single double bond on carbon 5 of the nucleus.
021243: Sterol produced in largest amount has a single double bond on carbon 7 of the nucleus.
021244: Sterol produced in largest amount has double bonds on carbons 5 and 7 of the nucleus.
021245: Sterol produced in largest amount has a single double bond on carbon 22 of the chain.
021246: Sterol produced in largest amount has a single double bond between carbon 24 and 28 on the chain.
021247: Sterol produced in largest amount is saturated.
021248: A 26 carbon sterol is > 2% of the total sterols.
021249: A 27 carbon sterol is > 2% of the total sterols.
021250: A 28 carbon sterol is > 2% of the total sterols.
021251: A 29 carbon sterol is > 2% of the total sterols.
021252: A 30 carbon sterol is > 2% of the total sterols.
021253: A δ-5 sterol is > 2% of the total sterols.
021254: A δ-7 sterol is > 2% of the total sterols.
021255: A δ-5,7 sterol is > 2% of the total sterols.
021256: A δ-5,8 sterol is > 2% of the total sterols.
021257: A δ-8 sterol is > 2% of the total sterols.
021258: A δ-22 sterol is > 2% of the total sterols.
021259: A 24-ethylidene (cis) sterol is > 2% of the total sterols.
021260: A 24-ethylidene (trans) sterol is > 2% of the total sterols.
021261: A 24-methylene sterol is > 2% of the total sterols.
021262: Methyl sterols are > 2% of the total sterols.
021263: The specific sterol present in largest quantity is cholesterol.
021264: The specific sterol present in largest quantity is brassicasterol.
021265: The specific sterol present in largest quantity is 24-methylene cholesterol.
021266: The specific sterol present in largest quantity is δ-8-ergosterol.
021267: The specific sterol present in largest quantity is ergosterol.
021268: The specific sterol present in largest quantity is δ-7,22-ergostadienol.

021269: The specific sterol present in largest quantity is fucosterol.
021270: Isofucosterol is the specific sterol present in the largest quantity.
021271: Clionasterol is the specific sterol present in the largest quantity.
021272: Poriferasterol is the specific sterol present in the largest quantity.
021273: Chondrillasterol is the specific sterol present in the largest quantity.
021274: 7-Dehydroporiferasterol is the specific sterol present in the largest quantity.
021001: Membranes contain lipid.
021002: A disaccharide residue is glycosidically linked to the 1-position of an α,β-diglyceride.
021003: Cell contains α-glucosylglucosyl diglyceride.
021004: Cell contains β-glucosylglucosyl diglyceride.
021005: Cell contains α-galactosylglucosyl diglyceride.
021006: Cell contains β-galactosylglucosyl diglyceride.
021007: Cell contains α-mannosylmannosyl diglyceride.
021008: Phosphatidyl-glycerol with amino acid ester linkages is present.
021009: Mycolic acids (C88 and higher) is present.
021010: C-15 branched acids are present.
021011: Cholesterol is extractable from cells (also see Section 5).
021012: Choline is bound with lipid-like substances in the cell.
021013: Ethanolamine is bound with lipid-like substances in the cell.

[NOTE 1: Items in the Table have the following forms:

Any amount of a 10 carbon saturated fatty acid moiety is cell associated and synthesized *de novo*.

A 10 carbon saturated fatty acid moiety is > 10% of the cell associated fatty acids (by weight) and synthesized *de novo*.

Any amount of a 10 carbon mono-unsaturated fatty acid moiety is cell associated and synthesized *de novo*.

A 10 carbon mono-unsaturated fatty acid moiety is > 10% of the cell associated fatty acids (by weight) and synthesized *de novo*.

Any amount of a 10 carbon di-unsaturated fatty acid moiety is cell associated and synthesized *de novo*.

A 10 carbon di-unsaturated fatty acid moiety is > 10% of the cell associated fatty acids (by weight) and synthesized *de novo*.

Section 21: Cell contents   143

Any amount of a 10 carbon methyl branched (iso-branch) fatty acid moiety is cell associated and synthesized *de novo*.

A 10 carbon methyl branched (iso-branch) fatty acid moiety is > 10% of the cell associated fatty acids (by weight) and synthesized *de novo*.

Any amount of a 10 carbon methyl branched (ante-iso-branch) fatty acid is cell associated and synthesized *de novo*.

A 10 carbon methyl branched (ante-iso-branch) fatty acid is > 10% of the cell associated fatty acids (by weight) and synthesized *de novo*.

Any amount of a 10 carbon hydroxy-substituted fatty acid moiety is cell associated and synthesized *de novo*.

A 10 carbon hydroxy-substituted fatty acid moiety is > 10% of the cell associated fatty acids (by weight) and synthesized *de novo*.]

NUMBER CARBONS	SATURATED ANY	SATURATED >10%	MONO-UNSATURATED ANY	MONO-UNSATURATED >10%	DI-UNSATURATED ANY	DI-UNSATURATED >10%
10	021014	021015	021016	021017	021018	021019
11	021026	021027	021028	021029	021030	021031
12	021038	021039	021040	021041	021042	021043
13	021050	021051	021052	021053	021054	021055
14	021062	021063	021064	021065	021066	021067
15	021074	021075	021076	021077	021078	021079
16	021086	021087	021088	021089	021090	021091
17	021098	021099	021100	021101	021102	021103
18	021110	021111	021112	021113	021114	021115
19	021122	021123	021124	021125	021126	021127
20	021134	021135	021136	021137	021138	021139
21	021146	021147	021148	021149	021150	021151
22	021158	021159	021160	021161	021162	021163
23	021170	021171	021172	021173	021174	021175
24	021182	021183	021184	021185	021186	021187

NUMBER CARBONS	METHYL BRANCHED ISO-BRANCH ANY	METHYL BRANCHED ISO-BRANCH >10%	METHYL BRANCHED ANTI-ISO-BRANCH ANY	METHYL BRANCHED ANTI-ISO-BRANCH >10%	HYDROXY SUBSTITUTED ANY	HYDROXY SUBSTITUTED >10%
10	021020	021021	021022	021023	021024	021025
11	021032	021033	021034	021035	021036	021037
12	021044	021045	021046	021047	021048	021049
13	021056	021057	021058	021059	021060	021061
14	021068	021069	021070	021071	021072	021073
15	021080	021081	021082	021083	021084	021085
16	021092	021093	021094	021095	021096	021097
17	021104	021105	021106	021107	021108	021109

144   Section 21:   Cell contents

          18   021116   021117 | 021118   021119 | 021120   021121
          19   021128   021129 | 021130   021131 | 021132   021133
          20   021140   021141 | 021142   021143 | 021144   021145
          21   021152   021153 | 021154   021155 | 021156   021157
          22   021164   021165 | 021166   021167 | 021168   021169
          23   021176   021177 | 021178   021179 | 021180   021181
          24   021188   021189 | 021190   021191 | 021192   021193

021194:   Any amount of a 17 carbon cyclopropane ring fatty
          acid moiety is cell associated and synthesized *de
          novo*.
021195:   A 17 carbon cyclopropane ring fatty acid moiety is
          > 10% of the cell associated fatty acids (by
          weight) and synthesized *de novo*.
021196:   Any amount of a 19 carbon cyclopropane ring fatty
          acid moiety is cell associated and synthesized *de
          novo*.
021197:   A 19 carbon cyclopropane ring fatty acid moiety is
          > 10% of the cell associated fatty acids (by
          weight) and synthesized *de novo*.
021198:   Any amount of a 16 carbon mono-unsaturated methyl
          branched (iso-branch) fatty acid moiety is cell
          associated and synthesized *de novo*.
021199:   A 16 carbon mono-unsaturated methyl branched (iso-
          branch) fatty acid moiety is > 10% of the cell
          associated fatty acids (by weight) and synthesized
          de novo.
021200:   A 17 carbon mono-unsaturated methyl branched (iso-
          branch) fatty acid moiety is cell associated and
          synthesized *de novo*.
021201:   A 17 carbon mono-unsaturated methyl branched (iso-
          branch) fatty acid moiety is > 10% of the cell
          associated fatty acids (by weight) and synthesized
          de novo.
021202:   Any amount of a 17 carbon mono-unsaturate methyl
          branched (ante-iso-branch) fatty acid moiety is
          cell associated and synthesized *de novo*.
021203:   A 17 carbon mono-unsaturated methyl branched (ante-
          iso-branch) fatty acid moiety is > 10% of the cell
          associated fatty acids (by weight) and synthesized
          de novo.

          CELL CONTENTS - INORGANIC

021308:   Ash is > 10% of cell dry weight.
021309:   Sodium ion is > 1.0% of cell dry weight.
021310:   Potassium ion is > 1.0% of cell dry weight.
021311:   Calcium ion is > 0.3% of cell dry weight.
021312:   Magnesium ion is > 0.4% of cell dry weight.

          CELL CONTENTS - HYDROCARBONS

021275:   Cell hydrocarbons are present.
021276:   More than 40% of cell hydrocarbons are of medium
          carbon chain length (C12-C20).
021277:   More than 40% of cell hydrocarbons are of long
          carbon chain length (C21-C35).

Section 21: Cell contents

021278: A 15 carbon saturated straight chain hydrocarbon is > 10% of the cellular hydrocarbons.
021279: A 17 carbon saturated straight chain hydrocarbon is > 10% of the cellular hydrocarbons.
021280: A 17 carbon mono-unsaturated straight chain hydrocarbon is > 10% of the cellular hydrocarbons.
021281: A 18 carbon saturated branched chain hydrocarbon is > 10% of the cellular hydrocarbons.
021282: A 21 carbon hexa-unsaturated straight chain hydrocarbon is > 10% of the cellular hydrocarbons.
021283: A 23 carbon saturated straight chain hydrocarbon is > 10% of the cellular hydrocarbons.
021284: A 25 carbon saturated straight chain hydrocarbon is > 10% of the cellular hydrocarbons.
021285: A 25 carbon mono-unsaturated straight chain hydrocarbon is > 10% of the cellular hydrocarbons.
021286: A 27 carbon saturated straight chain hydrocarbon is > 10% of the cellular hydrocarbons.
021287: A 27 carbon mono-unsaturated straight chain hydrocarbon is > 10% of the cellular hydrocarbons.

CELL CONTENTS - PROTEINS, PEPTIDES, AMINO ACIDS

021288: Protein is > 40% of cell dry weight.
021289: Glycine is > 5% of the cell amino acids.
021290: Alanine is > 5% of the cell amino acids.
021291: Valine is > 5% of the cell amino acids.
021292: Leucine is > 5% of the cell amino acids.
021293: Isoleucine is > 5% of the cell amino acids.
021294: Serine is > 5% of the cell amino acids.
021295: Threonine is > 5% of the cell amino acids.
021296: Cysteine is > 5% of the cell amino acids.
021297: Cystine is > 5% of the cell amino acids.
021298: Methionine is > 5% of the cell amino acids.
021299: Aspartic acid is > 5% of the cell amino acids.
021300: Glutamic acid is > 5% of the cell amino acids.
021301: Lysine is > 5% of the cell amino acids.
021302: Arginine is > 5% of the cell amino acids.
021303: Histidine is > 5% of the cell amino acids.
021304: Phenylalanine is > 5% of the cell amino acids.
021305: Tyrosine is > 5% of the cell amino acids.
021306: Proline is > 5% of the cell amino acids.
021307: Hydroxyproline is > 5% of the cell amino acids.

SECTION 22: LYSIS

[NOTE 1: Features concerning autolysins are listed in Section 23.]

022001: Majority of cells in a culture lyse spontaneously within 2 weeks.
022002: Cells lyse in distilled water.
022003: Magnesium ions protect against lysis in distilled water.
022004: Divalent cations protect against lysis in distilled water.
022005: Sodium ions protect against lysis in distilled water.
022006: Monovalent cations protect against lysis in distilled water.
022007: Cells are lysed by 10 mg lysozyme per 2-3 g wet weight of cells.
022008: Cells are lysed by 2% (final concentration) of sodium lauryl sulfate (sodium dodecyl sulfate).
022009: Cells are lysed by 10% bile.
022010: Cells are lysed by 40% bile.
022018: Cells are lysed by saponin.
022019: Cells are lysed by lysostaphin.
022011: Cells are lysed by osmotic shock (reduction of at least 10-fold in tonicity).
022012: Cells are lysed by freezing and thawing.
022013: Cells are easily lysed by sonic vibration within 5 min.
022014: Cells are lysed by bacteriophage.
022015: Sphaeroplasts are formed in osmotically protective media, e.g., 20% sucrose.
022016: Incorporation of glycine (0.1M) in media results in enlarged cells and L forms.
022017: Majority of cells are lysed in 5% NaCl after 4 hrs.

SECTION 23: CELL SURFACE (WALL OR MEMBRANE)

CELL SURFACE APPEARANCE

023143: Cell surface is smooth when viewed at the limit of light microscope resolution.
023144: Cell surface is striate (lined) when viewed at the limit of light microscope resolution.
023145: Cell surface is reticulate.
023146: Cell surface is punctate (having minute dots or depressions in the surface).
023147: Cell surface is areolate (divided into many open spaces or chambers).
023148: Cell surface is verrucose (papillae or knobs on the surface).

CELL WALL

023001: Cell wall is present.
023002: In electron micrographs of thin-sections of organisms the cell wall appears as a layer homogeneous in density, 100-800 angstroms thick, extracellular to the plasma membrane.
023003: In electron micrographs of thin-sections of organisms the cell wall appears as a triple-layered structure, 60-100 angstroms thick (outer membrane) separated by 1 or more layers of variable electron density from the plasma membrane.
023004: In electron micrographs of thin-sections of organisms the cell wall appears as a triple-layered structure, 60-100 angstroms thick (outer membrane) separated by 1 or more layers of variable electron density from the plasma membrane, but showing in addition 1 or more layers of definite order of structure extracellular to the outer membrane.
023005: Electron micrographs of negatively stained, freeze-etched or shadowed organisms show a smooth surface structure.
023006: Electron micrographs of negatively stained, freeze-etched or shadowed organisms show a wrinkled surface structure.
023007: Electron micrographs of negatively stained, freeze-etched or shadowed organisms show a regular surface pattern of linear striations.
023008: Electron micrographs of negatively stained, freeze-etched or shadowed organisms show a surface polygonal array of spherical sub-units.
023009: Electron micrographs of negatively stained, freeze-etched or shadowed organisms show a surface criss-cross pattern of fibers, often in pairs.
023010: Cell wall contains peptidoglycan (mucopeptide).
023011: Peptidoglycan represents <10% of the mass of the cell wall.
023012: Peptidoglycan represents >10% of the mass of the cell wall.
023013: The back-bone structure of the cell wall peptido-

glycan consists of alternating β-1,4-linked N-acetylglucosamine and N-acetylmuramic acid residues.

023014: N-acetylmuramic acid residues in the peptidoglycan bear O-acetyl substituents.

023015: The carboxyl groups of the N-acetylmuramic acid residues are substituted with peptide sub-units of the sequence -L-Ala-D-Glu-(diamino)amino acid-D-Ala.

023016: The carboxyl groups of the N-acetylmuramic acid residues are substituted with peptide sub-units of the sequence -L-Ala-D-Glu-α-NH_2 with the glutamic acid residue also linked in the gamma position with diamino amino-D-Ala acid sequence.

023017: The carboxyl groups of the N-acetyl muramic acid residues are substituted with peptide sub-units of the sequence -L-Ala-D-Glu linked in the alpha position of glutamic acid with glycine, and also in the gamma position of glutamic acid with a diamino amino D-Ala acid residue.

023018: The carboxyl groups of the N-acetylmuramic acid residues are substituted with peptide sub-units of the sequence -Gly-D-Glu (with the gamma position of glutamic acid linked to a HomoSer-D-Ala sequence).

023019: The carboxyl groups of the N-acetylmuramic acid residues are substituted with peptide sub-units of the sequence -L-Ser-D-Glu (the gamma position of glutamic acid being linked with a diamino-D-Ala amino acid sequence).

023020: The carboxyl groups of the N-acetylmuramic acid residues are substituted with peptide sub-units of the sequence -L-Ala-Thr-3-hydroxy-Glu(diamino)amino acid-D-Ala.

023134: The diamino-amino acid in the peptide sub-units is DAP.

023021: The diamino-amino acid in the peptide sub-units is DL-DAP.

023022: The diamino-amino acid in the peptide sub-units is LL-DAP.

023023: The diamino-amino acid in the peptide sub-units is DD-DAP.

023024: The diamino-amino acid in the peptide sub-units is L-lysine.

023025: The diamino-amino acid in the peptide sub-units is 2,4-diamino-butyric acid.

023026: The diamino-amino acid in the peptide sub-units is 2,6-diamino-3-hydroxypimelic acid.

023027: The diamino-amino acid in the peptide sub-units is L-ornithine.

023028: The diamino-amino acid in the peptide sub-units is hydroxylysine.

023029: The carboxyl groups of the N-acetylmuramic acid residues are substituted with peptide sub-units of a sequence other than that defined by characters 23015-23020.

Section 23: Cell surface   149

023030:   More than 50% of the peptide sub-units of the
          peptidoglycan are cross-linked.
023031:   Less than 50% of the peptide sub-units of the
          peptidoglycan are cross-linked.
023032:   Cross-linking bridges extend from an amino-group of
          the diamino-amino acid of one peptide sub-unit to
          the C-terminal D-Ala carboxyl of another peptide
          sub-unit.
023033:   Cross-linking bridges extend from the α-carboxyl
          group of the glutamic acid residue of one peptide
          sub-unit to the C-terminal D-Ala carboxyl of
          another peptide sub-unit.
023034:   Cross-linking bridges are direct and do not involve
          additional amino acids.
023035:   Cross-linking bridges are indirect and composed of
          short peptides of glycine and/or L-amino acids.
023036:   Cross-linking bridges are indirect and composed of
          a single D-amino acid.
023037:   Cross-linking bridges are indirect and composed of
          diamino-amino acid.
023038:   Cross-linking bridges are indirect and composed of
          several peptides, joined "head to tail," each
          having the same amino acid sequence as the peptide
          sub-units.
023039:   Cross-linking bridges are indirect and of a struc-
          ture not described by characters 023035 - 023038.
023040:   Teichoic acid is present in the cell wall.
023041:   Wall teichoic acid is a ribitol phosphate polymer.
023042:   Wall teichoic acid is a glycerol phosphate polymer.
023043:   Wall teichoic acid has glycosidic substituents.
023044:   Glycosidic substituents of wall teichoic acid are
          hexoses.
023045:   Glycosidic substituents of wall teichoic acid are
          hexosamines.
023046:   Glycosidic substituents of wall teichoic acid are
          alpha in configuration.
023047:   Glycosidic substituents of wall teichoic acid are
          beta in configuration.
023048:   Wall teichoic acid bears D-alanyl ester sub-
          stituents.
023049:   Wall teichoic acid bears acyl substituents other
          than D-alanine.
023050:   Phosphodiester linkages of wall teichoic acid
          involve alditol primary hydroxyl groups.
023051:   Phosphodiester linkages of wall teichoic acid
          involve primary and secondary hydroxyl groups of
          alditol residues.
023052:   Wall teichoic acids are group precipitinogens.
023053:   Serological activity of wall teichoic acid is
          specific to glycosidic substituents.
023054:   Serological activity of wall teichoic acid is
          specific to acyl substituents.
023055:   Serological activity of wall teichoic acid is
          specific to the alditol phosphate back-bone.
023056:   Wall teichoic acid is implicated as a bacteriophage
          receptor.

023057: Membrane glycerol teichoic acids are present in organism.
023058: Membrane teichoic acid is associated with lipid (lipoteichoic acid).
023059: Membrane teichoic acid has glycosidic substituents.
023060: Glycosidic substituents of membrane teichoic acid are hexoses.
023061: Glycosidic substituents of membrane teichoic acid are hexosamines.
023062: Glycosidic substituents of membrane teichoic acid are oligosaccharides.
023063: Glycosidic substituents of membrane teichoic acid are alpha in configuration.
023064: Glycosidic substituents of membrane teichoic acid are beta in configuration.
023065: Membrane teichoic acids bear D-alanyl ester substituents.
023066: Membrane teichoic acids bear acyl substituents other than D-alanine.
023067: Substitution of membrane glycerol teichoic acid is qualitatively identical to wall teichoic acid.
023068: Membrane teichoic acids are group precipitinogens.
023069: Serological specificity of membrane teichoic acid is specific to its glycosidic substituents.
023070: Serological specificity of membrane teichoic acid is specific to its acyl substituents.
023071: Serological specificity of membrane teichoic acid is specific to its alditol phosphate back-bone.
023072: Cell wall contains polysaccharides.
023073: Cell wall polysaccharide is group-specific precipitinogen.
023074: Cell wall polysaccharide is type-specific precipitinogen.
023075: Cell wall polysaccharide is implicated as a bacteriophage receptor.
023076: Cell wall polysaccharide contains hexose residues.
023077: Cell wall polysaccharide contains hexosamine residues.
023078: Cell wall polysaccharide contains pentose residues.
023079: Cell wall polysaccharide contains uronic acid residues.
023080: Cell wall polysaccharide contains deoxysugar residues.
023123: Cell wall polysaccharide contains 6-deoxyhexose residues.
023135: Cell wall polysaccharide contains 6-deoxytalose residues.
023124: Cell wall polysaccharide contains rhamnose residues.
023125: Cell wall polysaccharide contains galactose residues to < 2% of the cell wall dry weight.
023126: Cell wall polysaccharide contains galactose residues to > 2% of the cell wall dry weight.
023127: Cell wall polysaccharide contains glucose residues to < 2% of the cell wall dry weight.

Section 23: Cell surface

023128: Cell wall polysaccharide contains glucose residues to > 2% of the cell wall dry weight.
023129: Cell wall polysaccharide contains glucosamine residues.
023130: Cell wall polysaccharide contains hexosamine residues other than glucosamine.
023131: Cell wall polysaccharide contains fucose residues.
023132: Cell wall polysaccharide contains xylose residues.
023133: Carbohydrate content of the cell wall is > 20%.
023081: Cell wall polysaccharide contains alditolphosphate residues and is of the type - $(\text{sugar})x\text{-alditol-PO}_4$ --.
023082: Cell wall polysaccharide contains alditol phosphate residues and is of the type - $(\text{sugar-PO}_4\text{-alditol-PO}_4)x$ --.
023083: Cell wall polysaccharide bears alditol phosphate residues as side chain substituents.
023084: Cell wall polysaccharide contains phosphodiester linkages between sugar residues.
023085: Glycosidic linkages in cell wall polysaccharide are predominantly alpha in configuration.
023086: Glycosidic linkages in cell wall polysaccharide are predominantly beta in configuration.
023087: Lipid present in cell wall (particularly phospholipid).
023088: Total lipid of cell wall <3% dry weight.
023089: Lipopolysaccharides present in cell wall.
023090: Lipopolysaccharide can be extracted as an endotoxic complex with protein.
023091: Polysaccharide moiety of lipopolysaccharide contains 2-keto-3-deoxyoctonic acid (KDO).
023092: Polysaccharide of lipopolysaccharide contains heptose.
023093: Polysaccharide of lipopolysaccharide contains hexoses.
023094: Polysaccharide moiety of lipopolysaccharide contains 6-deoxyhexoses.
023095: Polysaccharide moiety of lipopolysaccharide contains 3,6-dideoxyhexoses.
023096: Polysaccharide moiety of lipopolysaccharide contains hexosamines.
023097: Polysaccharide moiety of lipopolysaccharide contains 6-deoxyhexosamines.
023098: Polysaccharide moiety of lipopolysaccharide contains dideoxyhexosamines.
023099: Lipopolysaccharide shows R-antigen specificity (rough variant).
023100: Lipopolysaccharide shows O-antigen specificity (smooth variant).
023101: Lipid component of lipopolysaccharide is <50% dry weight.
023102: Cell wall contains waxes (glycolipids to >5% of the dry weight).
023103: Cell wall contains lipoprotein.
023104: Cell wall contains protein not associated with lipid.
023105: Cell wall protein is antigenic.

023106: Cell wall protein shows type-specific antigenicity.
023107: Cell wall protein is associated with virulence.
023108: Cell wall protein appears to have "structural" role.
023109: Cell wall protein has enzymic activity.
023110: Cell walls contain autolysins.
023111: Autolysin has endo-N-acetylmuramidase activity.
023112: Autolysin has endo-N-acetylglucosaminidase activity.
023113: Autolysin has N-acetylmuramyl-L-alanine amidase activity.
023114: Autolysin has endopeptidase activity.
023115: Lytic enzymes convert whole organisms into protoplasts.
023116: Lytic enzymes convert whole organisms into sphaeroplasts.
023117: Cell wall synthesis is inhibited by penicillins.
023118: Cell wall synthesis is inhibited by cephalosporins.
023119: Cell wall synthesis is inhibited by vancomycin.
023120: Cell wall synthesis is inhibited by ristocetin.
023121: Cell wall synthesis is inhibited by bacitracin.
023122: Cell wall synthesis is inhibited by D-cycloserine or O-carbamyl-D-serine.

SURFACE STRUCTURE - OTHER THAN CELL WALLS

023136: Cell has pellicular alveolus.
023137: Cell has argyrome system (silverline system).
023138: Cell has perilemma.
023139: Cell has postoral suture.
023140: Cell has preoral suture.
023141: Cell has ribbed wall.
023142: Cell has suture lines.

SECTION 24: METABOLIC REACTIONS

CATABOLIC END PRODUCTS

[NOTE 1: The items in the Table have the following forms:

Acetic acid is produced from complex basal medium (eg., peptone-yeast extract broth).

Acetic acid is produced from D-fructose.

Acetic acid is produced from D-glucose.

Acetic acid is produced from lactose.

Acetic acid is produced from D,L-lactate.

Acetic acid is produced from pyruvate.

Acetic acid is produced from L-threonine.]

PRODUCTS \ SUBSTRATES	COMPLEX BASAL MEDIUM	D-FRUCTOSE	D-GLUCOSE	LACTOSE
FATTY ACIDS				
Acetic acid	024353	024389	024016	024216
Acetic acid without carbon dioxide	024354	024390	024017	024217
Butyric acid	024355	024391	024018	024218
iso-Butyric acid	024356	024392	024019	024219
Caproic acid	024357	024393	024020	024220
iso-Caproic acid	024358	024394	024021	024221
Caprylic acid	024359	024395	024022	024222
Formic acid	024360	024396	024023	024223
Fumaric acid	024361	024397	024211	024224
Oenanthic (Heptanoic) acid	024362	024398	024024	024225
Propionic acid	024363	024399	024025	024226
Valeric acid	024364	024400	024026	024227
iso-Valeric acid	024365	024401	024027	024228
KETO- AND HYDROXY- ACIDS				
Any form of lactic acid	024366	024402	024257	024258
Lactic acid (D-)	024367	024403	024028	024229
Lactic acid (L+)	024368	024404	024029	024230
Lactic acid (DL)	024369	024405	024030	024231

α-Ketobutyric acid	024432	024431	024430	024429
Gluconic acid	024370	024406	024031	024232
2-Ketogluconic acid	024371	024407	024032	024233
Pyruvic acid	024456	024457	024458	024459

DICARBOXYLIC ACID

Succinic acid	024372	024408	024033	024234

ALCOHOLS AND KETONE

Acetone	024373	024409	024034	024235
Acetylmethylcarbinol	024374	024410	024035	024236
2,3-Butanediol	024375	024411	024036	024237
Butanol	024376	024412	024037	024238
iso-Butanol	024377	024413	024038	024239
Ethanol	024378	024414	024039	024240
Glycerol	024379	024415	024040	024241
Hexanol	024380	024416	024041	024242
Pentanol	024381	024417	024042	024243
iso-Pentanol	024435	024436	024437	024438
Propanol	024382	024418	024043	024244
iso-Propanol	024383	024419	024044	024245

GASES

Carbon dioxide	024384	024420	024045	024246
Hydrogen gas	024385	024421	024046	024247

	SUBSTRATES	D,L-LACTATE	PYRUVATE	L-THREONINE
PRODUCTS				

FATTY ACIDS

Acetic acid	024050	024281	024317
Acetic acid without carbon dioxide	024259	024282	024318
Butyric acid	024051	024283	024111
iso-Butyric acid	024260	024284	024320
Caproic acid	024052	024285	024321
iso-Caproic acid	024261	024286	024322
Caprylic acid	024262	024287	024323
Formic acid	024263	024288	024324
Fumaric acid	024264	024289	024325
Oenanthic (Heptanoic) acid	024265	024290	024326
Propionic acid	024055	024291	024113

Section 24: Metabolic reactions   155

```
Valeric acid                  024057    024292    024328
iso-Valeric acid              024266    024293    024329
```

KETO- AND HYDROXY- ACIDS

```
Any form of lactic acid                 024294    024330
Lactic acid (D-)                        024295    024331
Lactic acid (L+)                        024296    024332
Lactic acid (DL)                        024297    024333
α-Ketobutyric acid            024428    024427    024112
Gluconic acid                           024298    024334
2-Ketogluconic acid                     024299    024335
Pyruvic acid                  024460              024461
```

DICARBOXYLIC ACID

```
Succinic acid                 024056    024300    024336
```

ALCOHOLS AND KETONES

```
Acetone                       024267    024301    024337
Acetylmethylcarbinol          024268    024302    024338
2,3-Butanediol                024269    024303    024339
Butanol                       024270    024304    024340
iso-Butanol                   024271    024305    024341
Ethanol                       024272    024306    024342
Glycerol                      024273    024307    024343
Hexanol                       024274    024308    024344
Pentanol                      024275    024309    024345
iso-Pentanol                  024439    024440    024441
Propanol                      024276    024310    024346
iso-Propanol                  024277    024311    024347
```

GASES

```
Carbon dioxide                024053    024312    024319
Hydrogen gas                  024054    024313    024327
```

EXTRACELLULAR POLYSACCHARIDES

024001: Cellulose is produced.
024252: Extracellular starch-like materials are produced.
024002: Extracellular dextrans are synthesized from
        sucrose.
024003: Extracellular levans are synthesized from sucrose.
013343: Extracellular glycoproteins are produced.

COMPLEX ORGANIC SUBSTANCES

024004: Agar is hydrolyzed (liquefied).
024005: Carrageenin is degraded.
024006: Sterile wedges of carrot are softened.
024007: Casein is hydrolyzed (peptonized).
024008: Collagen is degraded.
024009: Gelatin is hydrolyzed (liquefied).
024010: Keratin is degraded.

Section 24: Metabolic reactions

024011: Pectin is hydrolyzed.
024012: Sterile wedges of potato are softened.

COMPLEX BASAL MEDIUM CATABOLISM (E.G., PEPTONE YEAST EXTRACT BROTH)

024350: Complex basal medium is catabolized.
024351: Complex basal medium is catabolized aerobically.
024352: Complex basal medium is catabolized anaerobically.

FRUCTOSE CATABOLISM

024386: D-Fructose is catabolized.
024387: D-Fructose is catabolized aerobically.
024388: D-Fructose is catabolized anaerobically.
024124: D-Fructose is reduced to mannitol.

GLUCOSE CATABOLISM

024013: D-Glucose is catabolized.
024014: D-Glucose is catabolized aerobically.
024015: D-Glucose is catabolized anaerobically.
024253: In media containing D-Glucose plus an organic nitrogen source (eg., peptones, amino acids, etc.) an alkaline reaction occurs.
024254: An alkaline reaction occurs in media containing D-Glucose plus an inorganic nitrogen source.

LACTOSE CATABOLISM

024213: Lactose is catabolized.
024214: Lactose is catabolized aerobically.
024215: Lactose is catabolized anaerobically.
024125: Lactose is oxidized to 3-ketolactose.

LACTIC ACID CATABOLISM

024047: Lactate is catabolized.
024048: Lactate is catabolized aerobically.
024049: Lactate is catabolized anaerobically.

PYRUVIC ACID CATABOLISM

024278: Pyruvate is catabolized.
024279: Pyruvate is catabolized aerobically.
024280: Pyruvate is catabolized anaerobically.
024067: Pyruvate is decomposed, yielding acetate, hydrogen and carbon dioxide.

THREONINE CATABOLISM

024314: L-Threonine is catabolized.
024315: L-Threonine is catabolized aerobically.
024316: L-Threonine is catabolized anaerobically.

Section 24: Metabolic reactions

SYNTHESIS OF SPECIFIC ORGANIC COMPOUNDS

024424: Cells produce niacin which diffuses into the medium.

CATABOLISM OF ORGANIC COMPOUNDS

-ACIDS

024453: Acetate yields carbon dioxide.
024058: Acrylate reduction yields propionate.
024059: α-Crotonic acid is utilized.
024060: Formate is decomposed to hydrogen and carbon dioxide.
024061: Glutaric acid salts is utilized.
024452: Gluconate yields acetate.
024062: Gluconate yields reducing compounds (i.e., containing carbonyl groups).
024063: Gluconate yields 2-ketogluconate.
024064: Gluconate yields 5-ketogluconate.
024065: Gluconate yields 2,5-ketogluconate.
024066: α-Ketobutyrate yields glycine.
024068: Succinate yields propionate.
024069: Succinate yields carbon dioxide.

-ALCOHOLS (INCLUDING AROMATIC)

024070: Acetone is reduced to isopropyl alcohol.
024071: Catechol is cleaved by meta-cleavage.
024072: Catechol is cleaved by ortho-cleavage.
024073: Ethanol yields acetic acid.
024074: Ethanol is reduced to caproic acid.
024075: Glycerol yields dihydroxyacetone.
024076: Glycerol yields acrolein.
024451: D-Mannitol yields acid anaerobically.
024077: Phenol is oxidized.
024078: Protocatechuate is cleaved by meta-cleavage.
024079: Protocatechuate is cleaved by ortho-cleavage.

-AMINO ACIDS AND PEPTONES

024080: Acid is produced from carbohydrate-free media containing amino acids or peptones.
024081: Gas is produced from carbohydrate-free media containing amino acids or peptones.
024082: The Stickland reaction occurs in which mutual oxidation-reduction between 2 amino acids (one acting as a hydrogen donor and one acting as a hydrogen receptor) results in the formation of a keto acid, a fatty acid, and ammonia.
024083: Alanine yields acetate.
024084: Alanine yields propionate.
024454: α-Aminolevulinate yields porphyrins.
024212: L-Arginine utilization results in basic endproducts (medium becomes alkaline).
024085: Arginine yields citrulline.
024433: Arginine yields ornithine.

024434: Arginine yields putrescine.
024086: Aspartate yields acetate.
024087: Aspartate yields ethanol.
024088: Aspartate yields lactate.
024089: Citrulline yields ornithine.
024090: Glycine yields acetate.
024091: Glutamate yields acetate.
024092: Glutamate yields butyrate.
024093: Glutamate yields hydrogen.
024094: Histidine yields acetate.
024095: Histidine yields butyrate.
024096: Histidine yields formamide.
024097: Histidine yields formic acid.
024098: Histidine yields molecular hydrogen.
024099: Histidine yields lactate.
024100: Lysine yields acetate.
024101: Lysine yields butyrate.
024447: Phenylalanine yields phenylpyruvate.
024102: Serine yields acetate.
024103: Serine yields butyrate.
024104: Serine yields ethanol.
024105: Serine yields formate.
024106: Serine yields molecular hydrogen.
024107: Serine yields lactate.
024108: Serine yields propionate.
024109: Serine yields pyruvate.
024110: Serine yields valerate.
024114: Tryptophan yields indole.
024115: Tryptophan yields skatol.
024116: Tryptophan yields methylindole.
024117: Tryptophan yields pyruvate.
024118: Tyrosine yields phenol.
024119: Tyrosine yields p-cresol.
024120: Tyrosine yields p-hydroxyphenylacetic acid.
024121: Tyrosine yields p-hydroxyphenyllactic acid.
024209: Hippuric acid hydrolysis yields benzoic acid.

-CARBOHYDRATES

024122: D-Arabinose is oxidized to D-arabonic acid.
024123: L-Arabinose is oxidized to L-arabonic acid.
024126: D-Ribose is oxidized to D-ribonic acid.
024127: D-Xylose is oxidized to D-xylonic acid.

-MISCELLANEOUS

024426: Paraminosalicylate (PAS) is degraded.
024445: Salicylate is degraded.
024128: Methane is produced.
024129: Trimethylamine oxide is reduced to trimethylamine.
024130: Uric acid salts are utilized.
024131: Alkylamines are oxidized.
024132: Indole is oxidized to indigotin.

Section 24: Metabolic reactions

INORGANIC PRODUCTS OR SUBSTRATES

-AMMONIA, NITRATE, NITRITE, NITROGEN

024133: Ammonium ion is oxidized.
024134: Ammonium thiocyanate serves as a sole carbon and nitrogen source for growth and is oxidized to ammonium sulfate and carbon dioxide.
024135: Ammonia is produced.
024136: Ammonia is produced from peptones.
024137: Nitrite is oxidized.
024138: Nitrate is reduced.
024139: Nitrate is reduced to nitrite.
024442: Nitrate is reduced to nitrogen gas.
024449: Nitrate is reduced to nitric oxide.
024450: Nitrate is reduced to ammonia.
024210: Nitrite is reduced.
024140: Nitrite is reduced to nitrogen gas.
024448: Nitrite is reduced to nitrous oxide.
024141: Nitrite is reduced to ammonia.
024142: Nitrite is reduced to hydroxylamine.
024143: Molecular nitrogen is utilized.

-SULFUR COMPOUNDS

024144: Hydrogen sulfide is oxidized.
024145: Thiosulfate is oxidized.
024146: Tetrathionate is oxidized.
024147: Trithionate is oxidized.
024148: Hydrogen sulfide is produced.
024149: Thiosulfate is reduced to hydrogen sulfide.
024150: Sulfate is reduced to hydrogen sulfide.
024151: Sulfate is reduced to sulfite.
024152: Sulfite is reduced to hydrogen sulfide.
024153: Elementary sulfur is reduced to hydrogen sulfide.
024154: Hydrogen sulfide is produced from cysteine.
024155: Hydrogen sulfide is produced from cystine.
024156: Hydrogen sulfide is produced from reduced glutathione.
024157: Hydrogen sulfide is produced from oxidized glutathione.
024158: Hydrogen sulfide is produced from methionine.
024251: Hydrogen sulfide is produced from peptones.
024159: Elementary sulfur is oxidized.
024160: Sulfite is oxidized.
024161: Tetrathionate is reduced.

-OXYGEN, CARBON DIOXIDE, HYDROGEN, MISCELLANEOUS

024162: Molecular oxygen is utilized.
024163: Molecular oxygen is produced.
024455: Hydrogen peroxide is produced.
024164: Hydrogen peroxide is decomposed.
024425: Hydrogen peroxide decomposing activity is resistant to 68 C for 20 min.
024165: Molecular hydrogen is oxidized.

024166: Molecular hydrogen is oxidized to hydrogen peroxide.
024167: Molecular hydrogen is oxidized to water.
024168: Molecular hydrogen is produced.
024169: Hydrogen gas is specifically produced from the nitrogenous constituents of peptones.
024170: Carbon dioxide is produced.
024171: Carbon monoxide is oxidized to carbon dioxide.
024172: Carbon dioxide is specifically produced from the nitrogenous constituents of peptones.
024173: Ferrous iron is oxidized.
024174: Manganous compounds are oxidized in the absence of citrate.
024175: Tellurite is reduced.
024176: Selenite is reduced.
024177: Methane is oxidized to carbon dioxide and water.
024178: Methanol is oxidized to carbon dioxide and water.
024446: Ethanol is oxidized to carbon dioxide and water.

DYE REDUCTION AND INDICATOR TESTS

024179: Benzylviologen is reduced.
024348: Acid is produced in skim milk.
024349: Skim milk is coagulated.
024422: Gas is produced in skim milk.
024180: Litmus milk is acid.
024249: Litmus milk changes from acid to alkaline.
024181: Litmus milk is coagulated.
024182: Litmus milk is alkaline.
024250: Litmus milk changes from alkaline to acid.
024183: Litmus milk is peptonized.
024423: Gas is produced in skim miilk.
024184: Litmus is reduced.
024443: Cholera red reaction is positive (the addition of sulfuric acid to a nitrate peptone broth results in a red colorization).
024185: Methyl red test is positive.
024206: Benzidine test is positive.
024186: Methylene blue is reduced.
024462: Neutral red is reduced.
024255: Resazurin is reduced.
024187: Resazurin is reduced to resorufin.
024188: Resorufin is reduced.
024189: One or more sulfonphthalein indicators is reduced (bromthymol blue, phenol red, etc.).
024190: Tetrazolium dyes are reduced.
024191: Voges-Proskauer test is positive (also see feature number 24035).
024248: Kovacs' oxidase test is positive (smear from colony turns dark purple with tetramethylparaphenylenediamine dihydrochloride).
024444: Colony develops a blue coloration immediately upon application of an alcoholic solution of guaicum (gum guaiac in ethyl alcohol).

Section 24: Metabolic reactions

TOXIN PRODUCTION AND HEMOLYSIS

024192: A neurotropic toxin is produced.
024193: A diffusible toxin (exotoxin) is produced.
024194: A toxin is liberated into the medium only if cell autolysis occurs.
024195: An enterotoxin is produced.
024196: A hemolytic toxin is produced.
024256: A clostridial alpha-toxin is produced (Nagler reaction).
024197: Cells are hemolytic.
024198: Sheep blood hemolysis is alpha.
024199: Sheep blood hemolysis is beta.
024200: Ox blood hemolysis is alpha.
024201: Ox blood hemolysis is beta.
024202: Horse blood hemolysis is alpha.
024203: Horse blood hemolysis is beta.
024204: Human blood hemolysis is alpha.
024205: Human blood hemolysis is beta.
024207: Rabbit blood hemolysis is alpha.
024208: Rabbit blood hemolysis is beta.

SECTION 25: CARBOHYDRATE METABOLISM

[NOTE 1: For all feature numbers concerning D-Glucose, see also Section 24.]

[NOTE 2: The items in the Table have the following forms:

Dihydroxyacetone is utilized.

Dihydroxyacetone is oxidized.

Dihydroxyacetone is reduced.

Acid is produced from Dihydroxyacetone.

Gas is produced from Dihydroxyacetone.

Dihydroxyacetone can be used as the sole source of carbon.

Aesculin is hydrolyzed.]

SUBSTRATE	UTILIZED	OXIDIZED	REDUCED	ACID
At least 1 carbohydrate	025001	025059	025117	025175
MONOSACCHARIDES				
-TRIOSES				
Dihydroxyacetone	025002	025060	025118	025176
Glyceraldehyde	025003	025061	025119	025177
-TETROSES				
D-Erythrose	025381	025382	025383	025384
L-Erythrulose	025004	025062	025120	025178
D-Threose	025005	025063	025121	025179
-PENTOSES				
D-Arabinose	025006	025064	025122	025180
L-Arabinose	025007	025065	025123	025181
Deoxyribose	025374	025375	025376	025377
D-Lyxose	025008	025066	025124	025182
L-Lyxose	025009	025067	025125	025183
D-Ribose	025010	025068	025126	025184
D-Ribulose	025011	025069	025127	025185
D-Xylose	025012	025070	025128	025186
D-Xylulose	025013	025071	025129	025187

Section 25: Carbohydrate metabolism

-METHYL PENTOSES

D-Fucose	025014	025072	025130	025188
L-Fucose	025015	025073	025131	025189
D-Rhamnose	025016	025074	025132	025190
L-Rhamnose	025017	025075	025133	025191

-HEXOSES

D-Allose	025018	025076	025134	025192
D-Fructose	025019	025077	025135	025193
D-Galactose	025020	025078	025136	025194
D-Glucose	025021	025079	025137	025195
D-Mannose	025022	025080	025138	025196
L-Sorbose	025023	025081	025139	025197

-HEPTOSES

α-D-Galaheptose	025024	025082	025140	025198
β-D-Galaheptose	025025	025083	025141	025199
α-D-Glucoheptose	025026	025084	025142	025200
Sedoheptulose	025027	025085	025143	025201

-GLYCOSIDES

At least 1 glycoside

Aesculin	025028	025086	025144	025202
Amygdalin	025029	025087	025145	025203
Arbutin	025030	025088	025146	025204
Coniferin	025031	025089	025147	025205
α-Methyl-D-Glucoside	025033	025091	025149	025207
β-Methyl-D-Glucoside	025034	025092	025150	025208
α-Methyl-D-Mannoside	025035	025093	025151	025209
α-Methyl-D-Xyloside	025032	025090	025148	025206
Salicin	025036	025094	025152	025210

DISACCHARIDES

Cellobiose	025037	025095	025153	025211
Lactose	025038	025096	025154	025212
Maltose	025039	025097	025155	025213
Melibiose	025040	025098	025156	025214
Sucrose	025041	025099	025157	025215
Trehalose	025042	025100	025158	025216

TRISACCHARIDES

D-Melezitose	025043	025101	025159	025217
Raffinose	025044	025102	025160	025218

POLYSACCHARIDES

-PENTOSANS

Xylan	025045	025103	025161	025219

Section 25: Carbohydrate metabolism

```
            -HEXOSANS

Cellulose                       025046   025104   025162   025220
Chitin                          025047   025105   025163   025221
Dextran                         025048   025106   025164   025222
Dextrin                         025049   025107   025165   025223
Glycogen                        025050   025108   025166   025224
Inulin                          025051   025109   025167   025225
Starch                          025052   025110   025168   025226

            -OTHER POLYSACCHARIDES

At least 1 polysaccharide
Alginic Acid                    025053   025111   025169   025227
Bacterial Polysaccharide        025054   025112   025170   025228
Gum Arabic                      025055   025113   025171   025229
Hemicellulose                   025056   025114   025172   025230
Lignocellulose                  025057   025115   025173   025231
Mannan                          025058   025116   025174   025232
```

```
                                                           H
                                                           Y
                                                           D
                                                           R
                                              C   S        O
                                              A   O        L
                                          S   R   U        Y
                                   G      O   B   R        Z
                                   A      L   O   C        E
SUBSTRATE                          S      E   N   E        D

At least 1 carbohydrate          025233       025291

    MONOSACCHARIDES
      -TRIOSES

Dihydroxyacetone                 025234       025292
Glyceraldehyde                   025235       025293

      -TETROSES

D-Erythrose                      025385       025386
L-Erythrulose                    025236       025294
D-Threose                        025237       025295

      -PENTOSES

D-Arabinose                      025238       025296
L-Arabinose                      025239       025297
Deoxyribose                      025378       025379
D-Lyxose                         025240       025298
L-Lyxose                         025241       025299
D-Ribose                         025242       025300
D-Ribulose                       025243       025301
D-Xylose                         025244       025302
D-Xylulose                       025245       025303
```

Section 25: Carbohydrate metabolism   165

                     -METHYL PENTOSES

D-Fucose   025246   025304
L-Fucose   025247   025305
D-Rhamnose   025248   025306
L-Rhamnose   025249   025307

            -HEXOSES

D-Allose   025250   025308
D-Fructose   025251   025309
D-Galactose   025252   025310
D-Glucose   025253   025311
D-Mannose   025254   025312
L-Sorbose   025255   025313

            -HEPTOSES

α-D-Galaheptose   025256   025314
β-D-Galaheptose   025257   025315
α-D-Glucoheptose   025258   025316
Sedoheptulose   025259   025317

            -GLYCOSIDES

At least 1 glycoside   025364
Aesculin   025260   025318   025365
Amygdalin   025261   025319   025366
Arbutin   025262   025320   025367
Coniferin   025263   025321   025368
α-Methyl-D-Glucoside   025265   025323   025370
β-Methyl-D-Glucoside   025266   025324   025371
α-Methyl-D-Mannoside   025267   025325   025372
α-Methyl-D-Xyloside   025264   025322   025369
Salicin   025268   025326   025373

       DISACCHARIDES

Cellobiose   025269   025327
Lactose   025270   025328
Maltose   025271   025329
Melibiose   025272   025330
Sucrose   025273   025331
Trehalose   025274   025332

       TRISACCHARIDES

D-Melezitose   025275   025333
Raffinose   025276   025334

       POLYSACCHARIDES

            -PENTOSANS

Xylan   025277   025335   025350

Section 25: Carbohydrate metabolism

-HEXOSANS

Cellulose	025278	025336	025351
Chitin	025279	025337	025352
Dextran	025280	025338	025353
Dextrin	025281	025339	025354
Glycogen	025282	025340	025355
Inulin	025283	025341	025356
Starch	025284	025342	025357

-OTHER POLYSACCHARIDES

At least 1 polysaccharide			025349
Alginic Acid	025285	025343	025358
Bacterial Polysaccharide	025286	025344	025359
Gum Arabic	025287	025345	025360
Hemicellulose	025288	025346	025361
Lignocellulose	025289	025347	025362
Mannan	025290	025348	025363

025380: An alkaline reaction results from growth in basal medium used for carbohydrate utilization or fermentation tests.

SECTION 26: ALCOHOL METABOLISM

[NOTE 1: The items in the Table have the following forms:

Allyl Alcohol is utilized.

Allyl Alcohol is oxidized.

Allyl Alcohol is reduced.

Acid is produced from Allyl Alcohol.

Gas is produced from Allyl Alcohol.

Allyl Alcohol can be used as the sole source of carbon.]

SUBSTRATE	UTILIZED	OXIDIZED	REDUCED
At least 1 alcohol	026001	026103	026205
ALIPHATIC ALCOHOLS			
-MONOHYDRIC ALCOHOLS			
Allyl Alcohol	026002	026104	026206
1-Butanol	026003	026105	026207
2-Butanol	026004	026106	026208
Ethanol	026005	026107	026209
Geraniol(3,7-dimethyl-2,6-octadiene-1-ol)	026006	026108	026210
1-Hexanol	026007	026109	026211
1-Hexadecanol	026631	026632	026633
Methanol	026008	026110	026212
3-Methyl-1-butanol	026009	026111	026213
2-Methyl-1-Propanol	026010	026112	026214
2-Methyl-2-propanol	026011	026113	026215
1-Octanol	026012	026114	026216
1-Pentanol	026013	026115	026217
1-Propanol	026014	026116	026218
2-Propanol	026015	026117	026219
-DIHYDRIC ALCOHOLS			
2-Amino-2-Ethyl-1,3-Propanediol	026016	026118	026220
1,4-Butanediol	026017	026119	026221
DL-1,3-Butanediol	026018	026120	026222
D(-)2,3-Butanediol	026019	026121	026223
meso-2,3-Butanediol	026020	026122	026224
L(+)2,3-Butanediol	026021	026123	026225

Section 26: Alcohol metabolism

2-Butene-1,4-diol	026022	026124	026226
1,4-Butynediol	026023	026125	026227
Diethylene glycol	026024	026126	026228
Dipropylene Glycol	026025	026127	026229
1,2-Ethanediol	026026	026128	026230
1,7-Heptanediol	026027	026129	026231
1,6-Hexanediol	026028	026130	026232
2,5-Hexanediol	026029	026131	026233
D(+)3,4-Hexanediol	026030	026132	026234
meso-3,4-Hexanediol	026031	026133	026235
2-Nitro-2-Ethyl-1,3-Propanediol	026032	026134	026236
1,3-Pentanediol	026033	026135	026237
1,5-Pentanediol	026034	026136	026238
Polyethylene Glycol (-200)	026035	026137	026239
Polyethylene Glycol (-300)	026036	026138	026240
Polyethylene Glycol (-400)	026037	026139	026241
Polyethylene Glycol (-1500)	026038	026140	026242
D(-)1,2-Propanediol	026039	026141	026243
1,3-Propanediol	026040	026142	026244
Thiodiethylene Glycol	026041	026143	026245
Triethylene Glycol	026042	026144	026246

-TRIHYDRIC ALCOHOLS

1,2,4-Butanetriol	026043	026145	026247
1,2,6-Hexanetriol	026044	026146	026248
1,2,3-Propanetriol (Glycerol)	026045	026147	026249

-TETRAHYDRIC ALCOHOLS

Erythritol	026046	026148	026250
meso-Erythritol	026047	026149	026251
Pentaerythritol	026048	026150	026252

-PENTAHYDRIC ALCOHOLS

Adonitol	026049	026151	026253
iso-adonitol	026050	026152	026254
meso-Arabitol	026051	026153	026255
D-Arabitol	026052	026154	026256
L-Arabitol	026053	026155	026257
meso-Ribitol	026054	026156	026258
meso-Xylitol	026055	026157	026259

-HEXAHYDRIC ALCOHOLS

meso-Allitol	026056	026158	026260
Dulcitol	026057	026159	026261
L-Fucitol	026058	026160	026262
meso-Galactitol	026059	026161	026263
D-Glucitol	026060	026162	026264
L-Glucitol	026061	026163	026265
D-Iditol	026062	026164	026266
L-Iditol	026063	026165	026267
Lactositol	026064	026166	026268
D-Mannitol	026065	026167	026269
L-Mannitol	026066	026168	026270

Section 26: Alcohol metabolism   169

```
L-Rhamnitol                             026067      026169      026271
D-Sorbitol                              026068      026170      026272
L-Sorbitol                              026069      026171      026273

    -POLYHYDRIC ALCOHOLS

D-Glycero-D-Galactoheptitol             026070      026172      026274
meso-Glycero-Guloheptitol               026071      026173      026275
Perseitol                               026072      026174      026276
Polygalitol                             026073      026175      026277
Primulitol                              026074      026176      026278

    ALCOHOLS DERIVED FROM CYCLIC PARAFFINS

Cyclohexanol                            026075      026177      026279
Cycloheptanol                           026076      026178      026280
Cyclooctanol                            026077      026179      026281
Cyclopentanol                           026078      026180      026282
meso-Inositol                           026079      026181      026283
iso-Inositol                            026080      026182      026284
Pinitol                                 026082      026184      026286
Quercitol                               026083      026185      026287

    AROMATIC ALCOHOLS

    -MONOHYDRIC ALCOHOLS

Anisyl Alcohol                          026084      026186      026288
Benzyl Alcohol                          026085      026187      026289
Cinnamyl Alcohol                        026086      026188      026290
Coniferyl Alcohol                       026087      026189      026291
m-Cresol                                026088      026190      026292
o-Cresol                                026613      026614      026615
p-Cresol                                026619      026620      026621
Phenol                                  026089      026191      026293
2-Phenyl-ethanol                        026625      026626      026627

    -DIHYDRIC ALCOHOLS

Aesculetin                              026090      026192      026294
Catechol                                026091      026193      026295
Phenylethanediol                        026081      026183      026285
Resorcinol                              026092      026194      026296
Saligenin                               026093      026195      026297

    ALCOHOLS WITH ADDITIONAL ETHER-LINKED GROUPS

2(2-Butoxy-Ethoxy) Ethanol              026094      026196      026298
Diethylene Glycol
        Monoethyl Ether                 026095      026197      026299
2-Ethoxy Ethanol                        026096      026198      026300
2(2-Ethoxy-Ethoxy) Ethanol              026097      026199      026301
2(2-Methoxy-Ethoxy) Ethanol             026098      026200      026302
Ethylene Glycol Monoethyl Ether         026099      026201      026303
```

Section 26: Alcohol metabolism

KETOALCOHOLS

Substrate			
Acetoin	026100	026202	026304
D(-) Ethylpropionyl Carbinol	026101	026203	026305
L(+)Ethylpropionyl Carbinol	026102	026204	026306

SUBSTRATE	ACID	GAS	CARBOSE LOENE	SOURCE
At least 1 alcohol	026307	026409	026511	

ALIPHATIC ALCOHOLS

-MONOHYDRIC ALCOHOLS

Allyl Alcohol	026308	026410	026512
1-Butanol	026309	026411	026513
2-Butanol	026310	026412	026514
Ethanol	026311	026413	026515
Geraniol(3,7-dimethyl-2,6-octadiene-1-ol)	026312	026414	026516
1-Hexanol	026313	026415	026517
1-Hexadecanol	026634	026635	026636
Methanol	026314	026416	026518
3-Methyl-1-butanol	026315	026417	026519
2-Methyl-1-Propanol	026316	026418	026520
2-Methyl-2-propanol	026317	026419	026521
1-Octanol	026318	026420	026522
1-Pentanol	026319	026421	026523
1-Propanol	026320	026422	026524
2-Propanol	026321	026423	026525

-DIHYDRIC ALCOHOLS

2-Amino-2-Ethyl-1,3-Propanediol	026322	026424	026526
1,4-Butanediol	026323	026425	026527
DL-1,3-Butanediol	026324	026426	026528
D(-)2,3-Butanediol	026325	026427	026529
meso-2,3-Butanediol	026326	026428	026530
L(+)2,3-Butanediol	026327	026429	026531
2-Butene-1,4-diol	026328	026430	026532
1,4-Butynediol	026329	026431	026533
Diethylene Glycol	026330	026432	026534
Dipropylene Glycol	026331	026433	026535
1,2-Ethanediol	026332	026434	026536
1,7-Heptanediol	026333	026435	026537
1,6-Hexanediol	026334	026436	026538
2,5-Hexanediol	026335	026437	026539
D(+)3,4-Hexanediol	026336	026438	026540
meso-3,4-Hexanediol	026337	026439	026541

Section 26: Alcohol metabolism

2-Nitro-2-Ethyl-1,3-Propanediol	026338	026440	026542
1,3-Pentanediol	026339	026441	026543
1,5-Pentanediol	026340	026442	026544
Polyethylene Glycol (-200)	026341	026443	026545
Polyethylene Glycol (-300)	026342	026444	026546
Polyethylene Glycol (-400)	026343	026445	026547
Polyethylene Glycol (-1500)	026344	026446	026548
D(-)1,2-Propanediol	026345	026447	026549
1,3-Propanediol	026346	026448	026550
Thiodiethylene Glycol	026347	026449	026551
Triethylene Glycol	026348	026450	026552

-TRIHYDRIC ALCOHOLS

1,2,4-Butanetriol	026349	026451	026553
1,2,6-Hexanetriol	026350	026452	026554
1,2,3-Propanetriol (Glycerol)	026351	026453	026555

-TETRAHYDRIC ALCOHOLS

Erythritol	026352	026454	026556
meso-Erythritol	026353	026455	026557
Pentaerythritol	026354	026456	026558

-PENTAHYDRIC ALCOHOLS

Adonitol	026355	026457	026559
iso-adonitol	026356	026458	026560
meso-Arabitol	026357	026459	026561
D-Arabitol	026358	026460	026562
L-Arabitol	026359	026461	026563
meso-Ribitol	026360	026462	026564
meso-Xylitol	026361	026463	026565

-HEXAHYDRIC ALCOHOLS

meso-Allitol	026362	026464	026566
Dulcitol	026363	026465	026567
L-Fucitol	026364	026466	026568
meso-Galactitol	026365	026467	026569
D-Glucitol	026366	026468	026570
L-Glucitol	026367	026469	026571
D-Iditol	026368	026470	026572
L-Iditol	026369	026471	026573
Lactositol	026370	026472	026574
D-Mannitol	026371	026473	026575
L-Mannitol	026372	026474	026576
L-Rhamnitol	026373	026475	026577
D-Sorbitol	026374	026476	026578
L-Sorbitol	026375	026477	026579

-POLYHYDRIC ALCOHOLS

D-Glycero-D-Galactoheptitol	026376	026478	026580
meso-Glycero-Guloheptitol	026377	026479	026581
Perseitol	026378	026480	026582

Section 26: Alcohol metabolism

Polygalitol	026379	026481	026583
Primulitol	026380	026482	026584

ALCOHOLS DERIVED FROM CYCLIC PARAFFINS

Cyclohexanol	026381	026483	026585
Cycloheptanol	026382	026484	026586
Cyclooctanol	026383	026485	026587
Cyclopentanol	026384	026486	026588
meso-Inositol	026385	026487	026589
iso-Inositol	026386	026488	026590
Pinitol	026388	026490	026592
Quercitol	026389	026491	026593

AROMATIC ALCOHOLS

-MONOHYDRIC ALCOHOLS

Anisyl Alcohol	026390	026492	026594
Benzyl Alcohol	026391	026493	026595
Cinnamyl Alcohol	026392	026494	026596
Coniferyl Alcohol	026393	026495	026597
m-Cresol	026394	026496	026598
o-Cresol	026616	026617	026618
p-Cresol	026622	026623	026624
Phenol	026395	026497	026599
2-Phenyl-ethanol	026628	026629	026630

-DIHYDRIC ALCOHOLS

Aesculetin	026396	026498	026600
Catechol	026397	026499	026601
Phenylethanediol	026387	026489	026591
Resorcinol	026398	026500	026602
Saligenin	026399	026501	026603

ALCOHOLS WITH ADDITIONAL ETHER-LINKED GROUPS

2(2-Butoxy-Ethoxy) Ethanol	026400	026502	026604
Diethylene Glycol Monoethyl Ether	026401	026503	026605
2-Ethoxy Ethanol	026402	026504	026606
2(2-Ethoxy-Ethoxy) Ethanol	026403	026505	026607
2(2-Methoxy-Ethoxy) Ethanol	026404	026506	026608
Ethylene Glycol Monoethyl Ether	026405	026507	026609

KETOALCOHOLS

Acetoin	026406	026508	026610
D(-) Ethylpropionyl Carbinol	026407	026509	026611
L(+)Ethylpropionyl Carbinol	026408	026510	026612

SECTION 27: ALDEHYDE METABOLISM

[NOTE 1: The items in the Table have the following forms:

At least 1 aldehyde is utilized.

At least 1 aldehyde is oxidized.

At least 1 aldehyde is reduced.

Acid is produced from at least 1 aldehyde.

Gas is produced from at least 1 aldehyde.

At least 1 aldehyde can be used as the sole source of carbon.]

SUBSTRATE	UTILIZED	OXIDIZED	REDUCED
At least 1 aldehyde	027001	027008	027013
Acetaldehyde	027002	027009	027014
n-Butyraldehyde	027003	027010	027015
Formaldehyde	027004	027041	027016
o-Hydroxybenzaldehyde (salicylaldehyde)	027043	027044	027045
Hydroxypyruvic aldehyde	027005	027042	027017
Propionaldehyde	027006	027011	027018
Pyruvaldehyde	027007	027012	027019

SUBSTRATE	ACID	GAS	CARBON SOURCE
At least 1 aldehyde	027020	027027	027034
Acetaldehyde	027021	027028	027035
n-Butyraldehyde	027022	027029	027036
Formaldehyde	027023	027030	027037
o-Hydroxybenzaldehyde (salicylaldehyde)	027046	027047	027048
Hydroxypyruvic aldehyde	027024	027031	027038
Propionaldehyde	027025	027032	027039
Pyruvaldehyde	027026	027033	027040

SECTION 28: CARBOXYLIC ACID OR ESTER METABOLISM

[NOTE 1: The term "acid" covers acidic ions and salts.]

[NOTE 2: Ester hydrolysis questions follow the Table.]

[NOTE 3: The items in the Table have the following forms:

Acetic acid is utilized.

Acetic acid can be used as the sole source of carbon.

Acetic acid is oxidized.

Acetic acid is reduced.

Gas is produced from Acetic acid.

Acetic acid is decarboxylated.]

SUBSTRATE	UTILIZED	CARBON SOURCE	SOURCE OF URCE	OXIDIZED
At least 1 acid or ester	028001		028108	028215
SATURATED FATTY ACIDS				
-MONOCARBOXYLIC ACIDS				
Acetic acid	028002		028109	028216
Behenic acid (docosanoic acid)	028674		028675	028676
Butyric acid	028003		028110	028217
Capric acid (decanoic acid)	028650		028651	028652
Caproic acid (hexanoic acid)	028004		028111	028218
Caprylic acid (octanoic acid)	028005		028112	028219
Formic acid	028006		028113	028220
Isobutyric acid	028007		028114	028221
Isovaleric acid	028008		028115	028222
Lauric acid	028009		028116	028223
10-Methylhexadecanoic acid	028010		028117	028224
Myristic acid (tetradecanoic)	028686		028687	028688
Oenanthic acid (heptanoic acid)	028656		028657	028658
Palmitic acid	028011		028118	028225
Pelargonic acid (nonanoic)	028012		028119	028226
Pentadecanoic acid	028692		028693	028694
Propionic acid	028013		028120	028227
Stearic acid	028662		028663	028664
Tridecanoic acid	028014		028121	028228

Section 28: Carboxylic acid or ester metabolism

Undecanoic acid	028015	028122	028229
Valeric acid (pentanoic acid)	028016	028123	028230

-ESTER

Methyl formate	028017	028124	028231

-DICARBOXYLIC ACIDS

Adipic acid	028018	028125	028232
Azelaic acid	028019	028126	028233
Eicosanedioic acid	028020	028127	028234
Glutaric acid	028021	028128	028235
Malonic acid	028022	028129	028236
Oxalic acid	028023	028130	028237
Pimelic acid	028024	028131	028238
Sebacic acid	028025	028132	028239
Suberic acid	028026	028133	028240
Succinic acid	028027	028134	028241

UNSATURATED FATTY ACIDS

-MONOCARBOXYLIC ACIDS

Erucic acid	028028	028135	028242
Linoleic acid	028029	028136	028243
Linolenic acid	028030	028137	028244
cis-8-Octadecenoic acid	028031	028138	028245
trans-8-Octadecenoic acid	028032	028139	028246
cis-10-Octadecenoic acid	028033	028140	028247
8-Octadecynoic acid	028034	028141	028248
9-Octadecynoic (Stearolic) acid	028035	028142	028249
10-Octadecynoic acid	028036	028143	028250
Oleic acid	028037	028144	028251
Petroselaidic acid	028038	028145	028252
Petroselenic acid	028039	028146	028253

-ESTERS

Methyl arachidonate	028041	028148	028255
Ethyl oleate	028042	028149	028256
Methyl oleate	028043	028150	028257

-DICARBOXYLIC ACIDS

Citraconic acid	028044	028151	028258
Fumaric acid	028045	028152	028259
Itaconic acid	028046	028153	028260
Maleic acid	028047	028154	028261
Mesaconic acid	028048	028155	028262

-TRICARBOXYLIC ACIDS

Aconitic acid	028049	028156	028263
cis-aconitic acid	028050	028157	028264

Section 28: Carboxylic acid or ester metabolism

-HYDROXY-MONOCARBOXYLIC ACIDS

Glyceric acid	028051	028158	028265
DL-Glyceric acid	028052	028159	028266
Glycolic acid	028053	028160	028267
β-Hydroxybutyric acid	028054	028161	028268
Poly-β-hydroxybutyric acid	028055	028162	028269
D-Lactic acid	028056	028163	028270
DL-Lactic acid	028057	028164	028271
L-Lactic acid	028058	028165	028272
Ricinoleic acid	028040	028147	028254

-HYDROXY-DICARBOXYLIC ACIDS

D-Malic acid	028059	028166	028273
L-Malic acid	028060	028167	028274
Mucic acid	028061	028168	028275
D(-) Tartaric acid	028063	028170	028277
L(+) Tartaric acid	028064	028171	028278
DL-Tartaric acid	028644	028645	028646
meso-Tartaric acid	028065	028172	028279

-HYDROXY-TRICARBOXYLIC ACIDS

Citric acid	028066	028173	028280
Isocitric acid	028067	028174	028281

-KETO-MONOCARBOXYLIC ACIDS

Glyoxylic acid	028680	028681	028682
2-Ketogluconic acid	028068	028175	028282
5-Ketogluconic acid	028698	028699	028700
Levulinic acid	028069	028176	028283
β-Oxobutyric (Acetoacetic) acid	028070	028177	028284
Pyruvic acid	028071	028178	028285

-KETO-DICARBOXYLIC ACIDS

α-Ketoglutaric acid	028072	028179	028286
Oxaloacetic acid	028062	028169	028276

-KETO-TRICARBOXYLIC ACIDS

Oxalosuccinic acid	028073	028180	028287

-CYCLIC-MONOCARBOXYLIC ACIDS

Shikimic acid	028074	028181	028288
Cyclohexane carboxylic acid	028668	028669	028670

-CYCLIC-DICARBOXYLIC ACIDS

Phthalic acid	028075	028182	028289
Isophthalic acid	028076	028183	028290
Terephthalic acid	028077	028184	028291

Section 28: Carboxylic acid or ester metabolism 177

-AROMATIC ACIDS

Benzoic acid	028078	028185	028292
m-Hydroxybenzoic acid	028079	028186	028293
o-Hydroxybenzoic acid	028105	028212	028319
p-Hydroxybenzoic acid	028080	028187	028294
Benzoylformic acid	028081	028188	028295
2-Chlorophenoxyacetic acid	028082	028189	028296
4(2,4-Dichlorophenoxy)-butyric acid	028083	028190	028297
Mandelic acid	028084	028191	028298
D-Mandelic acid	028085	028192	028299
L-Mandelic acid	028086	028193	028300
Phenylacetic acid	028087	028194	028301
β-Phenylpropionic acid	028088	028195	028302
Quinic acid	028089	028196	028303
Salicylic acid	028090	028197	028304
Cinnamic acid	028091	028198	028305

ACIDS WITH OTHER GROUPS ATTACHED

2,2-Dichloropropionic acid	028092	028199	028306
2-Ethoxy Ethyl Acetic acid	028093	028200	028307
Ethoxyacetic acid	028094	028201	028308

SYNONYM

Valerianic acid [USE VALERIC #]	028095	028202	028309

SUGAR-DERIVED ACIDS

Arabonic acid	028096	028203	028310
Ascorbic acid	028097	028204	028311
Galactonic acid	028098	028205	028312
Galacturonic acid	028099	028206	028313
Gluconic acid	028100	028207	028314
D-Gluconic acid	028101	028208	028315
Glucuronic acid	028102	028209	028316
α-Methyl glucuronic acid	028103	028210	028317
Hydroxymethyl glutaric acid	028104	028211	028318
Pectin	028106	028213	028320
Saccharic acid	028107	028214	028321

Section 28: Carboxylic acid or ester metabolism

SUBSTRATE	REDUCED	GAS	DECARBOXYLATED
At least 1 acid or ester	028322	028429	028536

SATURATED FATTY ACIDS

-MONOCARBOXYLIC ACIDS

SUBSTRATE	REDUCED	GAS	DECARBOXYLATED
Acetic acid	028323	028430	028537
Behenic acid (docosanoic acid)	028677	028678	028679
Butyric acid	028324	028431	028538
Capric acid (decanoic acid)	028653	028654	028655
Caproic acid (hexanoic acid)	028325	028432	028539
Caprylic acid (octanoic acid)	028326	028433	028540
Formic acid	028327	028434	028541
Isobutyric acid	028328	028435	028542
Isovaleric acid	028329	028436	028543
Lauric acid	028330	028437	028544
10-Methylhexadecanoic acid	028331	028438	028545
Myristic acid (tetradecanoic)	028689	028690	028691
Oenanthic acid (heptanoic acid)	028659	028660	028661
Palmitic acid	028332	028439	028546
Pelargonic acid (nonanoic)	028333	028440	028547
Pentadecanoic acid	028695	028696	028697
Propionic acid	028334	028441	028548
Stearic acid	028665	028666	028667
Tridecanoic acid	028335	028442	028549
Undecanoic acid	028336	028443	028550
Valeric acid (pentanoic acid)	028337	028444	028551

-ESTER

SUBSTRATE	REDUCED	GAS	DECARBOXYLATED
Methyl formate	028338	028445	

-DICARBOXYLIC ACIDS

SUBSTRATE	REDUCED	GAS	DECARBOXYLATED
Adipic acid	028339	028446	028552
Azelaic acid	028340	028447	028553
Eicosanedioic acid	028341	028448	028554
Glutaric acid	028342	028449	028555
Malonic acid	028343	028450	028556
Oxalic acid	028344	028451	028557
Pimelic acid	028345	028452	028558
Sebacic acid	028346	028453	028559
Suberic acid	028347	028454	028560
Succinic acid	028348	028455	028561

Section 28: Carboxylic acid or ester metabolism

UNSATURATED FATTY ACIDS

-MONOCARBOXYLIC ACIDS

Erucic acid	028349	028456	028562
Linoleic acid	028350	028457	028563
Linolenic acid	028351	028458	028564
cis-8-Octadecenoic acid	028352	028459	028565
trans-8-Octadecenoic acid	028353	028460	028566
cis-10-Octadecenoic acid	028354	028461	028567
8-Octadecynoic acid	028355	028462	028568
9-Octadecynoic (Stearolic) acid	028356	028463	028569
10-Octadecynoic acid	028357	028464	028570
Oleic acid	028358	028465	028571
Petroselaidic acid	028359	028466	028572
Petroselenic acid	028360	028467	028573

-ESTERS

Methyl arachidonate	028362	028469	
Ethyl oleate	028363	028470	
Methyl oleate	028364	028471	

-DICARBOXYLIC ACIDS

Citraconic acid	028365	028472	028575
Fumaric acid	028366	028473	028576
Itaconic acid	028367	028474	028577
Maleic acid	028368	028475	028578
Mesaconic acid	028369	028476	028579

-TRICARBOXYLIC ACIDS

Aconitic acid	028370	028477	028580
cis-aconitic acid	028371	028478	028581

-HYDROXY-MONOCARBOXYLIC ACIDS

Glyceric acid	028372	028479	028582
DL-Glyceric acid	028373	028480	028583
Glycolic acid	028374	028481	028584
β-Hydroxybutyric acid	028375	028482	028585
Poly-β-hydroxybutyric acid	028376	028483	028586
D-Lactic acid	028377	028484	028587
DL-Lactic acid	028378	028485	028588
L-Lactic acid	028379	028486	028589
Ricinoleic acid	028361	028468	028574

-HYDROXY-DICARBOXYLIC ACIDS

D-Malic acid	028380	028487	028590
L-Malic acid	028381	028488	028591
Mucic acid	028382	028489	028592
D(-) Tartaric acid	028384	028491	028594
L(+) Tartaric acid	028385	028492	028595

Section 28: Carboxylic acid or ester metabolism

DL-Tartaric acid	028647	028648	028649
meso-Tartaric acid	028386	028493	028596

-HYDROXY-TRICARBOXYLIC ACIDS

Citric acid	028387	028494	028597
Isocitric acid	028388	028495	028598

-KETO-MONOCARBOXYLIC ACIDS

Glyoxylic acid	028683	028684	028685
2-Ketogluconic acid	028389	028496	028599
5-Ketogluconic acid	028701	028702	028703
Levulinic acid	028390	028497	028600
β-Oxobutyric (Acetoacetic) acid	028391	028498	028601
Pyruvic acid	028392	028499	028602

-KETO-DICARBOXYLIC ACIDS

α-Ketoglutaric acid	028393	028500	028603
Oxaloacetic acid	028383	028490	028593

-KETO-TRICARBOXYLIC ACIDS

Oxalosuccinic acid	028394	028501	028604

-CYCLIC-MONOCARBOXYLIC ACIDS

Shikimic acid	028395	028502	028605
Cyclohexane carboxylic acid	028671	028672	028673

-CYCLIC-DICARBOXYLIC ACIDS

Phthalic acid	028396	028503	028606
Isophthalic acid	028397	028504	028607
Terephthalic acid	028398	028505	028608

-AROMATIC ACIDS

Benzoic acid	028399	028506	028609
m-Hydroxybenzoic acid	028400	028507	028610
o-Hydroxybenzoic acid	028426	028636	028533
p-Hydroxybenzoic acid	028401	028508	028611
Benzoylformic acid	028402	028509	028612
2-Chlorophenoxyacetic acid	028403	028510	028613
4(2,4-Dichlorophenoxy)-butyric acid	028404	028511	028614
Mandelic acid	028405	028512	028615
D-Mandelic acid	028406	028513	028616
L-Mandelic acid	028407	028514	028617
Phenylacetic acid	028408	028515	028618
β-Phenylpropionic acid	028409	028516	028619
Quinic acid	028410	028517	028620
Salicylic acid	028411	028518	028621
Cinnamic acid	028412	028519	028622

Section 28: Carboxylic acid or ester metabolism

ACIDS WITH OTHER GROUPS ATTACHED

2,2-Dichloropropionic acid	028413	028520	028623
2-Ethoxy Ethyl Acetic acid	028414	028521	028624
Ethoxyacetic acid	028415	028522	028625

SYNONYM

Valerianic acid [USE VALERIC #]	028416	028523	028626

SUGAR-DERIVED ACIDS

Arabonic acid	028417	028524	028627
Ascorbic acid	028418	028525	028628
Galactonic acid	028419	028526	028629
Galacturonic acid	028420	028527	028630
Gluconic acid	028421	028528	028631
D-Gluconic acid	028422	028529	028632
Glucuronic acid	028423	028530	028633
α-Methyl glucuronic acid	028424	028531	028634
Hydroxymethyl glutaric acid	028425	028532	028635
Pectin	028427	028534	028637
Saccharic acid	028428	028535	028638

ESTER HYDROLYSIS

028639: At least 1 ester is hydrolyzed.
028640: Methyl formate is hydrolyzed.
028641: Methyl arachidonate is hydrolyzed.
028642: Ethyl oleate is hydrolyzed.
028643: Methyl oleate is hydrolyzed.

SECTION 29: AMINO ACID METABOLISM

[NOTE 1: The items in the Table have the following forms:

D-Alanine is utilized.

D-Alanine is required for growth.

D-Alanine can be used as the sole source of carbon.

D-Alanine can be used as the sole source of nitrogen.

D-Alanine can be used as the sole source of carbon and nitrogen (growth measurement).

D-Alanine is deaminated.

D-Alanine is decarboxylated.

D-Alanine is oxidized.

D-Alanine is reduced.

Gas is produced from D-Alanine.]

SUBSTRATE	UTILIZED	REQUIRED FOR GROWTH	CARBON SOURCE	SOURCE OF CARBON OR BOTH	NITROGEN SOURCE	SOURCE OF NITROGEN OR BOTH
At least 1 amino acid	029001	029052		029103	029154	
D-Alanine	029002	029053		029104	029155	
L-Alanine	029003	029054		029105	029156	
DL-Alanine	029552	029553		029511	029554	
β-Alanine	029004	029055		029106	029157	
m-Aminobenzoic Acid	029005	029056		029107	029158	
p-Aminobenzoic Acid	029006	029057		029108	029159	
DL-α-Aminobutyric Acid	029007	029058		029109	029160	
γ-Aminobutyric Acid	029008	029059		029110	029161	
γ-Aminocaproic Acid	029009	029060		029111	029162	
δ-Aminolevulinic Acid	029630	029631		029632	029633	
DL-α-Aminovaleric Acid	029010	029061		029112	029163	
δ-Aminovaleric Acid	029011	029062		029113	029164	
Anthranilic Acid	029012	029063		029114	029165	
L-Arginine	029013	029064		029115	029166	
DL-Arginine	029014	029065		029116	029167	
L-Asparagine	029015	029066		029117	029168	
L-Aspartic Acid	029016	029067		029118	029169	
Betaine	029017	029068		029119	029170	
DL-Carnitine	029620	029621		029622	029623	
L-Citrulline	029570	029571		029572	029573	

Section 29: Amino acid metabolism   183

DL-Citrulline	029018	029069	029120	029171
Creatine	029019	029070	029121	029172
L-Cysteine	029020	029071	029122	029173
L-Cystine	029021	029072	029123	029174
D-Glutamic Acid	029022	029073	029124	029175
L-Glutamic Acid	029023	029074	029125	029176
L-Glutamine	029522	029523	029524	029525
Glycine	029024	029075	029126	029177
Hippurate	029025	029076	029127	029178
L-Histidine	029026	029077	029128	029179
Indole-pyruvic Acid	029512	029513	029514	029515
Kynurenic Acid	029027	029078	029129	029180
L-Kynurenine	029028	029079	029130	029181
D-Leucine	029029	029080	029131	029182
L-Leucine	029030	029081	029132	029183
D-*iso*-Leucine	029031	029082	029133	029184
L-*iso*-Leucine	029032	029083	029134	029185
DL-*iso*-Leucine	029580	029581	029582	029583
DL-Norleucine	029033	029084	029135	029186
D-Lysine	029034	029085	029136	029187
L-Lysine	029035	029086	029137	029188
L-Methionine	029036	029087	029138	029189
Nicotinic acid	029532	029533	029534	029535
L-Ornithine	029037	029088	029139	029190
DL-Ornithine	029560	029561	029562	029563
D-Phenylalanine	029038	029089	029140	029191
L-Phenylalanine	029039	029090	029141	029192
DL-Phenylalanine	029600	029601	029602	029603
D-Proline	029040	029091	029142	029193
L-Proline	029041	029092	029143	029194
Sarcosine	029042	029093	029144	029195
D-Serine	029043	029094	029145	029196
L-Serine	029044	029095	029146	029197
DL-Serine	029542	029543	029544	029545
L-Threonine	029045	029096	029147	029198
DL-Threonine	029610	029611	029612	029613
D-Tryptophan	029046	029097	029148	029199
L-Tryptophan	029047	029098	029149	029200
D-Tyrosine	029048	029099	029150	029201
L-Tyrosine	029049	029100	029151	029202
D-Valine	029050	029101	029152	029203
L-Valine	029051	029102	029153	029204
DL-Valine	029575	029576	029577	029578

Section 29: Amino acid metabolism

SUBSTRATE	CASONDOLENE	SOURCE	DEAMINATED	DECARBOXYLATED	LATE-D
At least 1 amino acid	029205		029256		029307
D-Alanine	029206		029257		029308
L-Alanine	029207		029258		029309
DL-Alanine	029555		029556		029557
β-Alanine	029208		029259		029310
m-Aminobenzoic Acid	029209		029260		029311
p-Aminobenzoic Acid	029210		029261		029312
DL-α-Aminobutyric Acid	029211		029262		029313
γ-Aminobutyric Acid	029212		029263		029314
γ-Aminocapronic Acid	029213		029264		029315
δ-Aminolevulinic Acid	029634		029635		029636
DL-α-Aminovaleric Acid	029214		029265		029316
δ-Aminovaleric Acid	029215		029266		029317
Anthranilic Acid	029216		029267		029318
L-Arginine	029217		029268		029319
DL-Arginine	029218		029269		029320
L-Asparagine	029219		029270		029321
L-Aspartic Acid	029220		029271		029322
Betaine	029221		029272		029323
DL-Carnitine	029624		029625		029626
L-Citrulline	029574		029585		029586
DL-Citrulline	029222		029273		029324
Creatine	029223		029274		029325
L-Cysteine	029224		029275		029326
L-Cystine	029225		029276		029327
D-Glutamic Acid	029226		029277		029328
L-Glutamic Acid	029227		029278		029329
L-Glutamine	029526		029527		029528
Glycine	029228		029279		029330
Hippurate	029229		029280		029331
L-Histidine	029230		029281		029332
Indole-pyruvic Acid	029516		029517		029518
Kynurenic Acid	029231		029282		029333
L-Kynurenine	029232		029283		029334
D-Leucine	029233		029284		029335
L-Leucine	029234		029285		029336
D-iso-Leucine	029235		029286		029337
L-iso-Leucine	029236		029287		029338
DL-iso-Leucine	029584		029595		029596
DL-Norleucine	029237		029288		029339
D-Lysine	029238		029289		029340
L-Lysine	029239		029290		029341
L-Methionine	029240		029291		029342
Nicotinic acid	029536		029537		029538
L-Ornithine	029241		029292		029343
DL-Ornithine	029564		029565		029566

Section 29: Amino acid metabolism   185

D-Phenylalanine	029242	029293	029344
L-Phenylalanine	029243	029294	029345
DL-Phenylalanine	029604	029605	029606
D-Proline	029244	029295	029346
L-Proline	029245	029296	029347
Sarcosine	029246	029297	029348
D-Serine	029247	029298	029349
L-Serine	029248	029299	029350
DL-Serine	029546	029547	029548
L-Threonine	029249	029300	029351
DL-Threonine	029614	029615	029616
D-Tryptophan	029250	029301	029352
L-Tryptophan	029251	029302	029353
D-Tyrosine	029252	029303	029354
L-Tyrosine	029253	029304	029355
D-Valine	029254	029305	029356
L-Valine	029255	029306	029357
DL-Valine	029579	029590	029591

SUBSTRATE	OXIDIZED	REDUCED	GAS
At least 1 amino acid	029358	029409	029460
D-Alanine	029359	029410	029461
L-Alanine	029360	029411	029462
DL-Alanine	029558	029559	029640
β-Alanine	029361	029412	029463
m-Aminobenzoic Acid	029362	029413	029464
p-Aminobenzoic Acid	029363	029414	029465
DL-α-Aminobutyric Acid	029364	029415	029466
γ-Aminobutyric Acid	029365	029416	029467
γ-Aminocapronic Acid	029366	029417	029468
δ-Aminolevulinic Acid	029637	029638	029639
DL-α-Aminovaleric Acid	029367	029418	029469
δ-Aminovaleric Acid	029368	029419	029470
Anthranilic Acid	029369	029420	029471
L-Arginine	029370	029421	029472
DL-Arginine	029371	029422	029473
L-Asparagine	029372	029423	029474
L-Aspartic Acid	029373	029424	029475
Betaine	029374	029425	029476
DL-Carnitine	029627	029628	029629
L-Citrulline	029587	029588	029589
DL-Citrulline	029375	029426	029477
Creatine	029376	029427	029478
L-Cysteine	029377	029428	029479
L-Cystine	029378	029429	029480
D-Glutamic Acid	029379	029430	029481
L-Glutamic Acid	029380	029431	029482
L-Glutamine	029529	029530	029531

Section 29: Amino acid metabolism

Glycine	029381	029432	029483
Hippurate	029382	029433	029484
L-Histidine	029383	029434	029485
Indole-pyruvic Acid	029519	029520	029521
Kynurenic Acid	029384	029435	029486
L-Kynurenine	029385	029436	029487
D-Leucine	029386	029437	029488
L-Leucine	029387	029438	029489
D-*iso*-Leucine	029388	029439	029490
L-*iso*-Leucine	029389	029440	029491
DL-*iso*-Leucine	029597	029598	029599
DL-Norleucine	029390	029441	029492
D-Lysine	029391	029442	029493
L-Lysine	029392	029443	029494
L-Methionine	029393	029444	029495
Nicotinic acid	029539	029540	029541
L-Ornithine	029394	029445	029496
DL-Ornithine	029567	029568	029569
D-Phenylalanine	029395	029446	029497
L-Phenylalanine	029396	029447	029498
DL-Phenylalanine	029607	029608	029609
D-Proline	029397	029448	029499
L-Proline	029398	029449	029500
Sarcosine	029399	029450	029501
D-Serine	029400	029451	029502
L-Serine	029401	029452	029503
DL-Serine	029549	029550	029551
L-Threonine	029402	029453	029504
DL-Threonine	029617	029618	029619
D-Tryptophan	029403	029454	029505
L-Tryptophan	029404	029455	029506
D-Tyrosine	029405	029456	029507
L-Tyrosine	029406	029457	029508
D-Valine	029407	029458	029509
L-Valine	029408	029459	029510
DL-Valine	029592	029593	029594

187

SECTION 30: AMINE, AMIDE, LACTAM, PURINE, PYRIMIDINE METABOLISM

[NOTE 1: The items in the Table have the following forms:

Allylamine is utilized.

Allylamine is required for growth.

Allylamine can be used as the sole source of carbon.

Allylamine can be used as the sole source of nitrogen.

Allylamine can be used as the sole source of carbon and nitrogen.

Allylamine is deaminated.

Allylamine is oxidized.

Allylamine is reduced.]

SUBSTRATE	UTILIZED	REQUIRED GROWTH	CARBON SOURCE	NITROGEN SOURCE
At least 1 amine or amide or lactam	030001	030045	030089	030133
AMINE				
N-Acetylglucosamine	030377	030378	030379	030380
Allylamine	030002	030046	030090	030134
α-Amylamine	030003	030047	030091	030135
iso-Amylamine	030004	030048	030092	030136
Benzylamine	030005	030049	030093	030137
Butylamine	030006	030050	030094	030138
iso-Butylamine	030007	030051	030095	030139
Di-*iso*-Butylamine	030008	030052	030096	030140
Cadaverine	030009	030053	030097	030141
Ethylamine	030010	030054	030098	030142
Di-Ethylamine	030011	030055	030099	030143
Ethanolamine	030012	030056	030100	030144
Galactosamine	030013	030057	030101	030145
Glucosamine	030014	030058	030102	030146
Histamine	030015	030059	030103	030147
Methylamine	030016	030060	030104	030148
Tri-Methylamine	030017	030061	030105	030149

Section 30: Organic nitrogen metabolism

β-Methylbutylamine	030018	030062	030106	030150
Tri-Methylene Diamine	030353	030354	030355	030356
α-Methylglucosamine	030361	030362	030363	030364
β-Methylglucosamine	030369	030370	030371	030372
β-Phenylamine	030019	030063	030107	030151
Phenylethylamine	030020	030064	030108	030152
Propylamine	030021	030065	030109	030153
Di-Propylamine	030022	030066	030110	030154
iso-Propylamine	030023	030067	030111	030155
Tri-Propylamine	030024	030068	030112	030156

AMINE

Putrescine	030025	030069	030113	030157
Spermine	030026	030070	030114	030158
Spermidine	030027	030071	030115	030159
Taurine	030474	030475	030476	030477
Tryptamine	030028	030072	030116	030160
Tyramine	030029	030073	030117	030161

AMIDES

Acetamide	030030	030075	030119	030163
Acrylamide	030458	030459	030460	030461
Allantoin	030031	030074	030118	030162
Benzamide	030032	030076	030120	030164
Caproamide	030033	030077	030121	030165
Formamide	030434	030435	030436	030437
Iodoacetamide	030442	030443	030444	030445
Isobutyramide	030482	030483	030484	030485
Malonamide	030034	030078	030122	030166
Nicotinamide	030035	030079	030123	030167
iso-Nicotinamide	030036	030080	030124	030168
Oxamide	030466	030467	030468	030469
Propionamide	030037	030081	030125	030169
Pyrazinamide	030038	030082	030126	030170
Salicylamide	030039	030083	030127	030171
Succinamide	030040	030084	030128	030172
Valeramide	030450	030451	030452	030453

LACTAMS

ɣ-Butyrolactam	030041	030085	030129	030173
ε-Caprolactam	030042	030086	030130	030174
Pyroglutamic Acid	030043	030087	030131	030175
δ-Valerolactam	030044	030088	030132	030176

PURINES AND PYRIMIDINES

Adenine	030385	016119	030386	030387
Cytosine	030392	016120	030393	030394
Guanine	030399	016121	030400	030401
Hypoxanthine	030406	016122	030407	030408
Orotic acid	030490	016391	030491	030492
Thymine	030413	016123	030414	030415
Uracil	030420	016124	030421	030422

Section 30: Organic nitrogen metabolism

Uric acid		030497	016392	030498	030499
Xanthine		030427	016125	030428	030429

SUBSTRATE	CASE SOLE	SOURCE	DEAMINATED	OXIDIZED	REDUCED
At least 1 amine or amide or lactam		030177	030221	030265	030309
AMINE					
N-Acetylglucosamine		030381	030382	030383	030384
Allylamine		030178	030222	030266	030310
α-Amylamine		030179	030223	030267	030311
iso-Amylamine		030180	030224	030268	030312
Benzylamine		030181	030225	030269	030313
Butylamine		030182	030226	030270	030314
iso-Butylamine		030183	030227	030271	030315
Di-iso-Butylamine		030184	030228	030272	030316
Cadaverine		030185	030229	030273	030317
Ethylamine		030186	030230	030274	030318
Di-Ethylamine		030187	030231	030275	030319
Ethanolamine		030188	030232	030276	030320
Galactosamine		030189	030233	030277	030321
Glucosamine		030190	030234	030278	030322
Histamine		030191	030235	030279	030323
Methylamine		030192	030236	030280	030324
Tri-Methylamine		030193	030237	030281	030325
β-Methylbutylamine		030194	030238	030282	030326
Tri-Methylene Diamine		030357	030358	030359	030360
α-Methylglucosamine		030365	030366	030367	030368
β-Methylglucosamine		030373	030374	030375	030376
β-Phenylamine		030195	030239	030283	030327
Phenylethylamine		030196	030240	030284	030328
Propylamine		030197	030241	030285	030329
Di-Propylamine		030198	030242	030286	030330
iso-Propylamine		030199	030243	030287	030331
Tri-Propylamine		030200	030244	030288	030332
Putrescine		030201	030245	030289	030333
Spermine		030202	030246	030290	030334
Spermidine		030203	030247	030291	030335
Taurine		030478	030479	030480	030481
Tryptamine		030204	030248	030292	030336
Tyramine		030205	030249	030293	030337

AMIDES

Acetamide	030207	030250	030294	030338
Acrylamide	030462	030463	030464	030465
Allantoin	030206	030251	030295	030339
Benzamide	030208	030252	030296	030340
Caproamide	030209	030253	030297	030341
Formamide	030438	030439	030440	030441
Iodoacetamide	030446	030447	030448	030449
Isobutyramide	030486	030487	030488	030489
Malonamide	030210	030254	030298	030342
Nicotinamide	030211	030255	030299	030343
iso-Nicotinamide	030212	030256	030300	030344
Oxamide	030470	030471	030472	030473
Propionamide	030213	030257	030301	030345
Pyrazinamide	030214	030258	030302	030346
Salicylamide	030215	030259	030303	030347
Succinamide	030216	030260	030304	030348
Valeramide	030454	030455	030456	030457

LACTAMS

γ-Butyrolactam	030217	030261	030305	030349
ε-Caprolactam	030218	030262	030306	030350
Pyroglutamic Acid	030219	030263	030307	030351
δ-Valerolactam	030220	030264	030308	030352

PURINES AND PYRIMIDINES

Adenine	030388	030389	030390	030391
Cytosine	030395	030396	030397	030398
Guanine	030402	030403	030404	030405
Hypoxanthine	030409	030410	030411	030412
Orotic acid	030493	030494	030495	030496
Thymine	030416	030417	030418	030419
Uracil	030423	030424	030425	030426
Uric acid	030500	030501	030502	030503
Xanthine	030430	030431	030432	030433

030504: Putrescine or cadaverine is required for growth.

SECTION 31: HYDROCARBON AND KETONE METABOLISM

[NOTE 1: The items in the Table have the following forms.

At least 1 hydrocarbon can be used as the sole source of carbon.

At least 1 hydrocarbon is utilized.

At least 1 hydrocarbon is oxidized.

At least 1 hydrocarbon is reduced.

Acid is produced from at least 1 hydrocarbon.

Gas is produced from at least 1 hydrocarbon.]

SUBSTRATE	CARBON SOURCE	UTILIZED	OXIDIZED
HYDROCARBONS			
At least 1 hydrocarbon	031001	031099	031197
HYDROCARBONS - ALKANES (UNBRANCHED)			
n-Butane	031002	031100	031198
n-Decane	031003	031101	031199
n-Docosane	031004	031102	031200
n-Dodecane	031005	031103	031201
n-Dotriacontane	031613	031614	031615
n-Eicosane	031006	031104	031202
Ethane	031007	031105	031203
n-Heptadecane	031008	031106	031204
n-Heptane	031009	031107	031205
n-Hexadecane	031010	031108	031206
n-Hexane	031011	031109	031207
n-Hexatriacontane	031619	031620	031621
Methane	031012	031110	031208
n-Nonadecane	031013	031111	031209
n-Nonane	031014	031112	031210
n-Octacosane	031787	031788	031789
n-Octadecane	031015	031113	031211
n-Octane	031016	031114	031212
n-Pentadecane	031017	031115	031213
n-Pentane	031018	031116	031214
n-Propane	031019	031117	031215
n-Tetracosane	031763	031764	031765
n-Tetradecane	031020	031118	031216
n-Tridecane	031021	031119	031217

Section 31: Hydrocarbon and ketone metabolism

n-Triacontane	031607	031608	031609
n-Undecane	031022	031120	031218

HYDROCARBONS - ALKANES (BRANCHED)

Dimethyl Octane	031023	031121	031219
3-Ethyl Tetradecane	031024	031122	031220
2,2,4,4,6,8,8-Heptamethyl-Nonane	031637	031638	031639
2-Methyl Butane	031625	031626	031627
2-Methyl Heptane	031025	031123	031221
3-Methyl Heptane	031026	031124	031222
4-Methyl Heptane	031027	031125	031223
2-Methyl Hexane	031028	031126	031224
3-Methyl Hexane	031029	031127	031225
2-Methyl Pentadecane	031030	031128	031226
3-Methyl Pentadecane	031031	031129	031227
4-Methyl Pentadecane	031032	031130	031228
5-Methyl Pentadecane	031033	031131	031229
6-Methyl Pentadecane	031034	031132	031230
7-Methyl Pentadecane	031035	031133	031231
8-Methyl Pentadecane	031036	031134	031232
2-Methyl Undecane	031643	031644	031645
2,2,4,6,6-Pentamethylheptane	031775	031776	031777
Phytane	031595	031596	031597
Pristane (2,6,10,14-Tetramethylpentadecane)	031589	031590	031591
2,2,4-Trimethyl Pentane	031631	031632	031633

HYDROCARBONS - CYCLIC ALKANES

n-Butylcyclohexane	031037	031135	031233
Cyclohexane	031038	031136	031234
Decahydronaphthalene	031039	031137	031235
Dicyclohexyl	031685	031686	031687
Dimethylcyclohexane	031040	031138	031236
1,2-Dimethylcyclohexane	031661	031662	031663
1,4-Dimethylcyclohexane	031667	031668	031669
Ethylcyclohexane	031673	031674	031675
Isopropylcyclohexane	031835	031836	031837
Methylcyclohexane	031649	031650	031651
Methylcyclopentane	031655	031656	031657
Octocyclohexane	031679	031680	031681
Pentadecylcyclohexane	031601	031602	031603

HYDROCARBONS - ALKENES (UNBRANCHED)

1,11-Dodecadiene	031041	031139	031237
1-Dodecene	031042	031140	031238
Ethylene	031043	031141	031239
1-Heptene	031044	031142	031240
1-Hexadecene (hexadecylene)	031045	031143	031241
1-Octadecene	031046	031144	031242
1-Octene (n-caprylene)	031047	031145	031243
1-Pentadecene	031769	031770	031771
Propylene	031048	031146	031244

Section 31: Hydrocarbon and ketone metabolism   193

    HYDROCARBONS - ALKENES (BRANCHED)

2,2,4,6,6-Pentamethyl-3-heptene	031793	031794	031795

    HYDROCARBONS - CYCLIC ALKENES

Cyclic Olefines	031049	031147	031245
Pinene	031050	031148	031246

    HYDROCARBONS - ALKYNES

Acetylene	031051	031149	031247

    HYDROCARBONS - DERIVATIVES

1,2-Epoxyoctane	031052	031150	031248
Propylene oxide	031053	031151	031249

    HYDROCARBONS - MIXTURES

Asphalt	031054	031152	031250
Gasoline	031055	031153	031251
Kerosine	031056	031154	031252
Paraffin Oil	031057	031155	031253
Paraffin Wax	031058	031156	031254
Petroleum	031059	031157	031255

    HYDROCARBONS - AROMATIC

Acenaphthalene	031745	031746	031747
Anthracene	031060	031158	031256
Benzene	031061	031159	031257
n-Butylbenzene	031062	031160	031258
Chrysene (1,2-Benz-phenanthrene)	031817	031818	031819
p-Cymene	031063	031161	031259
2,3-Dimethylnaphthalene	031727	031728	031729
2,6-Dimethylnaphthalene	031733	031734	031735
Diphenylmethane	031709	031710	031711
n-Dodecylbenzene	031064	031162	031260
Ethylbenzene	031065	031163	031261
2-Ethylnaphthalene	031721	031722	031723
Mesitylene	031066	031164	031262
9-Methylanthracene	031751	031752	031753
2-Methylnaphthalene	031067	031165	031263
1-Methylnaphthalene	031068	031166	031264
1-Methylphenanthrene	031757	031758	031759
3-Methylphenanthrene	031069	031167	031265
9-Methylphenanthrene	031070	031168	031266
Naphthalene	031071	031169	031267
1-(α-Naphthyl)-Hendecane	031072	031170	031268
ω-Phenyldecane	031074	031172	031270
ω-Phenyldodecane	031075	031173	031271
ω-Phenyloctadecane	031077	031175	031273
Phenanthrene	031073	031171	031269

Section 31: Hydrocarbon and ketone metabolism

1-Phenyl-1-cyclohexene	031811	031812	031813
1-Phenyl-3,4-Dihydro-naphthalene	031799	031800	031801
3-Phenyleicosane	031076	031174	031272
1-Phenylheptane	031781	031782	031783
1-Phenylnaphthalene	031805	031806	031807
1-Phenyltridecane	031703	031704	031705
Pseudocumene	031078	031176	031274
Pyrene (Benzo-phenanthrene)	031823	031824	031825
Pyrogallol	031079	031177	031275
4-*tert*-Butylbenzene	031691	031692	031693
Toluene	031080	031178	031276
1,2,3,4-Tetrahydronaphthalene	031739	031740	031741
1,2,3,4-Tetramethylbenzene (Phrenitene)	031697	031698	031699
1,3,5-Triphenylbenzene	031715	031716	031717
Triphenylene (9,10-Benz-phenanthrene)	031829	031830	031831
Xylene	031081	031179	031277

HYDROCARBONS - AROMATIC DERIVATIVES

Cumyl Hydroperoxide	031082	031180	031278

KETONES

At least 1 ketone	031083	031181	031279

KETONES OF ALKANES (UNBRANCHED)

Acetone	031084	031182	031280
2-Butanone	031085	031183	031281
Dihydroxyacetone	031086	031184	031282
2-Octadecanone	031087	031185	031283
2-Pentanone	031088	031186	031284
2-Tridecanone	031089	031187	031285

KETONES OF CYCLIC ALKANES

Camphor	031090	031188	031286
Cyclohexane-1,2-dione	031091	031189	031287
Cyclohexane-1,3-dione	031092	031190	031288
Cyclohexanone	031093	031191	031289
Cyclooctanone	031094	031192	031290
Cyclopentnone	031095	031193	031291

KETONES OF AROMATIC HYDROCARBONS

Coumarin	031096	031194	031292
1,2-Naphthoquinone	031097	031195	031293
2-Naphthoquinone	031098	031196	031294

Section 31: Hydrocarbon and ketone metabolism   195

	REDUCED	ACID	GAS
SUBSTRATE			
HYDROCARBONS			
At least 1 hydrocarbon	031295	031393	031491
HYDROCARBONS - ALKANES (UNBRANCHED)			
n-Butane	031296	031394	031492
n-Decane	031297	031395	031493
n-Docosane	031298	031396	031494
n-Dodecane	031299	031397	031495
n-Dotriacontane	031616	031617	031618
n-Eicosane	031300	031398	031496
Ethane	031301	031399	031497
n-Heptadecane	031302	031400	031498
n-Heptane	031303	031401	031499
n-Hexadecane	031304	031402	031500
n-Hexane	031305	031403	031501
n-Hexatriacontane	031622	031623	031624
Methane	031306	031404	031502
n-Nonadecane	031307	031405	031503
n-Nonane	031308	031406	031504
n-Octacosane	031790	031791	031792
n-Octadecane	031309	031407	031505
n-Octane	031310	031408	031506
n-Pentadecane	031311	031409	031507
n-Pentane	031312	031410	031508
n-Propane	031313	031411	031509
n-Tetracosane	031766	031767	031768
n-Tetradecane	031314	031412	031510
n-Tridecane	031315	031413	031511
n-Triacontane	031610	031611	031612
n-Undecane	031316	031414	031512
HYDROCARBONS - ALKANES (BRANCHED)			
Dimethyl Octane	031317	031415	031513
3-Ethyl Tetradecane	031318	031416	031514
2,2,4,4,6,8,8-Heptamethyl-Nonane	031640	031641	031642
2-Methyl Butane	031628	031629	031630
2-Methyl Heptane	031319	031417	031515
3-Methyl Heptane	031320	031418	031516
4-Methyl Heptane	031321	031419	031517
2-Methyl Hexane	031322	031420	031518
3-Methyl Hexane	031323	031421	031519
2-Methyl Pentadecane	031324	031422	031520
3-Methyl Pentadecane	031325	031423	031521
4-Methyl Pentadecane	031326	031424	031522

5-Methyl Pentadecane	031327	031425	031523
6-Methyl Pentadecane	031328	031426	031524
7-Methyl Pentadecane	031329	031427	031525
8-Methyl Pentadecane	031330	031428	031526
2-Methyl Undecane	031646	031647	031648
2,2,4,6,6-Pentamethylheptane	031778	031779	031780
Phytane	031598	031599	031600
Pristane (2,6,10,14-Tetra-methylpentadecane)	031592	031593	031594
2,2,4-Trimethyl Pentane	031634	031635	031636

HYDROCARBONS - CYCLIC ALKANES

n-Butylcyclohexane	031331	031429	031527
Cyclohexane	031332	031430	031528
Decahydronaphthalene	031333	031431	031529
Dicyclohexyl	031688	031689	031690
Dimethylcyclohexane	031334	031432	031530
1,2-Dimethylcyclohexane	031664	031665	031666
1,4-Dimethylcyclohexane	031670	031671	031672
Ethylcyclohexane	031676	031677	031678
Isopropylcyclohexane	031838	031839	031840
Methylcyclohexane	031652	031653	031654
Methylcyclopentane	031658	031659	031660
Octocyclohexane	031682	031683	031684
Pentadecylcyclohexane	031604	031605	031606

HYDROCARBONS - ALKENES (UNBRANCHED)

1,11-Dodecadiene	031335	031433	031531
1-Dodecene	031336	031434	031532
Ethylene	031337	031435	031533
1-Heptene	031338	031436	031534
1-Hexadecene (hexadecylene)	031339	031437	031535
1-Octadecene	031340	031438	031536
1-Octene (n-caprylene)	031341	031439	031537
1-Pentadecene	031772	031773	031774
Propylene	031342	031440	031538

HYDROCARBONS - ALKENES (BRANCHED)

2,2,4,6,6-Pentamethyl-3-heptene	031796	031797	031798

HYDROCARBONS - CYCLIC ALKENES

Cyclic Olefines	031343	031441	031539
Pinene	031344	031442	031540

HYDROCARBONS - ALKYNES

Acetylene	031345	031443	031541

HYDROCARBONS - DERIVATIVES

1,2-Epoxyoctane	031346	031444	031542
Propylene oxide	031347	031445	031543

Section 31: Hydrocarbon and ketone metabolism

HYDROCARBONS - MIXTURES

Asphalt	031348	031446	031544
Gasoline	031349	031447	031545
Kerosine	031350	031448	031546
Paraffin Oil	031351	031449	031547
Paraffin Wax	031352	031450	031548
Petroleum	031353	031451	031549

HYDROCARBONS - AROMATIC

Acenaphthalene	031748	031749	031750
Anthracene	031354	031452	031550
Benzene	031355	031453	031551
n-Butylbenzene	031356	031454	031552
Chrysene (1,2-Benz-phenanthrene)	031820	031821	031822
p-Cymene	031357	031455	031553
2,3-Dimethylnaphthalene	031730	031731	031732
2,6-Dimethylnaphthalene	031736	031737	031738
Diphenylmethane	031712	031713	031714
n-Dodecylbenzene	031358	031456	031554
Ethylbenzene	031359	031457	031555
2-Ethylnaphthalene	031724	031725	031726
Mesitylene	031360	031458	031556
9-Methylanthracene	031754	031755	031756
2-Methylnaphthalene	031361	031459	031557
1-Methylnaphthalene	031362	031460	031558
1-Methylphenanthrene	031760	031761	031762
3-Methylphenanthrene	031363	031461	031559
9-Methylphenanthrene	031364	031462	031560
Naphthalene	031365	031463	031561
1-(α-Naphthyl)-Hendecane	031366	031464	031562
ω-Phenyldecane	031368	031466	031564
ω-Phenyldodecane	031369	031467	031565
ω-Phenyloctadecane	031371	031469	031567
Phenanthrene	031367	031465	031563
1-Phenyl-1-cyclohexene	031814	031815	031816
1-Phenyl-3,4-Dihydro-naphthalene	031802	031803	031804
3-Phenyleicosane	031370	031468	031566
1-Phenylheptane	031784	031785	031786
1-Phenylnaphthalene	031808	031809	031810
1-Phenyltridecane	031706	031707	031708
Pseudocumene	031372	031470	031568
Pyrene (Benzo-phenanthrene)	031826	031827	031828
Pyrogallol	031373	031471	031569
4-tert-Butylbenzene	031694	031695	031696
Toluene	031374	031472	031570
1,2,3,4-Tetrahydronaphthalene	031742	031743	031744
1,2,3,4-Tetramethylbenzene (Phrenitene)	031700	031701	031702
1,3,5-Triphenylbenzene	031718	031719	031720
Triphenylene (9,10-Benz-phenanthrene)	031832	031833	031834
Xylene	031375	031473	031571

Section 31: Hydrocarbon and ketone metabolism

HYDROCARBONS - AROMATIC DERIVATIVES

Cumyl Hydroperoxide	031376	031474	031572

KETONES

At least 1 ketone	031377	031475	031573
Acetone	031378	031476	031574
2-Butanone	031379	031477	031575
Dihydroxyacetone	031380	031478	031576
2-Octadecanone	031381	031479	031577
2-Pentanone	031382	031480	031578

KETONES OF ALKANES (UNBRANCHED)

2-Tridecanone	031383	031481	031579

KETONES OF CYCLIC ALKANES

Camphor	031384	031482	031580
Cyclohexane-1,2-dione	031385	031483	031581
Cyclohexane-1,3-dione	031386	031484	031582
Cyclohexanone	031387	031485	031583
Cyclooctanone	031388	031486	031584
Cyclopentnone	031389	031487	031585

KETONES OF AROMATIC HYDROCARBONS

Coumarin	031390	031488	031586
1,2-Naphthoquinone	031391	031489	031587
2-Naphthoquinone	031392	031490	031588

SECTION 32: FAT AND OIL METABOLISM

032001: At least 1 fat is hydrolyzed.
032002: Beef Tallow is hydrolyzed.
032003: Butter Fat is hydrolyzed.
032004: Coconut Oil is hydrolyzed.
032005: Corn Oil is hydrolyzed.
032006: Cottonseed Oil is hydrolyzed.
032007: Lard is hydrolyzed.
032008: Linseed Oil is hydrolyzed.
032009: Olive Oil is hydrolyzed.
032010: Tributyrin is hydrolyzed.
032011: Tricaprin is hydrolyzed.
032012: Tricaproin is hydrolyzed.
032013: Tricaprylin is hydrolyzed.
032014: Trilaurin is hydrolyzed.
032015: Trimyristin is hydrolyzed.
032016: Triolein is hydrolyzed.
032017: Tripalmitin is hydrolyzed.
032018: Tripropionin is hydrolyzed.
032019: Tristearin is hydrolyzed.
032020: Tween 20 is hydrolyzed.
032021: Tween 40 is hydrolyzed.
032022: Tween 60 is hydrolyzed.
032023: Tween 80 is hydrolyzed.

SECTION 33:   PRESERVATION OF STRAINS

033013: Strains may be stored at 22 C for 2 d.
033014: Strains may be stored at 22 C for 4 d.
033015: Strains may be stored at 22 C for 7 d.
033016: Strains may be stored at 22 C for 14 d.
033001: Strains may be stored at 25 C for 30 d.
033002: Strains may be stored at 5 C for 30 d.
033003: Agar slant cultures with a petrolatum overlay may be stored for 30 d.
033004: Strains may be stored in sterile soil (pH 6.0-8.0) for 30 d.
033017: Stock cultures are maintained aerobically.
033005: Strains require complete anaerobiosis during storage.
033006: Strains may be successfully lyophilized.
033018: Strain can be recovered after being lyophilized with serum.
033019: Strain can be recovered after being lyophilized with sucrose.
033020: Strain can be recovered after being lyophilized with skim milk.
033021: Strain may be recovered after being lyophilized with dimethylsulfoxide (DMSO).
033022: The storage atmosphere of lyophilized strain is vacuum.
033023: The storage atmosphere of lyophilized strain is argon.
033024: The storage atmosphere of lyophilized strain is nitrogen.
033025: Lyophilized strain is stored at 10 C.
033026: Lyophilized strain is stored at 4 C.
033027: Lyophilized strain is stored at below 0 C.
033028: Lyophilized strain is stored below -25 C.
033029: Lyophilized strain is stored below -50 C.
033007: Strains may be frozen at -20 C for 30 d.
033008: Strains may be frozen at -65 C (solid carbon dioxide) for 30 d.
033009: Strains may be frozen at -196 C (liquid nitrogen) for 30 d.
033030: Strain can be successfully recovered after freezing at cryogenic temperatures (below -100 C).
033031: Strain can be sucessfully recovered after freezing at low temperatures (-50 to -100 C).
033032: Strain can be recovered after being frozen in glycerol.
033033: Strain can be recovered after being frozen in dimethylsulfoxide (DMSO).
033034: Strain can be recovered after being frozen in host tissue.
033035: Strain can be recovered after being frozen with added sucrose.
033036: Percentage of cryoprotective agent (v/v) is 2.5%.
033037: Percentage of cryoprotective agent (v/v) is 5.0%.
033038: Percentage of cryoprotective agent (v/v) is 7.5%.
033039: Percentage of cryoprotective agent (v/v) is 10.0%.

Section 33: Preservation of strains

033040: Percentage of cryoprotective agent (v/v) is greater than 11.0%
033041: Equilibration time (time from addition of cryoprotective agent to time freezing program is started) is 0.
033042: Equilibration time is 15 min.
033043: Equilibration time is 30 min.
033044: Equilibration time is 45 min.
033045: Equilibration time is 60 min.
033046: Equilibration time is greater than 60 min.
033047: Equilibration temperature is room temperature.
033048: Equilibration temperature is 4 C.
033049: Freezing program is started at room temperature.
033050: Freezing program is started at 10 C.
033051: Freezing program is started at 4 C.
033052: Strain is frozen at the rate of 1 C/min.
033053: Strain is frozen at the rate of 2 C/min.
033054: Strain is frozen at the rate of 5 C/min.
033055: Strain is frozen at the rate of 10 C/min.
033056: Strain is directly plunged into liquid nitrogen from room temperature.
033057: After freezing program has lowered the temperature to -20 to -25 C, strain is plunged into liquid nitrogen.
033058: After freezing program has lowered the temperature to -26 to -30 C, strain is plunged into liquid nitrogen.
033059: After freezing program has lowered the temperature to -31 to -35 C, strain is plunged into liquid nitrogen.
033060: After freezing program has lowered the temperature to -36 to -40 C, strain is plunged into liquid nitrogen.
033061: After freezing program has lowered the temperature to -41 to -45 C, strain is plunged into liquid nitrogen.
033062: After freezing program has lowered the temperature to -46 to -50 C, strain is plunged into liquid nitrogen.
033063: After freezing program has lowered the temperature to -51 to -55 C, strain is plunged into liquid nitrogen.
033064: After freezing program has lowered the temperature to below -56 C, strain is plunged into liquid nitrogen.
033065: Strain is preserved as cysts.
033066: Strain is preserved as spores.
033067: Strain is preserved as vegetative cells.
033068: Strain can be stored dried.
033069: Strain can be stored only in broth or other liquid.
033070: Strain can be stored on slants under mineral oil.
033010: Strains survive only if maintained in appropriate animal hosts or in tissue cultures from these hosts.
033011: Strains survive storage in 15% glycerol at 0 C or below for 30 d.
033012: Strains survive 70% ethanol for 60 min.

SECTION 34: METABOLIC PATHWAYS AND ENZYMES

[NOTE 1: Enzyme names and enzyme numbers are taken from *Enzyme Nomenclature, Recommendations (1978) of the Nomenclature Committee of the International Union of Biochemistry*; Academic Press, New York. With few exceptions, the recommended names are listed first in Section 34, followed by "Other Names" or synonyms in a set of brackets, and/or finally followed by the official enzyme numbers in a set of parentheses.]

[NOTE.2: Enzymes common to two or more pathways are mentioned once only and are not repeated (although understood to be present) in the listing of enzymes in the second pathway.]

GLYCOLYSIS (EMP PATHWAY)

034001: Hexoses are catabolized, at least in part, to 2 moles of pyruvate (and beyond) by EMP pathway (demonstrated by isotope distribution studies, or equivalent specific test, NOT simply stoichiometry).
034002: Hexokinase [Heterophosphatase] (2.7.1.1) is present.
034003: A specific glucokinase (2.7.1.2) is present.
034004: A specific ketohexokinase (2.7.1.3) is present.
034005: Hexoses are phosphorylated by an enzyme not using adenosine-5'-triphosphate (ATP) as the phosphate donor.
034006: Glucosephosphate isomerase [Phosphohexose isomerase] (5.3.1.9) is present.
034007: 6-Phosphofructokinase [Phosphohexokinase] (2.7.1.11) is present.
034008: Fructose-bisphosphate aldolase [Zymohexase, Aldolase] (4.1.2.13) is present.
034009: Triosephosphate isomerase [Phosphotriose isomerase] (5.3.1.1) is present.
034010: Glyceraldehyde-phosphate dehydrogenase [Triosephosphate dehydrogenase] (1.2.1.12) is present.
034011: Phosphoglycerate kinase (2.7.2.3) is present.
034012: Phosphoglyceromutase [Glycerate phosphomutase] (2.7.5.3) is present.
034013: Enolase [Phosphopyruvate hydratase] (4.2.1.11) is present.
034014: Pyruvate kinase [Phosphoenolpyruvate kinase] (2.7.1.40) is present.

ENTNER-DOUDOROFF PATHWAY

034015: Hexoses are catabolized, at least in part, to equimolar quantities of glyceraldehyde-3-phosphate and pyruvate (and beyond) by the Entner-Doudoroff Pathway (demonstrated by isotope distribution

Section 34: Metabolic pathways and enzymes

studies, or equivalent specific test, NOT simply stoichiometry).
034016: Glucose-6-phosphate dehydrogenase [Zwischenferment] (1.1.1.49) is present.
034017: Gluconolactonase [Lactonase] (3.1.1.17) is present.
034018: Phospogluconate dehydratase (4.2.1.12) is present.
034019: Phospho-2-keto-3-deoxy-gluconate aldolase (4.1.2.14) is present.

HEXOSE MONOPHOSPHATE OR WARBURG-DICKENS SYSTEM

034020: Hexoses are catabolized, at least in part, to carbon dioxide and pyruvate (and beyond) by the HMP system (demonstrated by isotope distribution studies, or equivalent specific test, NOT simply stoichiometry).
034021: Pentoses are catabolized, at least in part, to pyruvate (and beyond) by the HMP system (demonstrated by isotope distribution studies, or equivalent specific test, NOT simply stoichiometry).
034022: Pentoses are catabolized, at least in part, to carbon dioxide and pyruvate (and beyond) by the HMP system (demonstrated by isotope distribution studies, or equivalent specific test, NOT simply stoichiometry).
034023: Xylose isomerase (5.3.1.5) is present.
034024: Xylulokinase (2.7.1.17) is present.
034025: L-Arabinose isomerase (5.3.1.4) is present.
034026: Ribulokinase (2.7.1.16) is present.
034027: L-Ribulosephosphate 4-epimerase (5.1.3.4) is present.
034028: Ribokinase (2.7.1.15) is present.
034029: Ribulosephosphate 3-epimerase [Phosphoribulose epimerase] (5.1.3.1) is present.
034030: Phosphogluconate dehydrogenase (decarboxylating) [Phosphogluconic acid dehydrogenase, 6-Phosphogluconic dehydrogenase, 6-Phosphogluconic acid carboxylase] (1.1.1.44) is present.
034031: Ribosephosphate isomerase [Phosphopentose isomerase, Phosphoriboseisomerase] (5.3.1.6) is present.
034033: Transketolase [Glycolaldehydetransferase] (2.2.1.1) is present.
034034: Transaldolase [Dihydroxyacetonetransferase] (2.2.1.2) is present.
034035: Fructose-bisphosphatase [Hexosediphosphatase] (3.1.3.11) is present.

PHOSPHOKETOLASE PATHWAYS

034036: Pentoses are catabolized, at least in part, to acetate and pyruvate (and beyond) by the pentose phosphoketolase pathway (demonstrated by isotope distribution studies, or equivalent specific test, NOT simply stoichiometry).
034037: Hexoses are catabolized, at least in part, to erythrose-4-phosphate and acetylphosphate (and

beyond) by the hexose phosphoketolase pathway (demonstrated by isotope distribution studies, or equivalent specific test, NOT simply stoichiometry).
034038: Phosphoketolase (4.1.2.9) is present.
034039: Fructose-6-phosphate phosphoketolase (4.1.2.22) is present.
034040: Acetate kinase [Acetokinase] (2.7.2.1) is present.

CITRIC ACID CYCLE

034041: Pyruvate is catabolized to carbon dioxide through the citric acid cycle (demonstrated by isotope distribution studies, or equivalent specific test, NOT simply stoichiometry).
034042: Pyruvate dehydrogenase (cytochrome) [Pyruvate dehydrogenase, Pyruvic dehydrogenase] (1.2.2.2) is present.
034043: Citrate (si)-synthase [Condensing enzyme] (4.1.3.7) is present.
034044: Aconitate hydratase [Aconitase] (4.2.1.3) is present.
034045: Isocitrate dehydrogenase (NAD^+) [Isocitric dehydrogenase] (1.1.1.41) is present.
034046: Isocitrate dehydrogenase ($NADP^+$) (1.1.1.42) is present.
034047: Oxoglutarate dehydrogenase [α-Ketoglutaric dehydrogenase] (1.2.4.2) is present.
034048: Dihydrolipoamide acetyltransferase [Lipoate acetyltransferase] (2.3.1.12) is present.
034049: Succinyl-CoA hydrolase [Succinyl-CoA acylase] (3.1.2.3) is present.
034050: Succinyl-CoA synthetase (GDP-forming) [Succinic thiokinase] (6.2.1.4) is present.
034051: Succinate dehydrogenase [Succinic dehydrogenase, Fumarate reductase, Fumaric hydrogenase] (1.3.99.1) is present.
034052: Fumarate hydratase [Fumarase] (4.2.1.2) is present.
034053: Malate dehydrogenase [Malic dehydrogenase] (1.1.1.37) is present.
034054: Malate synthase [Malate condensing enzyme, Glyoxylate transacetylase, Malate synthetase] (4.1.3.2) is present.

GLYOXYLATE CYCLE

034055: Isocitrate lyase [Isocitrase, Isocitritase, Isocitratase] (4.1.3.1) is present.
034056: Glyoxylate dehydrogenase (acylating) (1.2.1.17) is present.

AUTOTROPHIC CARBON DIOXIDE FIXATION

034057: Ribulosebisphosphate carboxylase [Carboxydismutase] (4.1.1.39) is present.
034059: Glyceraldehyde-phosphate dehydrogenase ($NADP^+$)

Section 34: Metabolic pathways and enzymes

(phosphorylating) [Triosephosphate dehydrogenase (NADP$^+$)] (1.2.1.13) is present.
034061: Phosphoribulokinase [Phosphopentokinase] (2.7.1.19) is present.

HETEROTROPHIC CARBON DIOXIDE FIXATION

034062: Phosphoenolpyruvate carboxylase (4.1.1.31) is present.
034063: Phosphoenolpyruvate carboxykinase (GTP) [Phosphoenolpyruvate carboxylase, Phosphopyruvate carboxylase] (4.1.1.32) is present.
034064: 2-Oxoglutarate synthase (1.2.7.3) is present.
034065: Acetyl-CoA carboxylase (6.4.1.2) is present.
034066: Propionyl-CoA carboxylase (ATP-hydrolysing) (6.4.1.3) is present.
034067: Methylcrotonoyl-CoA carboxylase (6.4.1.4) is present.
034068: Malate dehydrogenase (oxaloacetate-decarboxylating) ['Malic' enzyme, Pyruvic-malic carboxylase] (1.1.1.38) is present.
034069: Carbamate kinase (2.7.2.2) is present.
034070: Pyruvate carboxylase [Pyruvic carboxylase] (6.4.1.1) is present.

GENERAL REACTIONS INVOLVING PYRUVATE

034071: Methylmalonyl-CoA carboxyltransferase [Transcarboxylase] (2.1.3.1) is present.
034072: Pyruvate oxidase [Pyruvic oxidase] (1.2.3.3) is present.
034073: L-Lactate dehydrogenase [Lactic acid dehydrogenase] (1.1.1.27) is present.
034074: D-Lactate dehydrogenase [Lactic acid dehydrogenase] (1.1.1.28) is present.
034075: L-Lactate dehydrogenase (cytochrome) [Lactic acid dehydrogenase] (1.1.2.3) is present.
034076: D-Lactate dehydrogenase (cytochrome) [Lactic acid dehydrogenase] (1.1.2.4) is present.
034077: Pyruvate decarboxylase [α-Carboxylase, α-Ketoacid carboxylase] (4.1.1.1) is present.

PYRUVATE TO ALCOHOL PATHWAY

034078: Formate dehydrogenase [Formate hydrogenlyase] (1.2.1.2) is present.
034153: Formate dehydrogenase (cytochrome) (1.2.2.1) is present.
034154: Formate dehydrogenase (NADP$^+$) (1.2.1.43) is present.
034079: Phosphate acetyltransferase [Phosphotransacetylase, Phosphoacylase] (2.3.1.8) is present.
034080: Acetaldehyde dehydrogenase (acylating) (1.2.1.10) is present.
034081: Aldehyde dehydrogenase (1.2.1.3) is present.
034082: Alcohol dehydrogenase [Aldehyde reductase] (1.1.1.1) is present.

PYRUVATE TO BUTANEDIOL PATHWAY

034083: Pyruvate dehydrogenase (lipoamide) [Pyruvate dehydrogenase] (1.2.4.1) is present.
034084: Acetolactate decarboxylase (4.1.1.5) is present.
034085: D(-)-Butanediol dehydrogenase [Butyleneglycol dehydrogenase] (1.1.1.4) is present.
034086: Acetoin dehydrogenase [Diacetyl reductase] (1.1.1.5) is present.

PYRUVATE TO BUTYRATE PATHWAY

034087: Acetyl-CoA acetyltransferase [Thiolase, Acetoacetyl-CoA thiolase] (2.3.1.9) is present.
034088: 3-Hydroxybutyrate dehydrogenase (1.1.1.30) is present.
034089: Enoyl-CoA hydratase [Crotonase, Enoyl hydrase, Unsaturated acyl-CoA hydratase] (4.2.1.17) is present.
034090: Butyryl-CoA dehydrogenase [Butyryl dehydrogenase, Ethylene reductase, Unsaturated acyl-CoA reductase] (1.3.99.2) is present.
034091: Fatty acid-S-CoA transferase is present.

PYRUVATE TO ISOPROPANOL PATHWAY

034092: Acetoacetate decarboxylase (4.1.1.4) is present.
034093: Isopropanol dehydrogenase (NADP$^+$) (1.1.1.80) is present.

PROPIONIC ACID FERMENTATION

034094: Fumarate reductase (NADH) (1.3.1.6) is present.
034095: 3-Ketoacid CoA-transferase (2.8.3.5) is present.

POLYOL DEHYDROGENASES

034096: Mannitol dehydrogenase (1.1.1.67) is present.
034097: Mannitol dehydrogenase (cytochrome) (1.1.2.2) is present.
034098: Ribitol dehydrogenase (1.1.1.56) is present.
034099: D-Xyulose reductase (1.1.1.9) is present.
034100: D-Arabinitol dehydrogenase (1.1.1.11) is present.
034101: L-Arabinitol dehydrogenase (1.1.1.12) is present.
034102: L-Arabinitol dehydrogenase (ribulose-forming) (1.1.1.13) is present.

MISCELLANEOUS ENZYMES

034103: Alginase (3.2.1.16) is present. [Alginase (3.2.1.16) has been deleted from *Enzyme Nomenclature (1978)*.]
034104: Amidase [Acylamidase, Acylase] (3.5.1.4) is present.
034105: D-Amino-acid oxidase (1.4.3.3) acts non-specifically.

Section 34: Metabolic pathways and enzymes

034106: L-Amino-acid oxidase (1.4.3.2) acts non-specifically.
034107: α-Amylase [Diastase, Ptyalin, Glycogenase] (3.2.1.1) is present.
034145: An extracellular enzyme is present which decomposes starch to a crystalline dextrin.
034108: Arylsulphatase [Sulphatase] (3.1.6.1) is present.
034109: Catalase (1.11.1.6) is present (hydrogen peroxide decomposed by an enzyme containing a heme or porphyrin structural group).
034110: Catalase-pseudo (an enzyme not containing a heme structural unit) decomposes hydrogen peroxide to water and oxygen.
034111: Cellulase [Endo-1,4-β-glucanase] (3.2.1.4) is present.
034112: Chitinase [Chitodextrinase] (3.2.1.14) is present.
034113: Cytochrome oxidase (1.9.3.1 or 1.9.3.2) is produced.
034114: Coagulase is produced (blood plasma clotted).
034115: Deoxyribonuclease, a deoxyribonucleic acid digesting enzyme, is produced.
034116: Dextranase (3.2.1.11) is present.
034117: Dextransucrase [Sucrose 6-glucosyltransferase] (2.4.1.5) is present.
034118: Dextrin dextranase [Dextrin 6-glucosyltransferase] (2.4.1.2) is present.
034119: Amylo-1,6-glucosidase (3.2.1.33) is present.
034120: Ferredoxin is present.
034150: α-L-Fucosidase (3.2.1.51) is present.
034121: NAD$^+$ peroxidase (1.11.1.1) is present.
034122: Phosphorylase (2.4.1.1) is present.
034123: β-D-Galactosidase [Lactase] (3.2.1.23) is present.
034124: Exo-1,4-α-D-glucosidase [Glucoamylase, Amyloglucosidase, ɣ-Amylase, Lysosomal α-glucosidase, Acid maltase] (3.2.1.3) is present.
034148: α-D-Glucosidase [Maltase] (3.2.1.20) is present.
034125: β-D-Glucosidase [Gentiobiase, Cellobiase, Amygdalase] (3.2.1.21) is present.
034149: β-D-Glucuronidase (3.2.1.31) is present.
034126: Hyaluronidase (3.2.1.35 or 3.2.1.36) is produced.
034152: Carboxylesterase [Ali-esterase, B-esterase] (3.1.1.1) is produced.
034127: Inulinase [Inulase] (3.2.1.7) is present.
034128: Endo-1,3(4)-β-D-glucanase [Endo-1,3-β-D-glucanase, Laminarinase] (3.2.1.6) is present.
034129: Lecithinase is produced.
034130: Leucocidin is produced (lyses leucocytes).
034131: Levansucrase [Sucrose 6-fructosyltransferase] (2.4.1.10) is present.
034132: Methylmalonyl-CoA carboxyltransferase [Transcarboxylase] (2.1.3.1) is present.
034146: Acetyl-α-naphthylamine esterase [α-Esterase] is produced.
034147: Acetyl-β-naphthylamine esterase [α-Esterase] is produced.

034133: Neuraminidase [Sialidase] (3.2.1.18) is present.
034134: Pectinase [Polygalacturonase] (3.2.1.15) is present.
034135: Peroxidase (1.11.1.7) is present.
034136: Acid phosphatase [Acid phosphomonoesterase, Phosphomonoesterase, Glycerophosphatase] (3.1.3.2) is present.
034137: Alkaline phosphatase [Alkaline phosphomonoesterase, Phosphomonoesterase, Glycerophosphatase] (3.1.3.1) is produced.
034138: Plasmin [Fibrinase, Fibrinolysin] (3.4.21.7) is present.
034139: Ribonuclease (3.1.27.5 or 3.1.27.1) is produced.
034141: Rubredoxin is present.
034142: α,α-Trehalase (3.2.1.28) is present.
034143: Urease (3.5.1.5) is produced.
034144: Endo-1,4-β-D-xylanase (3.2.1.8) is present.
034151: Exo-1,4-β-D-xylosidase [Xylobiase, β-Xylosidase] (3.2.1.37) is present.

NUCLEOSIDASES AND RELATED ENZYMES - HYDROLYSING N-GLYCOSYL COMPOUNDS

[NOTE 3: These enzymes (3.2.2.1-3.2.14) catalyze the reactions an N-ribosylpurine + H_2O = a purine + D-ribose or an N-ribosylpyrimidine + H_2O = a pyrimidine + D-ribose.]

034155: A nucleosidase [Purine nucleosidase, N-ribosylpurine ribohydrolase] (3.2.2.1) is produced.
034156: Inosine nucleosidase [Inosinase, Inosine ribohydrolase] (3.2.2.2) is produced.
034157: Uridine nucleosidase [Uridine ribohydrolase] (3.2.2.3) is produced.
034158: AMP nucleosidase [AMP phosphoribohydrolase] (3.2.2.4) is produced.
034159: NAD^+ nucleosidase [NADase, DPNase, DPN hydrolase, NAD^+ glycohydrolase] (3.2.2.5) is produced. [Also catalyzes transfer of ADP ribose residues.]
034160: $NAD(P)^+$ nucleosidase [$NAD(P)^+$ glycohydrolase] (3.2.2.6) is produced. [Also catalyses transfer of ADPribose(P) residues.]
034161: Adenosine nucleosidase [Adenosinase, Adenosine ribohydrolase] (3.2.2.7) is produced. [Also acts on adenosine N-oxide.]
034162: N-Ribosylpyrimidine nucleosidase [Nucleoside ribohydrolase] (3.2.2.8) is produced. [Also hydrolyses purine ribonucleosides but at a slower rate.]
034163: Adenosylhomocysteine nucleosidase [S-Adenosyl-L-homocysteine homocysteinlribohydrolase] (3.2.2.9) is produced. [Also acts on 5'-methylthioadenosine to give adenine and 5-methylthioribose.]
034164: Pyrimidine-5'-nucleotide nucleosidase [Pyrimidine-5'-nucleotide phosphoribo(deoxyribo)hydrolase] (3.2.2.10) is produced. [Also acts on dUMP, dTMP and dCMP.]

Section 34: Metabolic pathways and enzymes

034165: Inosinate nucleosidase [5'-Inosinate phosphoribohydrolase] (3.2.2.12) is produced.
034166: 1-Methyladenosine nucleosidase [1-Methyladenosine ribohydrolase] (3.2.2.13) is produced.
034167: NMN nucleosidase [NMNase, Nicotinamidenucleotide phosphoribohydrolase] (3.2.2.14) is produced.

NUCLEOSIDASES AND RELATED ENZYMES - HYDROLYSING THIOETHER COMPOUNDS

034168: Adenosylhomocysteinase [S-Adenosyl-L-homocysteine hydrolase] (3.3.1.1) is produced and catalyzes the reaction: S-Adenosyl-L-homocysteine + H_2O = adenosine + L-homocysteine.
034169: Adenosylmethionine hydrolase [S-Adenosylmethionine cleaving enzyme, Methylmethionine-sulphonium-salt hydrolase, S-Adenosyl-L-methionine hydrolase] (3.3.1.2) is produced and catalyzes the reaction: S-Adenosyl-L-methionine + H_2O = methylthioadenosine + L-homoserine.
034170: Ribosylhomocysteinase [S-Ribosyl-L-homocysteine ribohydrolase] (3.3.1.3) is produced and catalyzes the reaction: S-Ribosyl-L-homocysteine + H_2O = ribose + L-homocysteine.

NUCLEOSIDASES AND RELATED ENZYMES - PENTOSYL TRANSFERASES

[NOTE 4: These enzymes (2.4.2.1 - 2.4.2.23) generally catalyze the reactions: Purine or pyrimidine nucleoside + orthophosphate or pyrophosphate = purine or pyrimidine + a ribose or ribosyl phosphate, ribosyl diphosphate, or ribosyl pyrophosphate.]

034171: Nucleoside ribosyltransferase [Nucleoside: purine (pyrimidine) ribosyltransferase] (2.4.2.5) is produced and catalyzes the reaction: D-Ribosyl-R + R' = D-ribosyl-R' + R.
034172: Nucleoside deoxyribosyltransferase [trans-N-Glucosidase, Nucleoside: purine (pyrimidine) deoxyribosyltransferase, nucleoside transdeoxyribosidase] (2.4.2.6) is produced and catalyzes the reaction: 2-Deoxy-D-ribosyl-R+R' = 2-deoxy-D-ribosyl-R' + R [R and R' represent various purines and pyrimidines.]
034173: Purine-nucleoside phosphorylase [Inosine phosphorylase, Purine-nucleoside:orthophosphate ribosyltransferase] (2.4.2.1) is produced. [Specificity uncertain; also catalyses ribosyltransferase reactions of the type catalyzed by EC 2.4.2.5.]
034174: Pyrimidine-nucleoside phosphorylase [Pyrimidine-nucleoside:orthophosphate ribosyltransrerase] (2.4.2.2) is produced.
034175: Uridine phosphorylase [Pyrimidine phosphorylase, Uridine:orthophosphate ribosyltransferase] (2.4.2.3) is produced.

034176: Thymidine phosphorylase [Pyrimidine phosphorylase, Thymidine:orthophosphate deoxyribosyltransferase] (2.4.2.4) is produced. [The enzyme may also catalyse reactions of the type catalyzed by EC 2.4.2.6.]
034177: Adenine phosphoribosyltransferase [AMP pyrophosphorylase, Transphosphoribosidase, AMP:pyrophosphate phosphoribosyltransferase] (2.4.2.7) is produced. [5-Amino-4-imidazolecarboxamide can replace adenine.]
034178: Hypoxanthine phosphoribosyltransferase [IMP pyrophosphorylase, Transphosphoribosidase, IMP:pyrophosphate phosphoribosyltransferase] (2.4.2.8) is produced. [Guanine and 6-mercaptopurine can replace hypoxanthine.]
034179: Uracil phosphoribosyltransferase [UMP pyrophosphorylase, UMP:pyrophosphate phosphoribosyltransferase] (2.4.2.9) is produced.
034180: Orotate phosphoribosyltransferase [Orotidylic acid phosphorylase, Orotidine-5'-phosphate pyrophosphorylase, Orotidine-5'-phosphate: pyrophosphate phosphoribosyltransferase] (2.4.2.10) is produced.
034181: Nicotinate phosphoribosyltransferase [Nicotinatenucleotide:pyrophosphate phosphoribosyltransferase] (2.4.2.11) is produced.
034182: Nicotinamide phosphoribosyltransferase [NMN pyrophosphorylase, Nicotinamidenucleotide: pyrophosphate phosphoribosyltransferase] (2.4.2.12) is produced.
034183: Guanosine phosphorylase [Guanosine:orthophosphate ribosyltransferase] (2.4.2.15) is produced. [Also acts on deoxyguanosine.]
034184: Urateribonucleotide phosphorylase [Urate-ribonucleotide:orthophosphate ribosyltransferase] (2.4.2.16) is produced.
034185: ATP phosphoribosyltransferase [Phosphoribosyl-ATP pyrophosphorylase, 1-(5'-Phosphoribosyl)- ATP:pyrophosphate phosphoribosyltransferase] (2.4.2.17) is produced.
034186: Nicotinatemononucleotide pyrophosphorylase (carboxylating) [Nicotinatenucleotide:pyrophosphate phosphoribosyltransferase (carboxylating)] (2.4.2.19) is produced and catalyzes the reaction: Nicotinate D-ribonucleotide + pyrophosphate + CO_2 = quinolinate + 5-phospho-α-D-ribose-1-diphosphate.
034187: Dioxotetrahydropyrimidine phosphoribosyltransferase [Dioxotetrahydro-pyrimidine ribonucleotide pyrophosphorylase, 2,4-Dioxotetra-hydropyrimidine nucleotide:pyrophosphate phosphoribosyltransferase] (2.4.2.20) is produced and catalyzes the reaction: A 2,4-dioxotetrahydropyrimidine D-ribonucleotide + pyrophosphate = a 2,4-dioxotetrahydropyrimidine + 5-phospho-α-D-ribose-1-diphosphate. [Acts (in the reverse direction) on uracil and other pyrimidines and pteridines containing a 2,4-diketo structure.]

Section 34: Metabolic pathways and enzymes 211

034188: Nicotinatenucleotide-dimethylbenzimidazole phosphoribosyltransferase (2.4.2.21) is produced and catalyzes the reaction: β-Nicotinate D-ribonucleotide + dimethyl-benzimidazole = nicotinate + 1-α-D-ribosyl-5,6-dimethylbenzimidazole 5'-phosphate. [Also acts on benzimidazole and the clostridial enzyme acts on adenine to form 7-α-D-ribosyl-adenine 5'-phosphate.]
034189: Xanthine phosphoribosyltransferase [5-Phospho-α-D-ribose-1-diphosphate:xanthine phosphoribosyltransferase] (2.4.2.22) is produced.
034190: Deoxyuridine phosphorylase [Deoxyuridine:orthophosphate deoxyribosyltransferase] (2.4.2.23) is produced.

NUCLEOSIDASES AND RELATED ENZYMES - TRANSFERRING A
PHOSPHORUS-CONTAINING GROUP-KINASES

[NOTE 5: These enzymes (2.7.1.20-2.7.1.73 and 2.7.1.76, 2.7.1.78 and 2.7.1.86) catalyze the general type of reaction: ATP + substrate = ADP + phosphorylated substrate as illustrated by 2.7.1.20.]

034191: Adenosine kinase [ATP-adenosine 5'-phosphotransferase] (2.7.1.20) is produced and catalyzes the reaction: ATP + adenosine = ADP + AMP. [2-Aminoadenosine can also act as acceptor.]
034192: Thymidine kinase [ATP-thymidine 5'-phosphotransferase] (2.7.1.21) is produced. [Deoxyuridine is also an acceptor, and dGTP is a donor.]
034193: Ribosylinicotinamide kinase [ATP:N-ribosyl-nicotinamide 5'-phosphotransferase] (2.7.1.22) is produced.
034194: Ribosylnicotinamide kinase [ATP:N-ribosyl-nicotinamide 5'-phosphotransferase] (2.7.1.22) is produced.
034195: NAD$^+$ kinase [ATP:NAD$^+$ 2'-phosphotransferase, DPN kinase] (2.7.1.23) is produced.
034196: Adenylylsulphate kinase [ATP:adenylylsulphate 3'-phosphotransferase] (2.7.1.25) is produced.
034197: Uridine kinase [ATP:uridine 5'-phosphotransferase] (2.7.1.48) is produced. [Cytidine can act as acceptor; GTP or ITP can act as donor.]
034198: Hydroxymethylpyrimidine kinase [ATP:2-methyl-4-amino-5-hydroxymethylpyrimidine 5-phosphotransferase] (2.7.1.49) is produced. [CTP, UTP, and GTP can act as a donor.]
034199: Inosine kinase [ATP:inosine 5'-phosphotransferase] (2.7.1.73) is produced.
034200: Deoxycytidine kinase [NTP:deoxycytidine 5'-phosphotransferase] (2.7.1.74) is produced and catalyzes the reaction: NTP + deoxycytidine = NDP + dCMP. [Cytosine arabinoside can act as acceptor; all natural nucleoside triphosphates (except dCTP) can act as donor.]

Section 34: Metabolic pathways and enzymes

034201: Deoxyadenosine kinase [ATP:deoxyadenosine 5'-phosphotransferase] (2.7.1.76) is produced. [Deoxyguanosine can also act as acceptor.]
034202: Nucleoside phosphotransferase [Nucleotide: 3'-deoxynucleoside 5'-phosphotransferase] (2.7.1.77) is produced and catalyzes the reaction: A nucleotide + 3'-deoxynucleoside = a nucleoside + 3'-deoxynucleoside 5'-monophosphate. [Phenyl-phosphate as well as 3'-and 5'-nucleotides can act as phosphate donor.]
034203: Polynucleotide 5'-hydroxylkinase [ATP:5'-dephosphopolynucleotide 5'-phosphotransferase] (2.7.1.78) is produced. [Also acts on 5'-dephospho-RNA and 3'-mononucleotides.]
034204: NADH kinase [ATP:NADH 2'-phosphotransferase] (2.7.1.86) is produced. [CTP, ITP, UTP, and GTP can also act as phosphate donors.]

NUCLEOSIDASES AND RELATED ENZYMES - PHOSPHOTRANSFERASES WITH A PHOSPHATE GROUP AS ACCEPTOR

[NOTE 6: The enzymes (2.7.4.3-2.7.4.14) differ functionally from the usual kinases in catalyzing the transfer of P from ATP or nucleoside triphosphate acting as donor to already phosphorylated acceptor compounds to yield ADP + nucleoside diphosphate. See examples of 2.7.4.3 and 2.7.4.4.]

034205: Adenylate kinase [ATP:AMP phosphotransferase, Myokinase] (2.7.4.3) is produced and catalyzes the reaction: ATP + AMP = ADP + ADP. [Inorganic triphosphate can also act as donor.]
034206: Nucleosidemonophosphate kinase [ATP:nucleosidemonophosphate phosphotransferase] (2.7.4.4) is produced and catalyzes the reaction: ATP + nucleoside monophosphate = ADP + nucleoside diphosphate. [Many nucleotides can act as acceptor; other nucleoside triphosphates can act instead of ATP.]
034207: Nucleosidediphosphate kinase [ATP:nucleoside-diphosphate phosphotransferase] (2.7.4.6) is produced. [Many nucleoside diphosphates can act as acceptor, while many ribo-and deoxyribonucleoside triphosphates can act as donor.]
034208: Phosphomethylpyrimidine kinase [ATP: 2-methyl-4-amino-5-phosphomethylpyrimidine phosphotransferase] (2.7.4.7) is produced.
034209: Guanylate kinase [ATP:(d)GMP phosphotransferase, Deoxyguanylate kinase] (2.7.4.8) is produced. [dGMP can also act as acceptor, and dATP can act as donor.]
034210: dTMP kinase [ATP:dTMP phosphotransferase] (2.7.4.9) is produced.
034211: Nucleosidetriphosphate-adenylate kinase [Nucleosidetriphosphate: AMP phosphotransferase] (2.7.4.10) is produced. [Many nucleoside triphosphates can act as donor.]

Section 34: Metabolic pathways and enzymes 213

034212: (Deoxy)adenylate kinase [ATP:(d)AMP phosphotransferase] (2.7.4.11) is produced. [AMP can also act as acceptor.]
034213: T2-induced deoxynucleotide kinase [ATP:(d)NMP phosphotransferase] (2.7.4.12) is produced. [dTMP and dAMP can act as acceptor; dATP can act as donor.]
034214: (Deoxy)nucleosidemonophosphate kinase [ATP;deoxynucleosidemonophosphate phosphotransferase] (2.7.4.13) is produced. [dATP can substitute for ATP.]
034215: Cytidylate kinase [ATP:CMP phosphotransferase, Deoxycytidylate kinase] (2.7.4.14) is produced. [UMP and dCMP can also act as acceptors.]

NUCLEOSIDASES AND RELATED ENZYMES - DIPHOSPHOTRANSFERASES

034216: Nucleotide pyrophosphokinase [ATP:nucleoside-5'-monophosphate pyrophosphotransferase] (2.7.6.4) is produced and catalyzes the reaction: ATP + nucleoside 5'-monophosphate = AMP + 5'-phosphonucleoside 3'-diphosphate. [Acts on 5'-mono-, di- and triphosphate derivatives of purine nucleosides.]

NUCLEOSIDASES AND RELATED ENZYMES - NUCLEOTIDYLTRANSFERASES

034217: NMN adenylyltransferase [ATP:NMN adenylyltransferase, NAD$^+$ + pyrophosphorylase] (2.7.7.1) is produced and catalyzes the reaction: ATP + nicotinamide ribonucleotide = pyrophosphate + NAD$^+$. [Nicotinate nucleotide can also act as acceptor. See also EC 2.7.7.18.]
034218: RNA nucleotidyltransferase [Nucleosidetriphosphate: RNA nucleotidyltransferase, RNA polymerase] (2.7.7.6) is produced and catalyzes the reaction: n Nucleoside triphosphate = n pyrophosphate + RNA. [Needs DNA as template. See also EC 2.7.7.19.]
034219: DNA nucleotidyltransferase [Deoxynucleosidetriphosphate: DNA deoxynucleotidyltransferase, DNA polymerase] (2.7.7.7) is produced and catalyzes the reaction: n Deoxynucleoside triphosphate = n-pyrophosphate + DNAn. [a DNA chain acts as a template, and the enzyme forms a complementary chain. Also acts on DNA-RNA hybrids.]
034220: Polyribonucleotide nucleotidyltransferase [Polyribonucleotide: orthophosphate nucleotidyltransferase, Polynucleotide phosphorylase] (2.7.7.8) is produced and catalyzes the reaction: RNA$n+1$ + orthophosphate = RNAn + a nucleoside diphosphate. [ADP, IDP, GDP, UDP and CDP can act as donor.]
034221: Polynucleotide adenylyltransferase [ATP:polynucleotide adenylyltransferase, NTP polymerase, RNA

214   Section 34: Metabolic pathways and enzymes

            adenylating enzyme] (2.7.7.19) is produced and
            catalyzes the reaction: n ATP + (nucleotide)m
            = n pyrophosphate + (nucleotide)m+n [Also
            acts slowly with CTP. The primer may be an RNA or
            DNA fragment or oligo(A) bearing a 3'-OH terminal
            group. See also EC 2.7.7.6.]
034222:   tRNA cytidylyltransferase [CTP:tRNA cytidyly-
            transferase, tRNA CCA-pyrophosphorylase]
            (2.7.7.21) is produced and catalyzes the reaction:
            CTP + tRNAn = pyrophosphate + tRNA n+1. [May
            be identical with EC2.7.7.25.]
034223:   tRNA adenylyltransferase [ATP:tRNA adenylyltrans-
            ferase, tRNA CCA-pyrophosphorylase] (2.7.7.25) is
            produced and catalyzes the reaction: ATP + tRNAn
            = pyrophosphate + tRNA n+1. [May be identical
            with EC 2.7.7.21.]
034224:   Nucleosidetriphosphate-hexose-1-phosphate
            nucleotidyltransferase [NTP:hexose-1-phosphate
            nucleotidyltransferase, NDPhexose pyrophos-
            phorylase] (2.7.7.28) is produced and catalyzes
            the reaction: Nucleoside triphosphate + hexose
            1-phosphate = pyrophosphate + NDPhexose.
            [Guanosine, inosine and adenosine diphosphate
            hexoses are substrates in the reverse reaction with
            either glucose or mannose as the sugar.]
034225:   DNA nucleotidylexotransferase [Nucleosidetriphos-
            phate:DNA deoxynucleotidylexo-transferase, Terminal
            deoxyribonucleotidyl transferase, Terminal addition
            enzyme] (2.7.7.31) is produced and catalyzes the
            reaction: n Deoxynucleoside triphosphate +
            (deoxynucleotide)m = n pyrophosphate +
            (deoxynucleotide)m+n. [Nucleoside may be ribo-
            or deoxyribo-, n must be greater than 3 and a
            3'-OH is required.]
034226:   Sugar-1-phosphate nucleotidyltransferase [NDP:
            sugar-1-phosphate nucleotidyltransferase, NDPsugar
            phosphorylase] (2.7.7.37) is produced and
            catalyzes the reaction: NDP + sugar 1-phosphate =
            orthophosphate + NDPsugar.
034227:   Gentamicin 2'-nucleotidyltransferase [NTP:genta-
            micin 2'-nucleotidyltransferase] (2.7.7.46) is
            produced and catalyzes the reaction: Nucleoside
            triphosphate + gentamicin = pyrophosphate +
            2'-nucleotidylgentamicin. [ATP, dATP, CTP, ITP
            and GTP can act as donors; kanamycin, tobramycin
            and sisomicin can also act as acceptors.]

    NUCLEOTIDASES AND RELATED ENZYMES - PHOSPHORIC MONOESTER
        HYDROLASES

[NOTE 7: These enzymes function as shown in the example
          of 3.1.3.31.]

034228:   5'-Nucleotidase [5'-Ribonucleotide phospho-
            hydrolase] (3.1.3.5) is produced. [Wide specifi-
            city for 5'-nucleotides.]

Section 34: Metabolic pathways and enzymes 215

034229: 3'-Nucleotidase [3'-ribonucleotide phosphohydrolase] (3.1.3.6) is produced. [Wide specificity for 3'-nucleotides.]

034230: Phosphoadenylate 3'-nucleotidase [Adenosine-3',5'-bisphosphate 3'-phosphohydrolase] (3.1.3.7) is produced. [Also acts on 3'-phosphoadenylylsulphate.]

034231: Nucleotidase [Nucleotide phosphohydrolase] (3.1.3.31) is produced and catalyzes the reaction: Nucleotide + H_2O = nucleoside + orthophosphate. [A wide specificity for 2'-, 3'- and 5'-nucleotides; also hydrolyses glycerol phosphate and 4-nitrophenyl phosphate.]

034232: Polynucleotide 3'-phosphatase [Polynucleotide 3'-phosphohydrolase, 2'(3')-Polynucleotidase] (3.1.3.32) is produced.

034233: Polynucleotide 5'-phosphatase [Polynucleotide 5'-phosphohydrolase, 5'-Polynucleotidase] (3.1.3.33) is produced. [Does not act on nucleoside monophosphates. Induced in *Escherichia coli* by T-even phages.]

034234: Deoxynucleotide 3'-phosphatase [Deoxyribonucleotide 3'-phosphohydrolase, 3'-Deoxynucleotidase] (3.1.3.34) is produced. [Also catalyses the selective removal of 3'-phosphate groups from DNA and oligodeoxynucleotides. Induced in *Escherichia coli* by T-even phages.]

034235: Thymidylate 5'-phosphatase [Thymidylate 5'-phosphohydrolase, Thymidylate 5'-nucleotidase] (3.1.3.35) is produced. [Also acts on 5'-methyl-dCMP and 5'-ribothymidylate but at lower rates.]

NUCLEOTIDASES AND RELATED ENZYMES - PHOSPHORIC DIESTER HYDROLASES

034236: Phosphodiesterase I [Oligonucleate 5'-nucleotidohydrolase, 5'-Exonuclease] (3.1.4.1) is produced and catalyzes the reaction: Hydrolytically removes 5'-nucleotides successively from the 3'-hydroxy termini of 3'-hydroxy-terminated oligonucleotides. [Low activity towards polynucleotides. A 3'-phosphate terminus on the substrate inhibits hydrolysis.]

034237: 2':3'-Cyclic-nucleotide 2'-phosphodiesterase [Nucleoside-2':3'-cyclic-phosphate 3'-nucleotidohydrolase] (3.1.4.16) is produced and catalyzes the reaction: Nucleoside 2':3'-cyclic phosphate + H_2O = nucleoside 3'-phosphate. [Also hydrolyses 3'-nucleoside monophosphates and bis-*p*-nitrophenyl phosphate, but not 3'-deoxynucleotides. Similar reactions are carried out by EC 3.1.4.8 and 3':5'-cyclic AMP, 3.1.4.22.]

034238: 3':5'-Cyclic-nucleotide phosphodiesterase [3':5'-Cyclic-nucleotide 5'-nucleotidohydrolase] (3.1.4.17) is produced and catalyzes the reaction: Nucleoside 3':5'-cyclic phosphate + H_2O = nucleoside 5'-phosphate. [Acts on 3':5'-cyclic

AMP, 3':5'-cyclic dAMP, 3':5'-cyclic IMP, 3':5'-cyclic GMP and 3':5'-cyclic CMP.]

034239: 2':3'-Cyclic-nucleotide 3'-phosphodiesterase [Nucleoside-2':3'-cyclic-phosphate 2'-nucleotidohydrolase] (3.1.4.37) is produced and catalyzes the reaction: Nucleoside 2':3'-cyclic phosphate + H_2O = nucleoside 2'-phosphate.

NUCLEOSIDASES AND RELATED ENZYMES - TRIPHOSPHORIC MONOESTER HYDROLASES

034240: dGTPase [dGTP triphosphohydrolase, Deoxy-GTPase], (3.1.5.1) is produced and catalyzes the reaction: dGTP + H_2O = deoxyguanosine + triphosphate. [Also acts on GTP.]

NUCLEOSIDASES AND RELATED ENZYMES - EXODEOXYRIBONUCLEASES PRODUCING 5'-PHOSPHOMONOESTERS

[NOTE 8: These enzymes all produce 5'-phosphomono-, di- or oligonucleotides by exonucleolytic cleavage in the 3' to 5' direction (3.1.11.1, 3.1.11.2) or the 5' to 3' direction (3.1.11.3, 3.1.11.4) or in either the 5' to 3' or 3' to 5' directions (3.1.11.5, 3.1.11.6).]

034241: Exodeoxyribonuclease I [*E. coli* exonuclease I. Similar enzymes: Mammalian DNase III, Exonuclease IV, T2 and T4 induced exodeoxyribonucleases] (3.1.11.1) is produced. [Preference for single stranded DNA. The *E. coli* enzyme hydrolyses glucosylated DNA. Formerly EC 3.1.4.25.]

034242: Exodeoxyribonuclease III [*E. coli* exonuclease III. Similar enzymes: *Haemophilus influenzae* exonuclease] (3.1.11.2) is produced. [Preference for double-stranded DNA. Has endonucleolytic activity near apurinic sites on DNA. Formerly EC 3.1.4.27.]

034243: Exodeoxyribonuclease (Lambda-induced) [Lambda exonuclease. Similar enzymes: T4, T5 and T7 exonucleases, Mammalian DNase IV] (3.1.11.3) is produced. [Preference for double-stranded DNA. Does not attack single-strand breaks. Formerly EC 3.1.4.28.]

034244: Exodeoxyribonuclease (Phage SP3-induced) [Phage SP3 DNase, DNA 5'-dinucleotidohydrolase] (3.1.11.4) is produced. [Preference for single-stranded DNA. Formerly EC 3.1.4.31.]

034245: Exodeoxyribonuclease V [*E. coli* exonuclease V. Similar enzyme: *H. influenzae* ATP-dependent DNase] (3.1.11.5) is produced. [Preference for double-stranded DNA. Possesses DNA-dependent ATPase activity. Acts endonucleolytically on single-stranded circular DNA.]

034246: Exodeoxyribonuclease VII [*E. coli* exonuclease VII. Similar to: *Micrococcus luteus* exonucle-

Section 34: Metabolic pathways and enzymes

ase] (3.1.11.6) is produced. [Preference for single-stranded DNA.]

NUCLEOSIDASES AND RELATED ENZYMES - EXORIBONUCLEASES PRODUCING 5'-PHOSPHOMONOESTERS

034247: Ribonuclease II [Exoribonuclease. Similar enzymes: *Lactobacillus plantarum* RNase, Mouse nuclear RNase, Oligoribonuclease of *E. coli*] (3.1.13.1) is produced and catalyzes the reaction: Exonucleolytic cleavage in the 3' to 5' direction. [Preference for single-stranded RNA. Formerly EC 3.1.4.20.]

034248: Exoribonuclease H (3.1.13.2) is produced and catalyzes the reaction: Exonucleolytic cleavage to 5'-phosphomonoester oligonucleotides in both 5' to 3' and 3' to 5' directions. [Attacks RNA in duplex with DNA strand. Found in certain oncorna viruses and animal cells.]

034249: Oligonucleotidase (3.1.13.3) is produced and catalyzes the reaction: Exonucleolytic cleavage of oligonucleotides to yield 5'-phosphomononucleotides. [Also hydrolyzes NAD^+ to NMN and AMP. Formerly EC 3.1.4.19.]

NUCLEOSIDASES AND RELATED ENZYMES - EXORIBONUCLEASES PRODUCING OTHER THAN 5'-PHOSPHOMONOESTERS

034250: Yeast ribonuclease [Similar enzyme: RNase U4] (3.1.14.1) is produced and catalyzes the reaction: Exonucleolytic cleavage to 3'-phosphomononucleotides.

NUCLEOSIDASES AND RELATED ENZYMES - EXONUCLEASES ACTIVE WITH EITHER RIBO- OR DEOXYRIBONUCLEIC ACIDS AND PRODUCING 5'-PHOSPHOMONOESTERS

034251: Venom exonuclease [Similar enzymes: Hog Kidney phosphodiesterase, *Lactobacillus* exonuclease] (3.1.15.1) is produced and catalyzes the reaction: Exonucleolytic cleavage in the 3' to 5' direction to yield 5'-phosphomonocleotides. [Preference for single-stranded substrate. The venom enzyme is also active with superhelical turns.]

NUCLEOSIDASES AND RELATED ENZYMES - EXONUCLEASES ACTIVE WITH EITHER RIBO- OR DEOXYRIBONUCLEIC ACIDS AND PRODUCING OTHER THAN 5'-PHOSPHOMONOESTERS

034252: Spleen exonuclease [3'-Exonuclease, Spleen phosphodiesterase. Similar enzymes: *Lactobacillus acidophilus*, *B. subtilis*, and Salmon testis nucleases.] (3.1.16.1) is produced and catalyzes the reaction: Exonucleolytic cleavage in the 5' to 3' direction to yield 3'-phosphomononucleotides. [Preference for single-stranded substrate. Formerly EC 3.1.4.18.]

NUCLEOSIDASES AND RELATED ENZYMES - ENDODEOXYRIBONUCLEASES PRODUCING 5'-PHOSPHOMONOESTERS

034253: Deoxyribonuclease I [Pancreatic DNase, DNase, thymonuclease. Similar enzymes: Streptococcal DNase (Streptodornase), T4 Endonuclease II, *E. coli* endonuclease I "Nicking" nuclease of calf thymus, Colicin E2 and E3] (3.1.21.1) is produced. [Preference for double-stranded DNA. Formerly EC 3.1.4.5.]

034254: Endodeoxyribonuclease IV (Phage T4-induced) [*E. coli* endonuclease IV. Similar enzymes: DNase V (mammalian), *Aspergillus sojae* DNase, *B. subtilis* endonuclease, T4 Endonuclease III, T7 Endoculease I, *Aspergillus* DNase K2, Vaccinia virus DNase VI, yeast DNase, *Chlorella* DNase] (3.1.21.2) is produced. [Preference for simple stranded DNA. Formerly EC 3.1.4.30. Enzymes 3.1.21.1 and 3.1.21.2 act in the endocucleolytic cleavage to 5'-phospho- and oligonucleotide and products.]

NUCLEOSIDASES AND RELATED ENZYMES - ENDODEOXYRIBONUCLEASES PRODUCING OTHER THAN 5'-PHOSPHOMONOESTERS

034255: Deoxyribonuclease II [DNase II, Pancreatic DNase II. Similar enzymes: Crab testes DNase, Snail DNase, Salmon testis DNase, Liver acid DNase, Human acid DNase of gastric mucosa and cervix] (3.1.22.1) is produced. [Preference for double-stranded DNA. Formerly EC 3.1.4.6. Enzymes 3.1.22.1, 3.1.22.2, and 3.1.22.3 act in the endonucleolytic cleavage to 3'-phosphomono- and oligonucleotides.]

034256: *Aspergillus* Deoxyribonuclease K1 [*Aspergillus* DNase K1] (3.1.22.2) is produced. [Preference for single-stranded DNA.]

034257: Endodeoxyribonuclease V [Similar enzymes from: Thymus endonuclease, *E. coli* endonuclease II, Human placenta endonuclease] (3.1.22.3) is produced.

NUCLEOSIDASES AND RELATED ENZYMES - SITE-SPECIFIC ENDODEOXYRIBONUCLEASES: CLEAVAGE IS SEQUENCE-SPECIFIC

034258: Endodeoxyribonuclease *Alu*I (3.1.23.1) is produced and catalyzes the reaction: AG↓CT. [Isolated from *Arthrobacter luteus*.]

034259: Endodeoxyribonuclease *Asu*I (3.1.23.2) is produced and catalyzes the reaction: G↓GNCC. [Isolated from *Anabaena subcylindrica*.]

034260: Endodeoxyribonuclease *Ava*I (3.1.23.3) is produced and catalyzes the reaction: C↓YCGRG [Isolated from *Anabaena variablilis*.]

Section 34: Metabolic pathways and enzymes   219

034261:   Endodeoxyribonuclease *Ava*II (3.1.23.4) is
          produced and catalyzes the reaction: G↓G(A or
          T)CC. [Isolated from *Anabaena variabilis*.]
034262:   Endodeoxyribonuclease *Bal*I (3.1.23.5) is produced
          and catalyzes the reaction: TGG↓CCA. [Isolated
          from *Brevibacterium albidum*.]
034263:   Endodeoxyribonuclease *Bam*HI [Similar enzymes
          from: *Bacillus amyloliquefaciens* and *Bacillus
          stearothermophilus* strains] (3.1.23.6) is
          produced and catalyzes the reaction: G↓GATCC.
          [Isolated from *Bacillus amyloliquefaciens* H]
034264:   Endodeoxyribonuclease *Bbv*I (3.1.23.7) is produced
          and catalyzes the reaction: GC(A or T)GC. [Iso-
          lated from *Bacillus brevis*.]
034265:   Endodeoxyribonuclease *Bcl*I (3.1.23.8) is produced
          and catalyzes the reaction: T↓GATCA. [Isolated
          from *Bacillus caldolyticus*.]
034266:   Endodeoxyribonuclease *Bgl*I (3.1.23.9) is produced
          and catalyzes the reaction: Reaction unknown.
          [Isolated from *Bacillus globigii*.]
034267:   Endodeoxyribonuclease *Bgl*II (3.1.23.10) is
          produced and catalyzes the reaction: A↓GATCT.
          [Isolated from *Bacillus globigii*.]
034268:   Endodeoxyribonuclease *Bpu*I (3.1.23.11) is
          produced and catalyzes the reaction: Reaction
          unknown. [Isolated from *Bacillus pumilus*.]
034269:   Endodeoxyribonuclease *Dpn*I (3.1.23.12) is
          produced and catalyzes the reaction: Gm6ATC.
          [Requires m6A at the site shown and will not cleave
          unmethylated sequence; cleavage site is unknown;
          isolated from pneumococcus.]
034270:   Endodeoxyribonuclease *Eco*RI (3.1.23.13) is
          produced and catalyzes the reaction: G↓AATTC.
          [Isolated from *E. coli* carrying a fi+ plasmid.]
034271:   Endodeoxyribonuclease *Eco*RII (3.1.23.14) is
          produced and catalyzes the reaction: ↓CC(A or
          T)GG. [Isolated from *E. coli* carrying a fi-
          plasmid.]
034272:   Endodeoxyribonuclease *Hae*I (3.1.23.15) is
          produced and catalyzes the reaction: (A or
          T)GG↓CC(A or T). [Isolated from *Haemophilus
          aegyptius*.]
034273:   Endodeoxyribonuclease *Hae*II. [Similar enzymes
          from: *Haemophilus influenzae* HI(*Hin*HI) and
          Neisseria gonorrhoea (*Ngo*I) (3.1.23.16) is
          produced and catalyzes the reaction: RGCGC↓Y.
034274:   Endodeoxyribonuclease *Hae*III [Endonuclease Z.
          Similar enzymes from: *Bacillus sphaericus
          Bacillus subtilis, Brevibacterium luteum,
          Providencia alcalifaciens, Haemophilus
          haemoglobinophilus* and *Streptococcus faecalis*
          strains] (3.1.23.17) is produced and catalyzes the
          reaction: GG↓CC. [Isolated from *Haemophilus
          aegyptius*. Cleaves duplex regions in single-
          stranded DNA.]
034275:   Endodeoxyribonuclease *Hga*I (3.1.23.18) is
          produced and catalyzes the reaction: (5'-3')

Section 34: Metabolic pathways and enzymes

GACGCNNNNN or (3'-5') CTGCGNNNNNNNNNN↓. [Isolated from *Haemophilus gallinarum*.]

034276: Endodeoxyribonuclease *Hha*I (3.1.23.19) is produced and catalyzes the reaction: GCG↓C. [Isolated from *Haemophilus haemolyticus*. Cleaves duplex regions in single-stranded DNA.]

034277: Endodeoxyribonuclease *Hind*II [Endonuclease R. Similar enzymes from: *Corynebacterium humiferum*, *Haemophilus influenzae* serotype C, *Moraxella nonliquefaciens* strains] (3.1.23.20) is produced and catalyzes the reaction: GTY↓RAC. [Isolated from *Haemophilus influenzae*, serotype d.]

034278: Endodeoxyribonuclease *Hind*III [Similar enzymes from: *Bordetella bronchiseptica*, *Corynebacterium humiferum* strains] (3.1.23.21) is produced and catalyzes the reaction: A↓AGCTT [Isolated from *Haemophilus influenzae*, serotype d.]

034279: Endodeoxyribonuclease *Hinf*I [Similar enzyme from: *Haemophilus haemolyticus* (*Hha*II)] (3.1.23.22) is produced and catalyzes the reaction: G↓ANTC. [Isolated from *Haemophilus influenzae*, serotype f.]

034280: Endodeoxyribonuclease *Hpa*I (3.1.23.23) is produced and catalyzes the reaction: GTT↓AAC. [Isolated from *Haemophilus parainfluenzae*.]

034281: Endodeoxyribonuclease *Hpa*II [Similar enzymes from: *Haemophilus aphrophilus* (*Hap*II), *Moraxella nonliquefaciens* (*Mno*I)] (3.1.23.24) is produced and catalyzes the reaction: C↓CGG. [Isolated from *Haemophilus parainfluenzae*.]

034282: Endodeoxyribonuclease *Hph*I (3.1.23.25) is produced and catalyzes the reaction: (5'-3') GGTGANNNNNNNN↓ or (3'-5') CCACTNNNNNNNN↓. [Isolated from *Haemophilus parahaemolyticus*.]

034283: Endodeoxyribonuclease *Kpn*I (3.1.23.26) is produced and catalyzes the reaction: GGTAC↓C. [Isolated from *Klebsiella pneumoniae*.]

034284: Endodeoxyribonuclease *Mbo*I [Similar enzymes from: *Diplococcus pneumoniae* (*Dpn*II), *Moraxella osloensis* (*Mos*I), *Staphylococcus aureus* (*Sau*3 Al)] (3.1.23.27) is produced and catalyzes the reaction: ↓GATC. [Isolated from *Moraxella bovis*.]

034285: Endodeoxyribonuclease *Mbo*II (3.1.23.28) is produced and catalyzes the reaction: (5'-3') GAAGANNNNNNNN↓ or (3'-5') CTTCTNNNNNNN↓. [Isolated from *Moraxella bovis*.]

034286: Endodeoxyribonuclease *Mnl*I (3.1.23.29) is produced and catalyzes the reaction: CCTC. [Cleavage site unknown but is a few nucleotides away from a recognition sequence. Isolated from *Moraxella nonliquefaciens*.]

034287: Endodeoxyribonuclease *Pfa*I (3.1.23.30) is produced and catalyzes the reaction: Unknown. [Isolated from *Pseudomonas facilis*.]

034288: Endodeoxyribonuclease *Pst*I [Similar enzymes from: *Bacillus subtilis* (*Bsu*247), *Streptomyces albus* (*Sal*PI), *Xanthomonas malvacearum*

Section 34: Metabolic pathways and enzymes

(*Xma*II)] (3.1.23.31) is produced and catalyzes the reaction: CTGCA↓G. [Isolated from *Providencia stuartii*.]

034289: Endodeoxyribonuclease *Pvu*I (3.1.23.32) is produced and catalyzes the reaction: Unknown. [Isolated from *Proteus vulgaris*.]

034290: Endodeoxyribonuclease *Pvu*II (3.1.23.33) is produced and catalyzes the reaction: CAG↓CTG. [Isolated from *Proteus vulgaris*.]

034291: Endodeoxyribonuclease *Sac*I [Similar enzymes from: *Streptomyces stanford* (*Sst*I)] (3.1.23.34) is produced and catalyzes the reaction: G↓AGCTC. [Isolated from *Streptomyces achromogenes*.]

034292: Endodeoxyribonuclease *Sac*II [Similar enzymes from: *Streptomyces stanford* (*Sst*II), *Thermopolyspora glauca* (*Tgl*I)] (3.1.23.35) is produced and catalyzes the reaction: CCGC↓GG. [Isolated from *Streptomyces achromogenes*.]

034293: Endodeoxyribonuclease *Sac*III [Similar enzymes from: *Streptomyces stanford* (*Sst*III)] (3.1.23.36) is produced and catalyzes the reaction: Unknown. [Isolated from *Streptomyces achromogenes*.]

034294: Endodeoxyribonuclease *Sal*I [Similar enzymes from: *Xanthomonas amaranthicola* (*Xam*I)] (3.1.23.37) is produced and catalyzes the reaction: G↓TCGAC. [Isolated from *Streptomyces albus* G.]

034295: Endodeoxyribonuclease *Sgr*I (3.1.23.38) is produced and catalyzes the reaction: Unknown. [Isolated from *Streptomyces griseus*.]

034296: Endodeoxyribonuclease *Taq*I (3.1.23.39) is produced and catalyzes the reaction: T↓CGA. [Isolated from *Thermus aquaticus*.]

034297: Endodeoxyribonuclease *Taq*II (3.1.23.40) is produced and catalyzes the reaction: Unknown. [Isolated from *Thermus aquaticus*.]

034298: Endodeoxyribonuclease *Xba*I (3.1.23.41) is produced and catalyzes the reaction: T↓CTAGA. [Isolated from *Xanthomonas badrii*.]

034299: Endodeoxyribonuclease *Xho*I [Similar enzymes: *Brevibacterium luteum* (*Blu*I), *Xanthomonas papavericola* (*Xpa*I)] (3.1.23.42) is produced and catalyzes the reaction: C↓TCGAG. [Isolated from *Xanthomonas holicola*.]

034300: Endodeoxyribonuclease *Xho*II (3.1.23.43) is produced and catalyzes the reaction: Unknown. [Isolated from *Xanthomonas holicola*.]

034301: Endodeoxyribonuclease *Xma*I [Similar enzyme from: *Serratia marcescens* (*Sma*I)] (3.1.23.44) is produced and catalyzes the reaction: C↓CCGGG. [Isolated from *Xanthomonas malvacearum*.]

034302: Endodeoxyribonuclease *Xni*I (3.1.23.45) is produced and catalyzes the reaction: Unknown [Isolated from *Xanthomonas nigromaculans*.]

Section 34: Metabolic pathways and enzymes

NUCLEOSIDASES AND RELATED ENZYMES - SITE-SPECIFIC ENDODEOXYRIBONUCLEASES - CLEAVAGE IS NOT SEQUENCE-SPECIFIC

[NOTE 9: These enzymes recognize a specific sequence of DNA although cleavage does not occur at that sequence. The precise reaction pattern of Enzymes 3.1.24.1 - 3.1.24.4 is unknown.]

034303: Endodeoxyribonuclease EcoB (3.1.24.1) is produced. [Absolute requirement for ATP (or dATP) and S-adenosyl-L-methionine. Cleaves randomly. ATPase activity in presence of DNA. Isolated from E. coli B.]

034304: Endodeoxyribonuclease EcoK (3.1.24.2) is produced. [Absolute requirement for ATP (or dATP) and S-adenosyl-L-methionine. Cleaves randomly. ATPase activity in presence of DNA. Isolated from E. coli K.]

034305: Endodeoxyribonuclease EcoP1 (3.1.24.3) is produced. [Requirement for S-adenosyl-L-methionine. Isolated from E. coli lysogenic for P10.]

034306: Endodeoxyribonuclease EcoP15 (3.1.24.4) is produced. [Requirement for S-adenosyl-L-methionine. Isolated from E. coli lysogenic for P15.]

NUCLEOSIDASES AND RELATED ENZYMES - SITE-SPECIFIC ENDODEOXYRIBONUCLEASES - SPECIFIC FOR ALTERED BASES

034307: Endodeoxyribonuclease (pyrimidine dimer) [Similar enzymes from: T4 endonuclease V, E. coli endonucleases III and V, Correndonuclease II] (3.1.25.1) is produced and catalyzes the reaction: Endonucleolytic cleavage near pyrimidine dimers to products with 5'-phosphate. [Acts on the damaged strand, 5' from the damaged site.]

034308: Endodeoxyribonuclease (apurinic or apyrimidinic) (3.1.25.2) is produced and catalyzes the reaction: Endonucleolytic cleavage near apurinic or apyrimidinic sites to products with 5'-phosphate. [Acts on damaged strand, 5' from the damaged site.]

NUCLEOSIDASS AND RELATED ENZYMES - ENDORIBONUCLEASES PRODUCING 5'-PHOSPHOMONOESTERS

034309: Ribonuclease (*Physarum polycephalum*) [Similar enzymes from: Pig liver nuclease, HeLa cell RNase, E. coli RNase, Bovine adrenal cortex RNase] (3.1.26.1) is produced.

034310: Ribonuclease alpha (3.1.26.2) is produced. [Specific for O-methylated RNA.]

034311: Ribonuclease III [Similar enzyme from: Calf thymus RNase] (3.1.26.3) is produced. [Specific for double-stranded RNA]

034312: Endoribonuclease H (calf thymus) [Similar enzymes from: E. coli, chicken embryo, human KB cells,

Section 34: Metabolic pathways and enzymes 223

rat liver, *Ustilago maydis*, human leukaemic cells, *Saccharomyces cerevisiae* (H2), and *Tetrahymena pyriformis*] (3.1.26.4] is produced. [Acts on RNA - DNA hybrids.]

034313: Ribonuclease P [Similar enzyme from: RNase NU from KB cells] (3.1.26.5) is produced. [Specificity for tRNA precursors.]

NUCLEOSIDASES AND RELATED ENZYMES - ENDORIBONUCLEASES PRODUCING OTHER THAN 5'-PHOSPHOMONOESTERS

[NOTE 10: These enzymes catalyze a generally two-stage endonucleoytic cleavage to 3'-phosphomononucleotides and oligonucleotides with 2', 3'-cycle intermediates.]

034314: Ribonuclease T2 [Ribonuclease II. Similar enzymes from: Plant RNase, *E. coli* RNase I, Rnase N2, Microbial RNase II] (3.1.27.1) is produced. [Formerly EC 2.7.7.17 and 3.1.4.23.]

034315: Ribonuclease (*Bacillus subtilis*) [Similar enzymes from: *Azotobacter agilis* RNase, *Proteus mirabilis* RNase] (3.1.27.2) is produced.

034316: Ribonuclease T1 [Guanyloribonuclease, *Aspergillus oryzae* ribonuclease, RNase N1 and N2. Similar enzymes: *N. crassa* RNase N1 and N2, *Ustilago sphaerogena* RNase, *Chalaropsis* Rnase, *B. subtilis* RNase, Microbial RNase I] (3.1.27.3) is produced. [Formerly EC 3.1.4.8.]

034317: Ribonuclease U2 [Similar enzymes: RNase U3, *Pleospora* RNase, *Trichoderma koningi* RNase III] (3.1.27.4) is produced.

034318: Ribonuclease (pancreatic) [RNase, RNase I, Pancreatic RNase, Ribonuclease I. Similar enzymes: Venom RNase, *Thiobacillus thioparus* RNase, *Xenopus laevis* RNase, *Rhizopus oligosporus* RNase] (3.1.27.5) is produced. [Formerly EC 2.7.7.16 and 3.1.4.22.]

034319: Ribonuclease (*Enterobacter*) (3.1.27.6) is produced. [Preference for cleavage at CpA. Homopolymers of A, U or G are not hydrolyzed.]

NUCLEOSIDASES AND RELATED ENZYMES - ENDONUCLEASES ACTIVE WITH EITHER RIBO- OR DEOXYRIBONUCLEIC ACIDS AND PRODUCING 5'-PHOSPHOMONOESTERS

034320: Endonuclease S1 (*Aspergillus*) [*Aspergillus* nuclease S1; Single stranded-nucleate endonuclease, Deoxyribonuclease S1. Similar enzymes: *N. crassa* nuclease 1423, Mung bean nuclease, *Penicillium citrium* nuclease P1] (3.1.30.1) is produced. [Preference for single-stranded substrate. Formerly EC 3.1.4.21.]

034321: Endonuclease (*Serratia marcescens*) [Similar enzymes: Silkworm and Potato nucleases, *Azotobacter* nuclease] (3.1.30.2) is produced. [Hydrolyses double- or single-stranded substrate. Formerly EC 3.1.4.9.]

NUCLEOSIDASES AND RELATED ENZYMES - ENDONUCLEASES ACTIVE WITH EITHER RIBO- OR DEOXYRIBONUCLEIC ACIDS AND PRODUCING OTHER THAN 5'-PHOSPHOMONOESTERS

034322: Micrococcal endonuclease [Similar enzymes: *Chlamydomonas* nuclease, Spleen phosphodiesterase Spleen endonuclease] (3.1.31.1) is produced and catalyzes the reaction: Endonucleolytic cleavage to 3'-phosphomono- and oligonucleotide end products. [Hydrolyses double- or single-stranded substrate. Formerly EC 3.1.4.7.]

NUCLEOSIDASES AND RELATED ENZYMES - PHOSPHATASES MAINLY ACTING ON NUCLEOSIDE PHOSPHATES

034323: Nucleosidediphosphatase [Nucleosidediphosphate phosphohydrolase] (3.6.1.6) is produced and catalyzes the reaction: A nucleoside diphosphate + H_2O = a nucleotide + orthophosphate. [Acts on IDP, GDP, UDP and also on D-ribose 5-diphosphate.]

034324: Nucleotide pyrophosphatase [Dinucleotide nucleotidohydrolase] (3.6.1.9) is produced and catalyzes the reaction: A dinucleotide + H_2O = 2 mononucleotides. [Substrates include NAD^+ $NADP^+$, FAD, CoA and also ATP and ADP.]

034325: Deoxycytidinetriphosphatase [dCTP nucleotidohydrolase, Deoxy-CTPase, dCTPase] (3.6.1.12) is produced and catalyzes the reaction: dCTP + H_2O = dCMP + pyrophosphate. [Also hydrolyses dCDP to dCMP and orthophosphate]

034326: ADPribose pyrophosphatase [ADPribose ribophosphohydrolase] (3.6.1.13) is produced and catalyzes the reaction: ADPribose + H_2O = AMP + D-ribose 5-phosphate.

034327: Adenosinetetraphosphatase [Adenosinetetraphosphate phosphohydrolase] (3.6.1.14) is produced and catalyzes the reaction: Adenosine 5'-tetraphosphate + H_2O = ATP + orthophosphate. [Also acts on inosine tetraphosphate and tripolyphosphate but shows little or no activity with other nucleotides or polyphosphates.]

034328: Nucleoside triphosphatase [Unspecific diphosphate phosphohydrolase] (3.6.1.15) is produced and catalyzes the reaction: NTP + H_2O = NDP + orthophosphate. [Also hydrolyses other nucleoside triphosphates, diphosphate, thiamin diphosphate and FAD.]

034329: m7G (5') pppN pyrophosphatase [7-Methylguanosine, 5'-triphosphoryl-5'-polynucleotide, 7-methylguanosine- 5'-phosphohydrolase] (3.6.1.30) is produced and catalyzes the reaction: 7-Methylguanosine 5'-triphosphoryl-5'-polynucleotide + H_2O = 7-methylguanosine 5'-phosphate + polynucleotide.

Section 34: Metabolic pathways and enzymes

NUCLEOSIDASES AND RELATED ENZYMES - CARBON-CARBON LYASES

034330: Deoxyribodipyrimidine photolyase [Deoxyribocyclobutadipyrimidine pyrimidinelyase, Photoreactivating enzyme, PR-enzyme] (4.1.99.3) is produced and catalyzes the reaction: Cyclobutadipyrimidine (in DNA) = 2 pyrimidine residues in (DNA). [Catalyzes reactivation by light of UV irradiated DNA.]

NUCLEOSIDASES AND RELATED ENZYMES - PHOSPHORUS-OXYGEN LYASES

[NOTE 11: The so-called 'nucleotidylcyclases' are included here, on the basis that pyrophosphate is eliminated from the nucleoside triphosphate.]

034331: Adenylate cyclase [ATP pyrophosphatelyase (cyclizing), Adenylyl cyclase, Adenyl cyclase] (4.6.1.1) is produced and catalyzes the reaction: ATP = 3':5'-cyclic AMP + pyrophosphate. [Also acts on dATP to form 3':5'-cyclic dAMP. Requires pyruvate.]

034332: Guanylate cyclase [GTP pyrophosphatelyase (cyclizing), Guanylyl cyclase, Guanyl cyclase] (4.6.1.2) is produced and catalyzes the reaction: GTP = 3':5'-cyclic GMP + pyrophosphate. [Also acts on ITP and deoxy-GTP.]

NUCLEOSIDASES AND RELATED ENZYMES - LIGASES FORMING AMINOACYL-TRNA AND RELATED COMPOUNDS

034333: Tyrosyl-tRNA synthetase [L-Tyrosine:tRNATyr ligase (AMP-forming)] (6.1.1.1) is produced and catalyzes the reaction: ATP + L-tyrosine + tRNATyr = AMP + pyrophosphate + L-tyrosyl-tRNATyr [Similar enzymes acylating a transfer RNA with the corresponding amino acids are numbered 6.1.1.1 - 6.1.1.22. The respective amino acids are the L enantiomorphs of tyrosine, tryptophan, threonine, leucine, isoleucine, lysine, alanine, valine, methionine, serine, aspartate, proline, cysteine, glutamate, glutamine, arginine, phenylalanine, histidine, and asparagine as well as D-alanine and glycine. The single example of Tyrosyl-tRNA synthetase (6.1.1.1) is given here. However, as the need arises, statements can be formulated and RKC numbers assigned for any of the other enzymes from 6.1.1.2 - 6.1.1.22.]

NUCLEOSIDASES AND RELATED ENZYMES - LIGASES FORMING PHOSPHORIC ESTER BONDS (RESTORING BROKEN PHOSPHODIESTER BONDS IN NUCLEIC ACIDS)

034334: Polydeoxyribonucleotide synthetase (ATP) [Poly-(deoxyribonucleotide): poly(deoxyribonucleotide) ligase(AMP-forming), Polynucleotide ligase, Sealase, DNA repair enzyme. DNA joinase, DNA ligase] (6.5.1.1) is produced and catalyzes the reaction: ATP + (deoxyribonucleotide)n + (deoxyribonucleotide)m= AMP + pyrophosphate + (deoxyribonucleotide) $n+m$. [Catalyses the formation of a phosphodiester at the site of a single-strand break in duplex DNA. RNA can also act as substrate to some extent.]

034335: Polydeoxyribonucleotide synthetase (NAD$^+$) [Poly-(deoxyribonucleotide): poly(deoxyribonucleotide) ligase(AMP-forming), NMN-forming) Polynucleotide ligase, (NAD$^+$) DNA repair enzyme, DNA joinase, DNA ligase](6.5.1.2) is produced and catalyzes the reaction: NAD$^+$ + (deoxyribonucleotide)n + (deoxyribonucleotide)m = AMP + NMN + (deoxyribonucleotide)$n+m$. [Catalyses the formation of a phosphodiester at the site of a single-strand break in duplex DNA. RNA can also act as substrate to some extent.]

034336: Polyribonucleotide synthetase(ATP) [Poly(ribo-nucleotide): poly(ribonucleotide) ligase (AMP-forming), RNA ligase] (6.5.1.3) is produced and catalyzes the reaction: ATP + (ribonucleotide)n + (ribonucleotide)m = AMP + pyrophosphate + (ribonucleotide)$n+m$.

ENZYMES ACTING ON PEPTIDE BONDS (PEPTIDE HYDROLASES)

The nomenclature of this large group of enzymes is difficult because the overall reaction of all peptidases and proteinases is essentially the same and substrate specificity in the ordinary sense is lacking. Many proteases attack nearly all denatured proteins and many native proteins. The two basic groups of these enzymes are the peptidases (expopeptidases: 3.4.11-17) and the proteinases (3.4.21-24). The peptidases are divided as follows: hydrolyzing single amino acids from the N-terminus (3.4.11) or the C-terminus (3.4.16-17) of the peptide chain; hydrolyzing dipeptide units from the N-terminus (3.4.14) or C-terminus (3.4.15). These, and further subdivisions of C-terminal enzymes are noted in the text. The proteinases are divided into subgroups based on their catalytic mechanism. Explanatory notes under the various subheadings are included in this document. In the following list of features many enzymes are omitted; only enzymes from microbiological sources or of microbiological significance are included.

ALPHA - AMINOACYLPEPTIDE HYDROLASES

[NOTE 12: These enzymes hydrolyse single amino acids from the N-terminus of the peptide chain according to the following general reaction: Aminoacyl peptide + H_2O = simple amino acid + peptide.]

Section 34: Metabolic pathways and enzymes

034337: Proline iminopeptidase [L-Prolylpeptide hydrolase] (3.4.11.5) is produced. [Formerly EC3.4.1.4.]
034338: Pyroglutamyl aminopeptidase [L-Pyroglutamyl-peptide hydrolase, Pyrrolidone-carboxylate peptidase] (3.4.11.8) is produced. [Removes pyroglutamate from various penultimate amino acid residues except L-proline. Occurs in *Pseudomonas*, *Bacillus subtilis*, rat liver.]
034339: Aminopeptidase P [Aminoacylprolylpeptide hydrolase, Aminoacylproline aminopeptidase] [3.4.11.9) is produced. [Releases any N-terminal amino acid only if adjacent to proline residue; requires Mn^{+2} Isolated from *E. coli*.]
034340: *Aeromonas proteolytica* aminopeptidase [α-Aminoacylpeptide hydrolase (*Aeromonas proteolytica*)] (3.4.11.10) is produced. [A zinc enzyme. Acts most rapidly on L-leucylpeptides, amide and B-naphthylamide. Does not cleave Glu-and Asp-bonds. Similar aminopeptidases were isolated from *E. coli* and *Staphylococcus thermophilus*.]
034341: Thermophilic aminopeptidase [α-Aminoacylpeptide hydrolase] (3.4.11.12) is produced. [Metalloenzymes of high temperature stability and of broad specificity, releasing all N-terminal amino acids, including arginine and lysine. Isolated from *Bacillus stearothermophilus*, *Talaromyces duponti*, *Mucor*.]

DIPEPTIDE HYDROLASES

034342: Dipeptidase [Dipeptide hydrolase] (3.4.13.11) is produced and catalyzes the reaction: Dipeptide + H_2O = 2 amino acids. [Many dipeptidases have comparable specificities. Found in *Mycobacterium phlei*, *E. coli*, *Streptococcus thermophilus*, yeast, and mammalian tissues.]

PEPTIDYLDIPEPTIDE HYDROLASES

034343: Dipeptidyl carboxypeptidase [Peptidyldipeptide hydrolase, Angiotensin converting enzyme. Peptidase P, Kinase II, Carboxycathepsin] (3.4.15.1) is produced and catalyzes the reaction: Polypeptidyldipeptide + H_2O = polypeptide + dipeptide. Also: Angiotensin I + H_2O = angiotensin II + His-Leu. [Cleaves C-terminal dipeptides from a variety of substrates. Enzyme isolated from *E. coli* cleaves N-blocked tripeptides, free tetra- and higher peptides, except those with a penultimate proline residue or those consisting of a chain of several glycine residues.]

SERINE CARBOXYPEPTIDASES

[NOTE 13: Optimum activity at acidic pH (4.5 - 6.0); contain a serine in their catalytic site; broad specificity; from microorganisms and many other plant and animal sources.]

034344: Serine carboxypeptidase [Peptidyl-L-amino acid hydrolase, carboxypeptidase C, lysosomal carboxypeptidase A, cathepsin A, lysomal carboxypeptidase B, cathepsin B2, cathepsin IV] (3.4.16.1) is produced and catalyzes the reaction: Peptidyl-L-amino acid + H_2O = peptide + L-amino acid (broad specificity). [Formerly EC 3.4.12.1 and 3.4.21.13.]

METALLO-CARBOXYPEPTIDASES

[NOTE 14: These enzymes release C-termimal amino acids.]

034345: Glycine carboxypeptidase [Peptidylglycine hydrolase, yeast carboxypeptidase] (3.4.17.4) is produced and catalyzes the reaction: Peptidylglycine + H_2O = peptide + glycine. [Also acts on peptides having a C-terminal L-leucine residue. Formerly, EC 3.4.2.3 and EC 3.4.12.8.]

034346: Alanine carboxypeptidase [Peptidyl-L-alanine hydrolase] (3.4.17.6) is produced and catalyzes the reaction: Peptidyl-L-alanine + H_2O = peptide + L-alanine. [Metallocarboxypeptidase from soil bacteria. Peptide can be replaced by a variety of pteroyl or acyl groups. Formerly EC 3.4.12.11.]

SERINE PROTEINASES

[NOTE 15: Trypsin (3.4.21.4) is a member of this group. Enzymes with trypsin like activity have been reported in some bacteria but there is no assurance that they really are trypsin.]

034347: *Myxobacter* α-lytic proteinase (3.4.21.12) is produced and catalyzes the reaction: Hydrolysis of proteins, especially bonds adjacent to L-alanine residues.

034348: Microbial serine proteinases (3.4.21.14) are produced and catalyze the reaction: Hydrolysis of proteins; and peptide amides. [There are many proteinases of many specificities listed under EC 3.4.21.14-33 from a wide variety of bacteria and fungi as listed below: Subtilisin; *Escherichia coli* periplasmic proteinase; *Aspergillus* alkaline proteinase; *Aspergillus* proteinase B from 10 species of *Aspergillus*; *Bacillus amyloliquefaciens* serine proteinase. [Formerly EC 3.4.21.15.]; *Tritirachium* alkaline proteinase [Proteinase K] hydrolyzing keratin and native proteins at the carboxyl of aromatic or hydrophobic amino acid residues; *Arthrobacter* and *Pseudomonas* serine proteinases of no clear specificity; extracellular proteinase from *Pseudomonas aeruginosa*; Thermomycolin [Thermomycolase] preferentially cleaving Ala-, Tyr-, Phe- found in the fungus *Malbranchea pulchella* var. *sulfurea*; Thermophilic *Streptomyces* serine proteinase hydrolyzing proteins in nonpolar sequences; thermophilic proteinase from *Streptomyces rectus*; *Candida lipolytica* serine proteinase.]

034349: *Alternaria* serine proteinase [Fungal proteinase] (3.4.21.16) is produced and catalyzes the reaction: Hydrolysis of proteins; no clear specificity.

034350: Staphylococcal serine proteinase (3.4.21.19) is produced and catalyzes the reaction: Preferential specificity: Glu-, Asp-. [Isolated from *Staphylococcus aureus*.]

THIOL PROTEINASES

034351: Asclepain [*Clostridium histolyticum* proteinase B] (3.4.22.7) is produced and catalyzes the reaction: No clear specificity. Specificity similar to that of papain.

034352: Clostripain [Clostridiopeptidase B, *Clostridium histolyticum* proteinase B] (3.4.22.8) is produced and catalyzes the reaction: Preferential cleavage: Arg-, Also Arg-Pro bond.

034353: Yeast proteinase B [Baker's yeast proteinase, Brewer's yeast proteinase] (3.4.22.9) is produced and catalyzes the reaction: No clear specificity. [A group of enzymes. Differ in pH optima. Baker's yeast Proteinase B, C, and brewer's yeast proteinase B ('peptidase B') belong here.]

034354: Streptococcal proteinase (3.4.22.10) is produced and catalyzes the reaction: The amino acid adjacent, at the amino terminal end, to the residue contributing the carbonyl to the susceptible peptic bond containing a bulky side chain. [Formed from inactive zymogen by proteolytic action. Formerly EC 3.4.4.18.]

034355: Staphylococcal thiol proteinase [Staphylococcal proteinase II] (3.4.22.13) is produced and catalyzes the reaction: Broad specificity on protein substrates.

MICROBIAL CARBOXYL (ACID) PROTEINASES

[NOTE 16: At least 14 fungal and protozoal proteinases of broad activity are listed under 3.4.23.6. Among these are Takadiastase, Trypsinogen kinase, Aspergillopeptidase, Penicillopepsin, Peptidase A, *Mucor* rennin, Baker's yeast proteinase A, and carboxyl proteinases from 3 species of *Aspergillus*, strains of *Penicillium*, *Rhizopus*, *Endothia*, *Mucor*, *Candida*, *Paecilomyces*, Sac-

charomyces, *Rhodotorula*, *Physarum*, and the
protozoa *Tetrahymena* and *Plasmodium*.]

034356: A microbial carboxyl proteinase (3.4.23.6) is produced.

METALLOPROTEINASES

034357: *Clostridium histolyticum* collagenase [Clostridiopeptidase A, collagenase A, Collagenase I] (3.4.24.3) is produced and catalyzes the reaction: Degrades helical regions of native collagen to small fragments. Preferential cleavage: -Gly in the sequence -Z-Pro-X-Gly-Pro-X. [Other forms with a broader specificity were isolated from the same source: collagenase B, collagenase II, pseudocollagenase. Collagenases of similar specificity have been isolated from *Streptomyces madurae* and *Bacteroides melaninogenicus*. Formerly EC 3.4.4.19. Collagenase B; formerly EC 3.4.99.5.]

034358: Microbial metalloproteinase (3.4.24.4) is produced. [The microbial metalloproteinases (3.4.24.4) generally are Zn and Ca proteins, but some may also be activated by Mg, Mn, or Fe. Preferential cleavages of bonds is often adjacent to a hydrophobic amino acid residue; sequences may contain valine, leucine, isoleucine, phenylanine, tyrosine or alanine. Some names given to this group of enzymes are: Clostridiopeptidase A, Collagenases A, B, I, II, Pseudocollagenase (*Clostridium*); Thermolysin (*Bacillus subtilis*); Protease III, Staphylokinase (*Staphylococcus aureus*); Coccus P(Proteinase) (*Serratia*); Pronase component (*Streptomyces griseus*); *P. roqueforti* protease II; Keratinases I, II, III (*Trichophyton*). Microbial metalloproteinases (3.4.24.4) are present in at least nine bacterial and four fungal genera and are doubtlessly more widespread. Should the need arise, specific coding of any protease feature of any strain can be accommodated in the RKC format.]

034359: *Achromobacter iophagus* collagenase (3.4.24.8) is produced and catalyzes the reaction: Degrades helical regions of native collagen to large fragments. [Extracellular Zn^{+2}-enzyme. In contrast to EC 3.4.24.3, degrades Z-Pro-Leu-Gly-Ala-D-Arg, X-Gly-Pro and X-Gly-Ala bonds in other proteins. Partially degraded forms of the enzyme are still active.]

034360: *Trichophyton schoenleinii* collagenase (3.4.24.9) is produced and catalyzes the reaction: Degrades native collagen in acid pH range. [Acid-stable fungal metalloenzyme.]

034361: *Trichophyton mentagrophytes* keratinase (3.4.24.10) is produced and catalyzes the reaction: Hydrolyses keratin; preferential cleavage at

Section 34: Metabolic pathways and enzymes

hydrophobic residues. [Keratinase I is extracellular; II and III cell-bound. Formerly EC 3.4.99.12.]

PROTEINASES OF UNKNOWN CATALYTIC MECHANISM

034362: *Streptomyces* alkalophilic keratinase (3.4.99.11) is produced and catalyzes the reaction: Hydrolyses keratin and poly-L-lysine; preferential cleavage: Ser-His, Leu-Val, Phe-Tyr, Lys-Ala. [Optimum activity at pH 13. Formerly EC 3.4.4.25.]

034363: *Penicillium notatum* extracellular proteinase (3.4.99.16) is produced. [Broad specificity for protein substrates.]

034364: Peptidoglycan endopeptidase [Glycylglycine endopeptidase, Endo-B-N-acetyl-glucosaminidase] (3.4.99.17) is produced and catalyzes the reaction: Hydrolyses pentaglycine cross-bridges at D-alanylglycine and glycylglycine linkages in the cell wall peptidoglycan. [From *Staphylococcus staphylolyticus*.]

034365: *Myxobacter* AL-I proteinase I (3.4.99.29) is produced and catalyzes the reaction: Preferential cleavage of bonds adjacent to a hydrophobic amino acid residue. [Inhibition studies suggest that mechanism of action is different from the four groups 3.4.21 - 3.4.24.]

034366: *Myxobacter* AL-I proteinase II (3.4.99.30) is produced and catalyzes the reaction: Preferential cleavage: -Lys.

SECTION 35: QUANTITATIVE ANTIBIOTIC SENSITIVITY
(DISC ZONE DIAMETERS; MINIMAL INHIBITORY CONCENTRATIONS)

[NOTE 1: SECTION 35 is to be used to record disc sensitivities for clinical applications (eg., the methods of Bauer et. al., 1966, commonly referred to as Kirby-Bauer or Bauer-Kirby Methods). Other tests of antimicrobial sensitivity or resistance should be coded under SECTION 19/40.]

[NOTE 2: Feature numbers 035001 through 035079 refer specifically to the Bauer-Kirby Methods. (See: Bauer, A. W., M. M. Kirby, J.C. Sherris and M. Turck. 1966. Antibiotic susceptibility and testing by a standardized single disc method. Am. J. Clin. Pathol. 45:493-496.) New feature numbers have been added to reflect the changes and additions as published by the National Committee for Clinical Laboratory Standards (NCCLS). (See: Performance Standards for Antimicrobial Disc Susceptibility Tests. Second edition; May, 1981; NCCLS, Vol. 1 No. 6; 771 E. Lancaster Avenue; Villanova, PA 19805.) (See also: Thornsberry, C. 1980. Disc agar diffusion susceptibility test. CDC 75-22 Rev. Jan. 1980. Bur. Labs. Bacteriol.; Clin. Lab. Branch; Antibiotics Invest. Section; CDC; Atlanta, GA 30333.]

[NOTE 3: The zone diameters are measured to the nearest mm (i.e., rounded on 0.5 mm if measurements are made with electro-optical devices).]

DISC ZONE DIAMETERS

035218: Amikacin, 30 µg disc, results in an inhibitory zone of 14 mm or less (resistant).
035219: Amikacin, 30 µg disc, results in an inhibitory zone of 15-16 mm (intermediate).
035220: Amikacin, 30 µg disc, results in an inhibitory zone of 17 mm or more (susceptible).
035001: Ampicillin, 10 µg disc, results in an inhibitory zone diameter of 20 mm or less (resistant for staphylococci).
035002: Ampicillin, 10 µg disc, results in an inhibitory zone diameter of 21 - 28 mm (intermediate for staphylococci).
035003: Ampicillin, 10 µg disc, results in an inhibitory zone diameter of 29 mm or more (susceptible for staphylococci).
035004: Ampicillin, 10 µg disc, results in an inhibitory zone diameter of 11 mm or less (resistant for enteric Gram negative, enterococci, and other organisms).
035005: Ampicillin, 10 µg disc, results in an inhibitory zone diameter of 12 - 13 mm (intermediate for enteric Gram negative and enterococci).

Section 35: Quantitative antibiotic sensitivity 233

035006: Ampicillin, 10 μg disc, results in an inhibitory zone diameter of 12 - 21 mm (intermediate for other organisms).
035007: Ampicillin, 10 μg disc, results in an inhibitory zone diameter of 14 mm or more (susceptible for enteric Gram negative and enterococci).
035008: Ampicillin, 10 μg disc, results in an inhibitory zone diameter of 22 mm or more (susceptible for other organisms).
035221: Bacitracin, 10 units disc, results in an inhibitory zone of 8 mm or less (resistant).
035222: Bacitracin, 10 units disc, results in an inhibitory zone of 9-12 mm (intermediate).
035223: Bacitracin, 10 units disc, results in an inhibitory zone of 13 mm or more (susceptible).
035009: Carbenicillin, 50 μg disc, results in an inhibitory zone diameter of 12 mm or less (resistant for *Pseudomonas* sp.).
035010: Carbenicillin, 50 μg disc, results in an inhibitory zone diameter of 13 - 14 mm (intermediate for *Pseudomonas* sp.).
035011: Carbenicillin, 50 μg disc, results in an inhibitory zone diameter of 15 mm or more (susceptible for *Pseudomonas* sp.).
035012: Carbenicillin, 50 μg disc, results in an inhibitory zone diameter of 17 mm or less (resistant for *Proteus* and *E. coli*).
035013: Carbenicillin, 50 μg disc, results in an inhibitory zone diameter of 18 - 22 mm (intermediate for *Proteus* and *E. coli*).
035014: Carbenicillin, 50 μg disc, results in an inhibitory zone diameter of 23 mm or more (susceptible for *Proteus* and *E. coli*).
035224: Carbenicillin, 100 μg disc, results in an inhibitory zone of 13 mm or less (resistant for *Pseudomonas* sp.).
035225: Carbenicillin, 100 μg disc, results in an inhibitory zone of 14-16 mm (susceptible for *Pseudomonas* sp.).
035226: Carbenicillin, 100 μg disc, results in an inhibitory zone of 17 mm or more (susceptible for *Pseudomonas* sp.).
035227: Carbenicillin, 100 μg disc, results in an inhibitory zone of 17 mm or less (resistant for *Enterobacteriaceae*).
035228: Carbenicillin, 100 μg disc, results in an inhibitory zone of 18-22 mm (intermediate for *Enterobacteriaceae*).
035229: Carbenicillin, 100 μg disc, results in an inhibitory zone of 23 mm or more (susceptible for *Enterobacteriaceae*).
035230: Cefamandole, 30μg disc, results in an inhibitory zone diameter of 14 mm or less (resistant).
035231: Cefamandole, 30μg disc, results in an inhibitory zone of 15-17mm (intermediate).
035232: Cefamandole, 30 μg disc, results in an inhibitory zone of 18 mm or more (susceptible).

035233: Cefotaxime, 30 µg disc, results in an inhibitory zone of 14 mm or less (resistant).
035234: Cefotaxime, 30 µg disc, results in an inhibitory zone of 15-22 mm (intermediate).
035235: Cefotaxime, 30 µg disc, results in an inhibitory zone of 23 mm or more (susceptible).
035236: Cefoxitin, 30 µg disc, results in an inhibitory zone of 14 mm or less (resistant).
035237: Cefoxitin, 30 µg disc, results in an inhibitory zone of 15-17 mm or less (intermediate).
035238: Cefoxitin, 30 µg disc, results in an inhibitory zone of 18 mm or more (susceptible).
035015: Cephaloridine, 30 µg disc, results in an inhibitory zone diameter of 11 mm or less (resistant).
035016: Cephaloridine, 30 µg disc, results in an inhibitory zone diameter of 12 - 15 mm (intermediate).
035017: Cephaloridine, 30 µg disc, results in an inhibitory zone diameter of 16 mm or more (susceptible).
035018: Cephalothin, 30 µg disc, results in an inhibitory zone diameter of 14 mm or less (resistant).
035019: Cephalothin, 30 µg disc, results in an inhibitory zone diameter of 15 - 17 mm (intermediate).
035020: Cephalothin, 30 µg disc, results in an inhibitory zone diameter of 18 mm or more (susceptible).
035021: Chloramphenicol, 30 µg disc, results in an inhibitory zone diameter of 12 mm or less (resistant).
035022: Chloramphenicol, 30 µg disc, results in an inhibitory zone diameter of 13 - 17 mm (intermediate).
035023: Chloramphenicol, 30 µg disc, results in an inhibitory zone diameter of 18 mm or more (susceptible).
035024: Clindamycin, 2 µg disc, results in an inhibitory zone diameter of 11 mm or less (resistant).
035025: Clindamycin, 2 µg disc, results in an inhibitory zone diameter of 12 - 15 mm (intermediate).
035026: Clindamycin, 2 µg disc, results in an inhibitory zone diameter of 16 mm or more (susceptible).
035239: Clindamycin, 2 µg disc, results in an inhibitory zone of 14 mm or less (resistant).
035240: Clindamycin, 2 µg disc, results in an inhibitory zone of 15-16 mm (intermediate).
035241: Clindamycin, 2 µg disc, results in an inhibitory zone of 17 mm or more (susceptible).
035027: Colistin, 10 µg disc, results in an inhibitory zone diameter of 8 mm or less (resistant).
035028: Colistin, 10 µg disc, results in an inhibitory zone diameter of 9 - 10 mm (intermediate).
035029: Colistin, 10 µg disc, results in an inhibitory zone diameter of 11 mm or more (susceptible).
035030: Erythromycin, 15 µg disc, results in an inhibitory zone diameter of 13 mm or less (resistant).
035031: Erythromycin, 15 µg disc, results in an in-

Section 35: Quantitative antibiotic sensitivity   235

           hibitory zone diameter of 14 - 17 mm (inter-
           mediate).
035032:   Erythromycin, 15 µg disc, results in an in-
           hibitory zone diameter of 18 mm or more (suscep-
           tible).
035033:   Gentamicin, 10 µg disc, results in an inhibitory
           zone diameter of 12 mm or less (resistant).
035034:   Gentamicin, 10 µg disc, results in an inhibitory
           zone diameter of 13 - 14 mm (intermediate).
035035:   Gentamicin, 10 µg disc, results in an inhibitory
           zone diameter of 15 mm or more (susceptible).
035036:   Kanamycin, 30 µg disc, results in an inhibitory
           zone diameter of 13 mm or less (resistant).
035037:   Kanamycin, 30 µg disc, results in an inhibitory
           zone diameter of 14 - 17 mm (intermediate).
035038:   Kanamycin, 30 µg disc, results in an inhibitory
           zone diameter of 18 mm or more (susceptible).
035039:   Lincomycin, 2 µg disc, results in an inhibitory
           zone diameter of 9 mm or less (resistant).
035040:   Lincomycin, 2 µg disc, results in an inhibitory
           zone diameter of 10 - 14 mm (intermediate).
035041:   Lincomycin, 2 µg disc, results in an inhibitory
           zone diameter of 15 mm or more (susceptible).
035042:   Methicillin, 5 µg disc, results in an inhibitory
           zone diameter of 9 mm or less (resistant).
035043:   Methicillin, 5 µg disc, results in an inhibitory
           zone diameter of 10 - 13 mm (intermediate).
035044:   Methicillin, 5 µg disc, results in an inhibitory
           zone diameter of 14 mm or more (susceptible).
035045:   Nafcillin or Oxacillin, 1 µg disc, results in an
           inhibitory zone diameter of 10 mm or less (resis-
           tant).
035046:   Nafcillin or Oxacillin, 1 µg disc, results in an
           inhibitory zone diameter of 11 - 12 mm (inter-
           mediate).
035047:   Nafcillin or Oxacillin, 1 µg disc, results in an
           inhibitory zone diameter of 13 mm or more (suscep-
           tible).
035048:   Nalidixic Acid, 30 µg disc, results in an in-
           hibitory zone diameter of 13 mm or less (resis-
           tant).
035049:   Nalidixic Acid, 30 µg disc, results in an in-
           hibitory zone diameter of 14 - 18 mm (inter-
           mediate).
035050:   Nalidixic Acid, 30 µg disc, results in an in-
           hibitory zone diameter of 19 mm or more (suscep-
           tible).
035051:   Neomycin, 30 µg disc, results in an inhibitory
           zone diameter of 12 mm or less (resistant).
035052:   Neomycin, 30 µg disc, results in an inhibitory
           zone diameter of 13 - 16 mm (intermediate).
035053:   Neomycin, 30 µg disc, results in an inhibitory
           zone diameter of 17 mm or more (susceptible).
035054:   Nitrofurantoin, 300 µg disc, results in an in-
           hibitory zone diameter of 14 mm or less (resis-
           tant).

236   Section 35:   Quantitative antibiotic sensitivity

035055: Nitrofurantoin, 300 µg disc, results in an inhibitory zone diameter of 15 - 18 mm (intermediate).
035056: Nitrofurantoin, 300 µg disc, results in an inhibitory zone diameter of 19 mm or more (susceptible).
035242: Nitrofurantoin, 300 µg disc, results in an inhibitory zone of 15-16 mm (intermediate).
035243: Nitrofurantoin, 300 µg disc, results in an inhibitory zone of 17 mm or more (susceptible).
035057: Penicillin G, 10 units disc, results in an inhibitory zone diameter of 20 mm or less (resistant for staphylococci).
035058: Penicillin G, 10 units disc, results in an inhibitory zone diameter of 21 - 28 mm (intermediate for staphylococci).
035059: Penicillin G, 10 units disc, results in an inhibitory zone diameter of 29 mm or more (susceptible for staphylococci).
035060: Penicillin G, 10 units disc, results in an inhibitory zone diameter of 11 mm or less (resistant for enteric Gram negative, enterococci, and other organisms).
035061: Penicillin G, 10 units disc, results in an inhibitory zone diameter of 12 - 13 mm (intermediate for enteric Gram negative and enterococci).
035062: Penicillin G, 10 units disc, results in an inhibitory zone diameter of 12 - 21 mm (intermediate for other organisms).
035063: Penicillin G, 10 units disc, results in an inhibitory zone diameter of 14 mm or more (susceptible for enteric Gram negative and enterococci).
035064: Penicillin G, 10 units disc, results in an inhibitory zone diameter of 22 mm or more (susceptible for other organisms).
035065: Polymyxin B, 300 units disc, results in an inhibitory zone diameter of 10 mm or less (resistant).
035066: Polymyxin B, 300 units disc, results in an inhibitory zone diameter of 11 - 14 mm (intermediate).
035067: Polymyxin B, 300 units disc, results in an inhibitory zone diameter of 15 mm or more (susceptible).
035244: Polymyxin B, 300 units disc, results in an inhibitory of zone of 8 mm or less (resistant).
035245: Polymyxin B, 300 units disc, results in an inhibitory zone of 9-11 mm (intermediate).
035246: Polymyxin B, 300 units disc, results in an inhibitory zone of 12 mm or more (susceptible).
035068: Streptomycin, 10 µg disc, results in an inhibitory zone diameter of 11 mm or less (resistant).
035069: Streptomycin, 10 µg disc, results in an inhibitory zone diameter of 12 - 14 mm (intermediate).

Section 35: Quantitative antibiotic sensitivity

035070: Streptomycin, 10 µg disc, results in an inhibitory zone diameter of 15 mm or more (susceptible).
035071: Sulfonamides, 300 µg disc, results in an inhibitory zone diameter of 12 mm or less (resistant).
035072: Sulfonamides, 300 µg disc, results in an inhibitory zone diameter of 13 - 16 mm (intermediate).
035073: Sulfonamides, 300 µg disc, results in an inhibitory zone diameter of 17 mm or more (susceptible).
035247: Sulfonamides, 250 µg disc, results in an inhibitory zone of 12 mm or less (resistant).
035248: Sulfonamides, 250 µg disc, results in an inhibitory zone of 13-16 mm (intermediate).
035249: Sulfonamides, 250 µg disc, results in an inhibitory zone of 17 or more (susceptible).
035074: Tetracycline, 30 µg disc, results in an inhibitory zone diameter of 14 mm or less (resistant).
035075: Tetracycline, 30 µg disc, results in an inhibitory zone diameter of 15 - 18 mm (intermediate).
035076: Tetracycline, 30 µg disc, results in an inhibitory zone diameter of 19 mm or more (susceptible).
035250: Tobramycin, 10 µg disc, results in an inhibitory zone diameter of 12 mm or less (resistant).
035251: Tobramycin, 10 µg disc, results in an inhibitory zone diameter of 13-14 mm (intermediate).
035252: Tobramycin, 10 µg disc, results in an inhibitory zone diameter of 15 mm or more (susceptible).
035253: Tricarcillin, 75 µg disc, results in an inhibitory zone diameter of 11 mm or less (resistant for *Pseudomonas aeruginosa*).
035254: Tricarcillin, 75 µg disc, results in an inhibitory zone diameter of 12-14 mm (intermediate for *Pseudomonas aeruginosa*).
035255: Tricarcillin, 75 µg disc, results in an inhibitory zone diameter of 15 mm or more (susceptible for *Pseudomonas aeruginosa*).
035256: Trimethoprim, 5µg disc, results in an inhibitory zone diameter of 10 mm or less (resistant).
035257: Trimethoprim, 5µg disc, results in an inhibitory zone diameter of 11-15 mm (intermediate).
035258: Trimethoprim, 5µg disc, results in an inhibitory zone of 16 mm or more (susceptible).
035259: Trimethoprim sulfamethoxazole, 1.25 µg of trimethoprim and 23.75 µg of sulfamethoxazole per disc, results in an inhibitory zone of 10 mm or less (resistant).
035260: Trimethoprim sulfamethoxazole, 1.25 µg of trimethoprim and 23.75 µg of sulfamethoxazole per disk, results in an inhibitory zone of 11-15 mm (intermediate).

Section 35: Quantitative antibiotic sensitivity

035261: Trimethoprim sulfamethoxazole, 1.25 µg of trimethoprim and 23.75 µg of sulfamethoxazole per disk, results in an inhibitory zone of 16 mm or more (susceptible).
035077: Vancomycin (Vancocin), 30 µg disc, results in an inhibitory zone diameter of 9 mm or less (resistant).
035078: Vancomycin (Vancocin), 30 µg disc, results in an inhibitory zone diameter of 10 - 11 mm (intermediate).
035079: Vancomycin (Vancocin), 30 µg disc, results in an inhibitory zone diameter of 12 mm or more (susceptible).

MINIMAL INHIBITORY CONCENTRATIONS (MIC)

035216: The Minimum Inhibitory Concentration (MIC) of Amikacin is _____ µg/ml.
035080: The Minimum Inhibitory Concentration (MIC) of Amoxicillin is _____ µg/ml.
035081: The Minimum Inhibitory Concentration (MIC) of Ampicillin is _____ µg/ml.
035082: The Minimum Inhibitory Concentration (MIC) of Bacitracin is _____ units/ml.
035083: The Minimum Inhibitory Concentration (MIC) of Capreomycin is _____ µg/ml.
035084: The Minimum Inhibitory Concentration (MIC) of Carbenicillin is _____ µg/ml.
035262: The Minimum Inhibitory Concentration (MIC) of Cefamandole is _____ µg/ml.
035263: The Minimum Inhibitory Concentration (MIC) of Cefotaxime is _____ µg/ml.
035085: The Minimum Inhibitory Concentration (MIC) of Cefoxitin is _____ µg/ml.
035086: The Minimum Inhibitory Concentration (MIC) of Celesticetin is _____ µg/ml.
035087: The Minimum Inhibitory Concentration (MIC) of Cephalexin is _____ µg/ml.
035088: The Minimum Inhibitory Concentration (MIC) of Cephaloglycine is _____ µg/ml.
035089: The Minimum Inhibitory Concentration (MIC) of Cephaloridine is _____ µg/ml.
035090: The Minimum Inhibitory Concentration (MIC) of Cephalothin is _____ µg/ml.
035091: The Minimum Inhibitory Concentration (MIC) of Cephapirin is _____ µg/ml.
035092: The Minimum Inhibitory Concentration (MIC) of Chloromycetin (Chloramphenicol) is _____ µg/ml.
035093: The Minimum Inhibitory Concentration (MIC) of Chlortetracycline (Aureomycin) is _____ µg/ml.
035094: The Minimum Inhibitory Concentration (MIC) of Clindamycin is _____ µg/ml.
035095: The Minimum Inhibitory Concentration (MIC) of Cloxacillin is _____ µg/ml.
035096: The Minimum Inhibitory Concentration (MIC) of Colistin Coly-mycin) is _____ µg/ml.

Section 35:   Quantitative antibiotic sensitivity   239

035097: The Minimum Inhibitory Concentration (MIC) of
         Dicloxacillin is _____ µg/ml.
035098: The Minimum Inhibitory Concentration (MIC) of
         Erythromycin (Ilotycin) is _____ µg/ml.
035099: The Minimum Inhibitory Concentration (MIC) of
         Fusidic Acid is _____ µg/ml.
035100: The Minimum Inhibitory Concentration (MIC) of
         Gentamicin is _____ µg/ml.
035101: The Minimum Inhibitory Concentration (MIC) of
         Kanamycin is _____ µg/ml.
035102: The Minimum Inhibitory Concentration (MIC) of
         Lincomycin is _____ µg/ml.
035103: The Minimum Inhibitory Concentration (MIC) of
         Methicillin is _____ µg/ml.
035104: The Minimum Inhibitory Concentration (MIC) of
         Metronidazole is _____ µg/ml.
035105: The Minimum Inhibitory Concentration (MIC) of
         Minocycline is _____ µg/ml.
035106: The Minimum Inhibitory Concentration (MIC) of
         Nafcillin is _____ µg/ml.
035107: The Minimum Inhibitory Concentration (MIC) of
         Nalidixic Acid is _____ µg/ml.
035108: The Minimum Inhibitory Concentration (MIC) of
         Neomycin (Mycifradin) is _____ µg/ml.
035109: The Minimum Inhibitory Concentration (MIC) of
         Nitrofurantoin (Furadantin/Macrodantin) is
         _____ µg/ml.
035110: The Minimum Inhibitory Concentration (MIC) of
         Novobiocin (Albamycin) is _____ µg/ml.
035111: The Minimum Inhibitory Concentration (MIC) of
         Oleandomycin is _____ µg/ml.
035112: The Minimum Inhibitory Concentration (MIC) of
         Oxacillin is _____ µg/ml.
035113: The Minimum Inhibitory Concentration (MIC) of
         Oxytetracycline Tetramycin, Terramycin) is
         _____ µg/ml.
035114: The Minimum Inhibitory Concentration (MIC) of
         Paromomycin is _____ µg/ml.
035115: The Minimum Inhibitory Concentration (MIC) of
         Penicillin G is _____ units/ml.
035116: The Minimum Inhibitory Concentration (MIC) of
         Phenethicillin is _____ µg/ml.
035117: The Minimum Inhibitory Concentration (MIC) of
         Phenoxymethyl Penicillin is _____ µg/ml.
035118: The Minimum Inhibitory Concentration (MIC) of
         Polymyxin B (Aerosporin) is _____ units/ml.
035119: The Minimum Inhibitory Concentration (MIC) of
         Rifampin (Rifampicin) is _____ µg/ml.
035120: The Minimum Inhibitory Concentration (MIC) of
         Spectinomycin is _____ µg/ml.
035121: The Minimum Inhibitory Concentration (MIC) of
         Streptomycin or Dihydroxystreptomycin is
         _____ µg/ml.
035122: The Minimum Inhibitory Concentration (MIC) of
         Subtilin is _____ µg/ml.
035123: The Minimum Inhibitory Concentration (MIC) of
         Sulfamethoxazole/Trimethoprim is _____ µg/ml.

035124: The Minimum Inhibitory Concentration (MIC) of
Sulfisoxazole (Gantrisin) is _____ µg/ml.
035217: The Minimum Inhibitory Concentration (MIC) of
Sulfonamides is _____ µg/ml.
035125: The Minimum Inhibitory Concentration (MIC) of
Tetracycline (Achromycin) is _____ µg/ml.
035126: The Minimum Inhibitory Concentration (MIC) of
Thiostrepton is _____ µg/ml.
035127: The Minimum Inhibitory Concentration (MIC) of
Tobramycin is _____ µg/ml.
035264: The Minimum Inhibitory Concentration (MIC) of
Tricarcillin is _____ µg/ml.
035265: The Minimum Inhibitory Concentration (MIC) of
Trimethoprim is _____ µg/ml.
035128: The Minimum Inhibitory Concentration (MIC) of
Vancomycin (Vancocin) is _____ µg/ml.

NUMERICAL INHIBITORY ZONE DIAMETERS

035266: Inhibitory Zone Diameter for Amikacin concentration
(disc) 30 µg is _____ mm.
035129: Inhibitory Zone Diameter for Ampicillin concentration (disc) 2 µg is _____ mm.
035130: Inhibitory Zone Diameter for Ampicillin concentration (disc) 10 µg is _____ mm.
035131: Inhibitory Zone Diameter for Ampicillin concentration (disc) 30 µg is _____ mm.
035132: Inhibitory Zone Diameter for Bacitracin concentration (disc) 0.02 unit is _____ mm.
035133: Inhibitory Zone Diameter for Bacitracin concentration (disc) .04 unit is _____ mm.
035134: Inhibitory Zone Diameter for Bacitracin concentration (disc) 0.4 unit is _____ mm.
035135: Inhibitory Zone Diameter for Bacitracin concentration (disc) 2 units is _____ mm.
035136: Inhibitory Zone Diameter for Bacitracin concentration (disc) 5 units is _____ mm.
035137: Inhibitory Zone Diameter for Bacitracin concentration (disc) 10 units is _____ mm.
035138: Inhibitory Zone Diameter for Carbenicillin concentration (disc) 50 µg is _____ mm.
035139: Inhibitory Zone Diameter for Carbenicillin concentration (disc) 100 µg is _____ mm.
035267: Inhibitory Zone Diameter for Cefamandole concentration (disc) 30 µg is _____ mm.
035268: Inhibitory Zone Diameter for Cefotaxime concentration (disc) 30 µg is _____ mm.
035269: Inhibitory Zone Diameter for Cefoxitin concentration (disc) 30 µg is _____ mm.
035140: Inhibitory Zone Diameter for Cephaloglycine concentration (disc) 30 µg is _____ mm.
035141: Inhibitory Zone Diameter for Cefoxithin concentration (disc) 30 µg is _____ mm.
035142: Inhibitory Zone Diameter for Cephaloridine concentration (disc) 30 µg is _____ mm.
035143: Inhibitory Zone Diameter for Cephalothin concentration (disc) 10 µg is _____ mm.

Section 35: Quantitative antibiotic sensitivity 241

035144: Inhibitory Zone Diameter for Cephalothin concentration (disc) 30 µg is _____ mm.
035145: Inhibitory Zone Diameter for Chloromycetin (Chloramphenicol) concentration (disc) 2.5 µg is _____ mm.
035146: Inhibitory Zone Diameter for Chloromycetin (Chloramphenicol) concentration (disc) 5 µg is _____ mm.
035147: Inhibitory Zone Diameter for Chloromycetin (Chloramphenicol) concentration (disc) 10 µg is _____ mm.
035148: Inhibitory Zone Diameter for Chloromycetin (Chloramphenicol) concentration (disc) 30 µg is _____ mm.
035149: Inhibitory Zone Diameter for Chlortetracycline (Aureomycin) concentration (disc) 5 µg is _____ mm.
035150: Inhibitory Zone Diameter for Chlortetracycline (Aureomycin) concentration (disc) 30 µg is _____ mm.
035151: Inhibitory Zone Diameter for Clindamycin concentration (disc) 2 µg is _____ mm.
035152: Inhibitory Zone Diameter for Colistin concentration (disc) 2 µg is _____ mm.
035153: Inhibitory Zone Diameter for Colistin concentration (disc) 10 µg is _____ mm.
035154: Inhibitory Zone Diameter for 2,4-diamino-6,7-diisopropyl-pteridine (O/129, Vibriostat) crystals on agar is _____ mm.
035155: Inhibitory Zone Diameter for 2,4-diamino-6,7-diisopropyl-pteridine (O/129, Vibriostat) concentration (disc) 20 µg is _____ mm.
035156: Inhibitory Zone Diameter for 2,4-diamino-6,7-diisopropyl-pteridine (O/129, Vibriostat) concentration (disc) 40 µg is _____ mm.
035157: Inhibitory Zone Diameter for Erythromycin (Ilotycin) concentration (disc) 2 µg is _____ mm.
035158: Inhibitory Zone Diameter for Erythromycin (Ilotycin) concentration (disc) 15 µg is _____ mm.
035159: Inhibitory Zone Diameter for Erythromycin (Ilotycin) concentration (disc) 60 µg is _____ mm.
035160: Inhibitory Zone Diameter for Gentamicin concentration (disc) 10 µg is _____ mm.
035161: Inhibitory Zone Diameter for Kanamycin concentration (disc) 5 µg is _____ mm.
035162: Inhibitory Zone Diameter for Kanamycin concentration (disc) 30 µg is _____ mm.
035163: Inhibitory Zone Diameter for Kanamycin concentration (disc) 1000 µg is _____ mm.
035164: Inhibitory Zone Diameter for Lincomycin concentration (disc) 2 µg is _____ mm.
035165: Inhibitory Zone Diameter for Mandelamine (Methene-

amine mandelate) concentration (disc) 3 *mg* is
_____ mm.
035166: Inhibitory Zone Diameter for Methicillin concentration (disc) 5 µg is _____ mm.
035167: Inhibitory Zone Diameter for Nafcillin concentration (disc) 1 µg is _____ mm.
035168: Inhibitory Zone Diameter for Nalidixic acid concentration (disc) 5 µg is _____ mm.
035169: Inhibitory Zone Diameter for Nalidixic acid concentration (disc) 30 µg is _____ mm.
035170: Inhibitory Zone Diameter for Neomycin (Mycifradin) concentration (disc) 5 µg is _____ mm.
035171: Inhibitory Zone Diameter for Neomycin (Mycifradin) concentration (disc) 30 µg is _____ mm.
035172: Inhibitory Zone Diameter for Nitrofurantoin concentration (disc) 30 µg is _____ mm.
035173: Inhibitory Zone Diameter for Nitrofurantoin concentration (disc) 100 µg is _____ mm.
035174: Inhibitory Zone Diameter for Nitrofurantoin concentration (disc) 300 µg is _____ mm.
035175: Inhibitory Zone Diameter for Novobiocin (Albamycin) concentration (disc) 5 µg is _____ mm.
035176: Inhibitory Zone Diameter for Novobiocin (Albamycin) concentration (disc) 10 µg is _____ mm.
035177: Inhibitory Zone Diameter for Novobiocin (Albamycin) concentration (disc) 30 µg is _____ mm.
035178: Inhibitory Zone Diameter for Oleandomycin concentration (disc) 15 µg is _____ mm.
035179: Inhibitory Zone Diameter for Oxacillin concentration (disc) 1 µg is _____ mm.
035180: Inhibitory Zone Diameter for Oxacillin concentration (disc) 2 µg is _____ mm.
035181: Inhibitory Zone Diameter for Oxytetracycline (Tetramycin, terramycin) concentration (disc) 2.5 µg is _____ mm.
035182: Inhibitory Zone Diameter for Oxytetracycline (Tetramycin, terramycin) concentration (disc) 5 µg is _____ mm.
035183: Inhibitory Zone Diameter for Oxytetracycline (Tetramycin, terramycin) concentration (disc) 30 µg is _____ mm.
035184: Inhibitory Zone Diameter for Penicillin G concentration (disc) 1 unit is _____ mm.
035185: Inhibitory Zone Diameter for Penicillin G concentration (disc) 2 units is _____ mm.
035186: Inhibitory Zone Diameter for Penicillin G concentration (disc) 2.5 units is _____ mm.
035187: Inhibitory Zone Diameter for Penicillin G concentration (disc) 5 units is _____ mm.
035188: Inhibitory Zone Diameter for Penicillin G concentration (disc) 10 units is _____ mm.
035189: Inhibitory Zone Diameter for Polymyxin B (Aerosporin) concentration (disc) 5 units is _____ mm.
035190: Inhibitory Zone Diameter for Polymyxin B (Aerosporin) concentration (disc) 30 units is _____ mm.

Section 35: Quantitative antibiotic sensitivity 243

035191: Inhibitory Zone Diameter for Polymyxin B (Aerosporin) concentration (disc) 50 units is _____ mm.
035192: Inhibitory Zone Diameter for Polymyxin B (Aerosporin) concentration (disc) 300 units is _____ mm.
035193: Inhibitory Zone Diameter for Puromycin concentration (disc) 200 µg is _____ mm.
035194: Inhibitory Zone Diameter for Rifampin (Rifampicin) concentration (disc) 15 µg is _____ mm.
035195: Inhibitory Zone Diameter for Streptomycin concentration (disc) 2.0 µg is _____ mm.
035196: Inhibitory Zone Diameter for Streptomycin concentration (disc) 2.5 µg is _____ mm.
035197: Inhibitory Zone Diameter for Streptomycin concentration (disc) 10 µg is _____ mm.
035198: Inhibitory Zone Diameter for Streptomycin concentration (disc) 30 µg is _____ mm.
035199: Inhibitory Zone Diameter for Streptomycin concentration (disc) 50 µg is _____ mm.
035200: Inhibitory Zone Diameter for Sulfadiazine concentration (disc) 1 mg is _____ mm.
035201: Inhibitory Zone Diameter for Sulfisoxazole (gantrisin) concentration (disc) 5 µg is _____ mm.
035202: Inhibitory Zone Diameter for Sulfisoxazole (gantrisin) concentration (disc) 0.1 mg is _____ mm.
035203: Inhibitory Zone Diameter for Sulfisoxazole (gantrisin) concentration (disc) 0.25 mg is _____ mm.
035270: Inhibitory Zone Diameter for Sulfonamides concentration (disc) 250 µg is _____mm.
035204: Inhibitory Zone Diameter for Sulfonamides concentration (disc) 300 µg is _____ mm.
035205: Inhibitory Zone Diameter for Tetracycline (Achromycin) concentration (disc) 2.5 µg is _____ mm.
035206: Inhibitory Zone Diameter for Tetracycline (Achromycin) concentration (disc) 5 µg is _____ mm.
035207: Inhibitory Zone Diameter for Tetracycline (Achromycin) concentration (disc) 10 µg is _____ mm.
035208: Inhibitory Zone Diameter for Tetracycline (Achromycin) concentration (disc) 30 µg is _____ mm.
035209: Inhibitory Zone Diameter for Tobramycin concentration (disc) 10 µg is _____ mm.
035271: Inhibitory Zone Diameter for Tricarcillin concentration (disc) 75 µg is _____mm.
035272: Inhibitory Zone Diameter for Trimethoprim concentration (disc) 10 µg is _____ mm.
035273: Inhibitory Zone Diameter for Trimethoprim sulfamethoxazole, 1.25 µg of trimethoprim and 23.75 µg of sulfamethoxazole concentration (disc) is _____mm.

Section 35: Quantitative antibiotic sensitivity

035210: Inhibitory Zone Diameter for Triple sulfa (sulfadiazine/sulfamethazine/sulfamerazine) concentration (disc) 0.1 mg is _____ mm.
035211: Inhibitory Zone Diameter for Triple sulfa (sulfadiazine/sulfamethazine/sulfamerazine) concentration (disc) 0.25 mg is _____ mm.
035212: Inhibitory Zone Diameter for Triple sulfa (sulfadiazine/sulfamethazine/sulfamerazine) concentration (disc) 1 mg is _____ mm.
035213: Inhibitory Zone Diameter for Vancomycin (Vancocin) concentration (disc) 5.0 µg is _____ mm.
035214: Inhibitory Zone Diameter for Vancomycin (Vancocin) concentration (disc) 7.5 µg is _____ mm.
035215: Inhibitory Zone Diameter for Vancomycin (Vancocin) concentration (disc) 30.0 µg is _____ mm.

SECTION 36: INTERNAL ORGANELLES

[NOTE 1: The terms in this section are used as defined in Corliss, J.O. (1979), The Ciliated Protozoa, 2nd ed., Pergamon Press, Oxford and as also defined by Bold, H.C. and M.J. Wynne (1978), Introduction to the Algae, Prentice-Hall, Inc., Englewood Cliffs, New Jersey.]

036001: Paramylum (paramylon) bodies are present.
036002: Two paramylum (paramylon) bodies are present.
036003: Three to 10 paramylum (paramylon) bodies are present.
036004: More than 10 paramylum (paramylon) bodies are present.
036005: Paramylum (paramylon) bodies are rod-like.
036006: Paramylum (paramylon) bodies are ovoid.
036007: Paramylum (paramylon) bodies are annular.
036008: Paramylum (paramylon) bodies are circular.
036009: Paramylum (paramylon) bodies are discoid.
036010: Paramylum (paramylon) bodies are granular.
036011: Statocyst (concrement vacuole) present.
036012: Contractile vacuole is present.
036013: One contractile vacuole is present.
036014: Two contractile vacuoles are present.
036015: Three contractile vacuoles are present.
036016: Four contractile vacuoles are present.
036017: Five contractile vacuoles are present.
036018: Six contractile vacuoles are present.
036019: Seven contractile vacuoles are present.
036020: Eight contractile vacuoles are present.
036021: Nine contractile vacuoles are present.
036022: Ten contractile vacuoles are present.
036023: More than 10 contractile vacuoles are present.
036024: Contractile vacuole is present posteriorly.
036025: Contractile vacuole is present anteriorly.
036026: Contractile vacuoles are centrally located.
036027: Contractile vacuoles are in a reservoir.
036028: Contractile vacuoles are present when placed in fresh water.
036029: Cell has cyrtos.
036030: Cell has cytoproct.
036031: Cell has pinocytotic vesicles.
036032: Cell has secretory ampulla.
036033: Cell has afferent canal (pulsating, nephridial, collecting, radial).
036034: Iodinophilous vacuole is present.
036035: Cell has chloroplasts (chromotophores).
036181: Chloroplast has single thylakoids.
036182: Chloroplast has thylakoids in groups of 2.
036183: Chloroplast has thylakoids in groups of 3.
036184: Chloroplast contains granum or grana (stack or stacks of thylakoids).
036036: Cell has 1 chloroplast (chromotophore).
036037: Cell has 2 chloroplasts (chromotophores).
036038: Cell has 3 chloroplasts (chromotophores).
036039: Cell has 4 chloroplasts (chromotophores).

Section 36: Internal organelles

036158: Cell contains 5 chloroplasts (chromatophores).
036159: Cell contains 6 chloroplasts (chromotophores).
036160: Cell contains > 6 chloroplasts (chromatophores).
036161: Cells have a variable number of chloroplasts (chromotophores).
036175: Chloroplast (chromotophore) is red.
036040: Chloroplast (chromotophore) is orange.
036042: Chloroplast (chromotophore) is yellow.
036041: Chloroplast (chromotophore) is green.
036177: Chloroplast (chromotophore) is blue.
036179: Chloroplast (chromotophore) is purple.
036043: Chloroplast (chromotophore) is brown.
036178: Chloroplast (chromotophore) is grey.
036180: Chloroplast (chromotophore) is black.
036162: Chloroplast (chromatophore) shape is campanulate (bell-shaped).
036044: Chloroplast (chromotophore) is discoidal.
036045: Chloroplast (chromotophore) is ovoid.
036046: Chloroplast (chromotophore) is band-form.
036047: Chloroplast (chromotophore) is cup-like.
036048: Chloroplast (chromotophore) is fusiform.
036049: Chloroplast (chromotophore) is diffuse.
036050: Chloroplast (chromotophore) branches (network).
036051: Chloroplast (chromotophore) is rod-shaped.
036052: Chloroplast (chromotophore) is sheet-like.
036053: Chloroplast (chromotophore) is lateral.
036054: Chloroplast (chromotophore) is plate-like.
036055: Chloroplast (chromotophore) is ovoid.
036163: Chloroplast (chromatophore) shape is stellar.
036164: Clorcplast (chromatophore) is elongate, i.e., length is greater than width.
036165: Chloroplast (chromatophore) surface is granular.
036166: Chloroplast (chromatophore) location in the cell is axial.
036167: Chloroplast (chromatophore) location in the cell is parietal.
036168: Chloroplast (chromatophore) location in the cell is lateral.
036169: Chloroplast (chromatophore) location in the cell is radial.
036170: Chloroplast (chromatophore) location is in the anterior (end that leads while swimming) half of the cell.
036171: Chloroplast (chromatophore) location is in the posterior (end that trails while swimming) half of the cell.
036172: Chloroplast (chromatophore) location in the cell is variable or scattered.
036173: Chloroplast (chromatophore) has 2 anterior chloroplast lobes.
036174: Chloroplast (chromatophore) has 4 anterior chloroplast lobes.
036056: Stigma is present.
036187: Stigma are placed in anterior (end that leads while swimming) half of the cell.
036057: Stigma is located near base of flagellum.
036188: Stigma is attached to the chloroplast.

Section 36: Internal organelles 247

036058: Stigma is red.
036186: Stigma is orange.
036059: Pyrenoid is present.
036060: Pyrenoid is spherical.
036061: Cells have 1 pyrenoid.
036062: Cells have 2 pyrenoids.
036063: Cells have 3 pyrenoids.
036064: Cells have 4 pyrenoids.
036065: Pyrenoid is surrounded by starch envelope.
036066: Pyrenoid is centrally located.
036067: Pyrenoid is posteriorly located.
036068: Mitochondrion is present.
036069: Mitochondrion is spherical.
036070: Mitochondrion is oval.
036071: Mitochondrion is irregularly shaped.
036072: More than 1 mitochondrion is present.
036073: Cell has chondriome.
036074: Cells have an axostyle.
036075: Parabasal body is present.
036076: Parabasal body is single.
036077: Parabasal body is double.
036078: Parabasal body is multiple.
036079: Parabasal body is pyriform.
036080: Parabasal body is rod-like.
036081: Parabasal body is curved.
036082: Parabasal body is bandiform.
036083: Parabasal body is collar-like.
036084: Parabasal body is spirally coiled.
036085: Parabasal thread is present.
036086: Blepharoplast is present.
036087: Blepharoplast is a small granule.
036088: Blepharoplast is ovoid.
036089: Blepharoplast is rod-shaped.
036090: Kinetoplast (DNA core in close proximity to basal body) is present.
036091: Kinetoplast (DNA core in close proximity to basal body) is disc-shaped.
036092: Kinetoplast (DNA core in close proximity to basal body) is circular.
036093: Kinetoplast (DNA core in close proximity to basal body) is oval.
036094: Kinetoplast (DNA core in close proximity to basal body) is ellipsoidal.
036095: Kinetoplast (DNA core in close proximity to basal body) is rod-shaped.
036096: Kinetoplast diameter is < 0.4 µm.
036097: Kinetoplast diameter is 0.4 µm.
036098: Kinetoplast diameter is 0.5 µm.
036099: Kinetoplast diameter is 0.6 µm.
036100: Kinetoplast diameter is 0.7 µm.
036101: Kinetoplast diameter is 0.8 µm
036102: Kinetoplast diameter is 0.9 µm.
036103: Kinetoplast diameter is 1.0 µm.
036104: Kinetoplast diameter is 1.1 µm.
036105: Kinetoplast diameter is 1.2 µm.
036106: Kinetoplast diameter is 1.3 µm.
036107: Kinetoplast diameter is 1.4 µm.

Section 36: Internal organelles

036108: Kinetoplast diameter is 1.5 µm.
036109: Kinetoplast diameter is greater than 1.5 µm.
036110: Cell has brood pouch (chamber) (marsupium).
036111: Cell has birth Pore.
036112: Cell has crystallocyst.
036113: Cell has fibrocysts.
036114: Cell has clathrocysts.
036115: Cell has cyrtocysts.
036116: Cell has pexicysts.
036117: Cell has rhabdocysts.
036185: Cell contains trichocysts.
036118: Cell has akontobolocyst (syn. spindle trichocyst).
036119: Cell has ampullocyst.
036120: Cell has cnidocyst.
036121: Cell has karyophore.
036122: Cell has Lamina Corticalis.
036123: Cell has Lieberkuhn's organelle.
036124: Cell has Lkm fiber.
036125: Cell has Km fiber.
036126: Cell has lysosomes.
036127: Cell has myonemes.
036128: Cell has nematodesma.
036129: Cell has cathetodesma.
036130: Cell has postciliodesma.
036131: Cell has skeletal plates.
036132: Cell has skeletal plaques.
036133: Cell has kinetodesma.
036134: Cell has ribbed wall.
036135: Cells have scales.
036136: Cells have calcareous discs.
036137: Cells have a skeleton composed of spines.
036156: The endalveolar skeletal elements are composed of calcium carbonate.
036157: The endalveolar skeletal elements are composed of cellulose.
036138: Cell has spongioplasm.
036139: Cell has epiplasm.
036140: Cell has phagoplasm.
036141: Cell has striated bands.
036142: Cell has postciliary microtubules (radial fibers).
036143: Cell has basal microtubules (basal fibers).
036144: Cell has subkinetal microtubules.
036145: Cell has transverse microtubules (transverse fibers).
036146: Cell has transverse fibrous spur.
036147: Cell has retractor fibers.
036148: Cell has trichite.
036149: Cell has tubulins.
036150: Cell has unit membrane.
036151: Cell has basal microtubules (basal fibers).
036152: Cell has PBB-complex (polar basal body-complex).
036153: Cell has paralabial organ.
036154: Cell has oral ribs.
036155: Endosymbiotes are present.

SECTION 37: NUCLEUS

[NOTE 1: "Nucleus" and "macronucleus" are equivalent terms, and are to be distinguished from micronucleus when the latter is present. For specific definitions of terms see Corliss, J.O. (1979), The Ciliated Protozoa, 2nd ed., Pergamon Press, Oxford.]

037001: One or more nuclei (macronuclei) are present.
037002: Single macronucleus is present.
037003: Two macronuclei are present.
037004: Cells are binucleate.
037005: Cells are multinucleated.
037006: Nucleus (macronucleus) is anterior.
037007: Nucleus (macronucleus) is central.
037008: Nucleus (macronucleus) is posterior.
037009: In addition to the nucleus (macronucleus) 1 micronucleus is present.
037010: In addition to the nucleus (macronucleus) 2 micronucleus are present.
037011: In addition to the nucleus (macronucleus) 3 micronuclei are present.
037012: In addition to the nucleus (macronucleus) 4 micronuclei are present.
037013: In addition to the nucleus (macronucleus) > 4 micronuclei are present.
037014: Nucleus (macronucleus) is oblong.
037015: Nucleus (macronucleus) is ellipsoid.
037016: Nucleus (macronucleus) is pyriform.
037017: Nucleus (macronucleus) is spherical.
037018: Nucleus (macronucleus) is rope-shaped.
037019: Nucleus (macronucleus) is lenticular.
037020: Nucleus (macronucleus) is globular.
037021: Nucleus (macronucleus) is bandiform.
037022: Nucleus (macronucleus) is U-shaped.
037023: Nucleus (macronucleus) is L-shaped.
037024: Nucleus (macronucleus) is T-shaped.
037025: Nucleus (macronucleus) is moniliform.
037094: Nucleus (macronucleus) is heteromeric.
037095: Nucleus (macronucleus) is homomeric.
037096: Nucleus (macronucleus) has a binnenkorper.
037026: Micronucleus is hemispherical.
037027: Micronucleus is crescent-shaped.
037028: Micronucleus is vesicular.
037029: Micronucleus is fusiform.
037030: Longest axis of nucleus (macronucleus) is <0.5 µm.
037031: Longest axis of nucleus (macronucleus) is 0.5-1.0 µm.
037032: Longest axis of nucleus (macronucleus) is 1.1-1.5 µm.
037033: Longest axis of nucleus (macronucleus) is 1.6-2.0 µm.
037034: Longest axis of nucleus (macronucleus) is 2.1-2.5 µm.

037035: Longest axis of nucleus (macronucleus) is 2.6-3.0 µm.
037036: Longest axis of nucleus (macronucleus) is 3.1-3.5 µm.
037037: Longest axis of nucleus (macronucleus) is 3.6-4.0 µm.
037038: Longest axis of nucleus (macronucleus) is 4.1-4.5 µm.
037039: Longest axis of nucleus (macronucleus) is 4.6-5.0 µm.
037040: Longest axis of nucleus (macronucleus) is 5.1-5.5 µm.
037041: Longest axis of nucleus (macronucleus) is 5.6-6.0 µm.
037042: Longest axis of nucleus (macronucleus) is 6.1-6.5 µm.
037043: Longest axis of nucleus (macronucleus) is 6.6-7.0 µm.
037044: Longest axis of nucleus (macronucleus) is 7.1-7.5 µm.
037045: Longest axis of nucleus (macronucleus) is 7.6-8.0 µm.
037046: Longest axis of nucleus (macronucleus) is 8.1-8.5 µm.
037047: Longest axis of nucleus (macronucleus) is 8.6-9.0 µm.
037048: Longest axis of nucleus (macronucleus) is 9.1-9.5 µm.
037049: Longest axis of nucleus (macronucleus) is 9.6-10.0 µm.
037050: Longest axis of nucleus (macronucleus) is 10.1-15.0 µm.
037051: Longest axis of nucleus (macronucleus) is 15.1-20.0 µm.
037052: Longest axis of nucleus (macronucleus) is 20 µm.
037053: Shortest axis of nucleus (macronucleus) is < 0.5 µm.
037054: Shortest axis of nucleus (macronucleus) is 0.5-1.0 µm.
037055: Shortest axis of nucleus (macronucleus) is 1.1-1.5 µm.
037056: Shortest axis of nucleus (macronucleus) is 1.6-2.0 µm.
037057: Shortest axis of nucleus (macronucleus) is 2.1-2.5 µm.
037058: Shortest axis of nucleus (macronucleus) is 2.6-3.0 µm.
037059: Shortest axis of nucleus (macronucleus) is 3.1-3.5 µm.
037060: Shortest axis of nucleus (macronucleus) is 3.6-4.0 µm.
037061: Shortest axis of nucleus (macronucleus) is 4.1-4.5 µm.
037062: Shortest axis of nucleus (macronucleus) is 4.6-5.0 µm.

Section 37: Nucleus

037063: Shortest axis of nucleus (macronucleus) is 5.1-5.5 µm.
037064: Shortest axis of nucleus (macronucleus) is 5.6-6.0 µm.
037065: Shortest axis of nucleus (macronucleus) is 6.1-6.5 µm.
037066: Shortest axis of nucleus (macronucleus) is 6.6-7.0 µm.
037067: Shortest axis of nucleus (macronucleus) is 7.1-7.5 µm.
037068: Shortest axis of nucleus (macronucleus) is 7.6-8.0 µm.
037069: Shortest axis of nucleus (macronucleus) is 8.1-8.5 µm.
037070: Shortest axis of nucleus (macronucleus) is 8.6-9.0 µm.
037071: Shortest axis of nucleus (macronucleus) is 9.1-9.5 µm.
037072: Shortest axis of nucleus (macronucleus) is 9.6-10.0µm.
037073: Shortest axis of nucleus (macronucleus) is 10.1-15.0 µm.
037074: Shortest axis of nucleus (macronucleus) is 15.1-20.0 µm.
037075: Shortest axis of nucleus (macronucleus) is 20 µm.
037076: Nuclear envelope is more complex than a double membrane.
037077: Nuclear membrane persists during mitosis.
037078: Nuclear membrane disintegrates by anaphase.
037079: Nuclear division occurs within the cyst.
037080: Nuclear division occurs during excystment.
037081: Endosome (nucleolus, karyosome) is present.
037082: Nucleus has numerous endosomal granules.
037083: Endosome is rounded.
037084: Endosome is central.
037085: Endosome is electron dense.
037097: Nuclear mitosis is endomitotic.
037098: Nuclear mitosis is epimitotic.
037099: During mitosis spindle fibers are extranuclear.
037100: During mitosis spindle fibers are endonuclear.
037101: During mitosis nuclear envelope partially differentiates.
037102: During mitosis nuclear envelope remains intact.
037103: During mitosis nuclear envelope completely differentiates.
037086: Nucleolus disintegrates during mitosis.
037087: Nucleolus divides into 2 polar masses during mitosis.
037088: Cell is a karyonide.
037089: Cell exhibits oligoploidy.
037090: Cell is diploid.
037091: Cell is haploid.
037092: Cell is heterokaryotic.
037093: Cell has heteromerous macronucleus.

SECTION 38:   NUCLEIC ACIDS

014046: Cells are prokaryotic.
014047: Cells are mesokaryotic.
014048: Cells are eukaryotic.
036068: Mitochondrion is present
036072: More than one mitochondrion is present.
037081: Endosome (nucleolus, karysome) is present.
037090: Cells are diploid (2N or twice the basic (haploid) number set of chromosomes).
037091: Cells are haploid (N or single basic set of chromosomes).
038001: Chromosomal DNA of cells is a single molecule.
038002: Chromosomal DNA of cells is in several or many chromosomes.
038003: Chromosomal DNA is complexed with histones.
038004: Extrachromosomal DNA occurs (plasmids or episomes).
038005: DNA replication begins at one point (site) and moves around the DNA circle.
038006: DNA replication begins at a number of sites simultaneously.
038007: DNA replication begins at multiple sites and moves in both directions from each site.

[NOTE 1: The B-form of DNA (Watson-Crick model) is the prevailing form in the living cell and exists as a right-handed helix in fibers at very high relative humidity of *ca* 92% and in solutions of low ionic strength.]

038008: The A-form of DNA is present.
038009: The B-form of DNA is present.
038010: The C-form of DNA is present.
038011: The D-form of DNA is present.
038012: The E-form of DNA is present.
038013: The Z-form of DNA (left-handed helix) is present.
038014: The DNA has 8 base pairs per turn.
038015: The DNA has 7 1/2 base pairs per turn.
038016: The DNA has 11 base pairs per turn.
038017: The DNA has 10 base pairs per turn.
038018: The DNA has 9 1/3 base pairs per turn.
038019: The DNA has 12 base pairs per turn.
038020: The DNA molecule lacks guanine.
038021: The DNA is supercoiled.
038022: The DNA is negatively supercoiled (underwound).
038023: The DNA is positively supercoiled (overwound).
038024: 5S RNA's fits a 4-helix secondary RNA structure with four base-paired regions.
038025: 5S RNAs fits a 5-helix secondary RNA structure.
038026: During cell division mitosis occurs.
037099: During mitosis spindle fibers are extranuclear.
037100: During mitosis spindle fibers are endonuclear.
037101: During mitosis nuclear envelope partially differentiates.
037102: During mitosis nuclear envelope remains intact.

Section 38:   Nucleic acids   253

037103: During mitosis nuclear envelope completely differentiates.
037097: Nuclear mitosis is endomitotic.
037098: Nuclear mitosis is epimitotic.
038027: During sexual reproduction meiosis does not occur.
038028: During sexual reproduction only portions of genetic complement are reassorted.
038029: During sexual reproduction a regular process of meiosis and reassortment of the whole chromosome complement occurs.
038030: Internal cell membranes or endoplasmic reticulum (a membrane network running through the cell) are present.
038031: Ribosomes form polysomes.
038032: Smooth endoplasmic reticulum (without attached ribosomes) is present.
038033: Rough endoplasmic reticulum (with attached ribosomes) is present.
038034: Mesosomes are associated with internal cell membranes.
038035: The respiratory system of the cell is part of the plasma membrane or mesosome.
038036: The respiratory system of the cell is in the mitochondria.
038037: Photosynthesis occurs other than in the chloroplasts.
036035: Cell has chloroplasts (chromatophores).
038038: Cellular ribosomes are 70 S.
038039: Cellular ribosomes are 80 S.
038040: Ribosomes of mitochondria are 70 S.
038041: Ribosomes of chloroplasts are 70 S.
038042: Microtubules are present.
038043: Microtubules are present in flagella.
038044: Microtubules are present in cilia.
038045: Microtubules are present in basal bodies.
038046: Microtubules are present in mitotic spindle apparatus.
038047: Microtubules are present in centrioles.
038048: Cells have a single long DNA molecule arranged in the general shape of a circle.
038049: Ribothymine in TψC loop in tRNA.
038050: Rifamycin inhibits mRNA synthesis.
038051: Rifamycin inhibits RNA synthesis in chloroplasts.
038052: Rifamycin inhibits RNA synthesis in mitochondria.
038053: Streptovaricin(s) inhibits RNA synthesis.
038054: Streptolydigins inhibits RNA synthesis.
038055: Amanitan inhibits mRNA synthesis in eukaryotes.
038056: Amanitan does not inhibit mRNA synthesis in prokaryotes.
038057: Actinomycin inhibits total RNA synthesis.

SECTION 101: *SALMONELLA*

[NOTE 1: Some microbiological products from different manufacturers may differ in certain respects. For example, *Salmonella* O Group B antisera produced by Bacto contains factors 4 and 5 while Group B produced by the Centers for Disease Control (CDC) contains either 4,5, and 12 or 4,5, and 27. Wellcome Group B contains single factors 4,5, or 27. Thus, where this occurs, it is necessary to assign unique code numbers for each manufacturer's product in order to avoid ambiguity and to provide full information.]

101020: Organism is monophasic.

O ANTIGENS - POLYVALENT ANTISERA - BACTO

101001: Organism reacts with *Salmonella* O pooled antisera containing somatic (O) Kauffmann-White Groups A,B,C,D,E,F, and G.
101002: Organism reacts with Bacto preparation 2975-32-9 containing somatic (O) Groups A,B,C,D,E,F,G,H, and I (Poly A - I) and Vi antibodies (part of Bacto Set 2975-32-9 and trade named Bacto MinESS Antisera Set II.
101003: Organism reacts with Bacto preparation 2264-47-2 containing (Poly A-I and Vi factors 1-16,19,22-25, 34 and Vi).
101004: Organism reacts with Bacto preparation 2534-47-6 (Poly A) containing somatic (O) Groups A,B,C,D,E1, E2,E3,E4, and L.
101005: Organism reacts with Bacto preparation 2535-47-5 (Poly B) containing somatic (O) Groups C1,C2,F,G, and H.
101006: Organism reacts with Bacto preparation 2536-47-4 (Poly C) containing somatic (O) Groups I,J,K,M,N, and O.
101007: Organism reacts with Bacto preparation 2537-47-3 (Poly D) containing somatic (O) Groups P,Q,R,S,T, and U.
101008: Organism reacts with Bacto preparation 2538-47-2 (Poly E) containing somatic (O) Groups V,W,X,Y, and Z.
101009: Organism reacts with Bacto preparation 2645-47-2 (Poly F) containing somatic (O) Groups 51-55.
101010: Organism reacts with Bacto preparation 2646-47-1 (Poly G) containing somatic (O) Groups 56-61.

O ANTIGENS - POLYVALENT ANTISERA - CDC

101066: Organism reacts with CDC *Salmonella* O Pooled Antisera 28a,28b.
101067: Organism reacts with CDC *Salmonella* O Pooled Antisera 28a,28c.

Section 101: *Salmonella* 255

101068: Organism reacts with CDC *Salmonella* O Pooled Antisera 40a,40b.
101069: Organism reacts with CDC *Salmonella* O Pooled Antisera 1,40a,40c.
101070: Organism reacts with CDC *Salmonella* O Pooled Antisera 43a,43b.
101071: Organism reacts with CDC *Salmonella* O Pooled Antisera 45a,45b.
101072: Organism reacts with CDC *Salmonella* O Pooled Antisera 45a,45c.
101073: Organism reacts with CDC *Salmonella* O Pooled Antisera 47ab,47ac.
101074: Organism reacts with CDC *Salmonella* O Pooled Antisera 48a,48b.
101075: Organism reacts with CDC *Salmonella* O Pooled Antisera 48a,48c.
101076: Organism reacts with CDC *Salmonella* O Pooled Antisera 50a,50b,50d.

O ANTIGENS - POLYVALENT ANTISERA - WELLCOME

101400: Organism reacts with Wellcome *Salmonella* O, POLY A-G Antisera.
101401: Organism reacts with Wellcome *Salmonella* O, POLY A-S Antisera.

SPICER-EDWARDS POLYVALENT H ANTISERA

101520: Organism reacts with Bacto Code 2265, Spicer-Edwards 1 *Salmonella* polyvalent H antisera. (Reacts with H antigens a,b,c,d,e,h, G complex, and i. The G Complex component of *Salmonella* H antisera Spicer-Edwards 1 and 4 reacts with antigens f,g; f,g,s; f,g,t; g,m; g,m,q; g,m,s; g,m,s,t; g,m,t; g,p; g,p,s; g,p,u; g,(p),z51; g,q; g,s,t; g,t; m,p,t,u and m,t.)
101521: Organism reacts with Bacto Code 2266, Spicer-Edwards 2 *Salmonella* polyvalent H antisera. (Reacts with H antigens a,b,c,k,r,y, and z29.)
101522: Organism reacts with Bacto Code 2267, Spicer-Edwards 3 *Salmonella* polyvalent H antisera. (Reacts with H antigens a,d,e,h,k,z,z4 complex, and z29.) The Z4 complex component reacts with z4,z23; z4,z24 and z4,z32.)
101523: Organism reacts with Bacto Code 2268, Spicer-Edwards 4 *Salmonella* polyvalent H antisera. (Reacts with H antigens b,d, G complex, k,r,z, and z10. The G Complex component of *Salmonella* H antisera Spicer-Edwards 1 and 4 reacts with antigens f,g; f,g,s; f,g,t; g,m; g,m,q; g,m,s; g,m,s,t; g,m,t; g,p; g,p,s; g,p,u; g,(p),z51; g,q; g,s,t; g,t; m,p,t,u and m,t.)
101524: Organism reacts with Bacto Code 2270, Spicer-Edwards EN Complex *Salmonella* polyvalaent H antisera. (Reacts with H antigens e,n,x and e,n,z15).

101525: Organism reacts with Bacto Code 2271, Spicer-Edwards L Complex *Salmonella* polyvalent H antisera. (Reacts with H antigens 1,v; 1,w; 1,z13; and 1,z28).
101526: Organism reacts with Bacto Code 2272, Spicer-Edwards 1 Complex *Salmonella* polyvalent H antisera. (Reacts with H antigens 1,2; 1,5; 1,6; and 1,7).

H ANTIGENS - WELLCOME

101530: Organism reacts with Wellcome *Salmonella* H, Phase 1 - 2 Antisera (1,2,5,6,7).
101531: Organism reacts with Wellcome *Salmonella* H, Phase 2 Antiserum.
101535: Organism reacts with Wellcome *Salmonella* H, Rapid Diagnostic 1 Antisera (b,d,E,r).
101536: Organism reacts with Wellcome *Salmonella* H, Rapid Diagnostic 2 Antisera (b,E,k,L).
101537: Organism reacts with Wellcome *Salmonella* H, Rapid Diagnostic 3 Antisera (d,E,G,k).
101540: Organism reacts with Wellcome *Salmonella* H, POLY E Antisera (eh,enx,enz15).
101541: Organism reacts with Wellcome *Salmonella* H, POLY G Antisera (gm,gp,qq,gst).
101542: Organism reacts with Wellcome *Salmonella* H, POLY L Antisera (lv,lw).

[NOTE 2: Within the limits of each manufacturer's antisera, somatic and flagellar antigens in the following Tables are consistent with *Salmonella* serovars in Le Minor (1984). (*=absorbed)]

[NOTE 3: The items in the Table pertain to a statement in the following format:

101410: Organism reacts with Wellcome *Salmonella* somatic (O) antisera Group A, Factor 2.]

Wellcome	CDC	Bacto	Group	Factor
101410		101202	A	2
	101050		A	1*,2,12*
	101051		B	4,5,12;4,12,27
	101052		B	4*,5,12*
	101053		B	4*,12*,27
101411		101204	B	4
	101101		B	4,5
101412		101205	B	5
101413		101207	C1	7
	101054		C1	6*,7
101414			C1	6,7
	101055		C2	8*,20
101415	101308	101208	C2	8
101416		101220	C3	20
101417		101209	D	9

Section 101: *Salmonella*

Wellcome	CDC	Bacto	Group	Factor
	101056		D1	9,12*
	101057		D2	9,46
101418			E	1,3,10 15,19,34
101419		101210	E1	10
101420		101215	E2	15
101421		101219	E4	19
		101234	E3	34
		101102	E	1,3,10, 15,19
	101058		E1	3,10
	101059		E2	3*,15
	101060		E3	3*,15*,34
	101061		E4	1*,3,19
101422	101311	101211	F	11
	101063		G1	13,22,36
101423		101103	G	13,22
	101062		G2	1*,13,23,37
	101064		H	6,14,24
		101104	H	14,24
	101065		H	1*,6*,14,25
		101214	H,C,K	14
101424			H,C	14
101427	101318	101218	K	18
101425	101316	101216	I	16
101426	101317	101217	J	17
101428	101321	101221	L	21
101429		101228	M	28
101430	101332	101230	N	30
101431	101336	101235	O	35
101432	101339	101238	P	38
101433	101340	101239	Q	39
101434		101240	R	40
101435	101344	101241	S	41
	101345	101242	T	42
		101243	U	43
	101348	101244	V	44
		101245	W	45
	101352	101246		46
		101247	X	47
		101248	Y	48
		101250	Z	50
		101201		1
		101212		12
		101222		22
		101223		23
		101224		24
		101225		25
		101227		27
		101236		36
		101237		37
	101362	101251		51
	101363	101252		52
	101364	101253		53

Section 101: *Salmonella*

Wellcome	CDC	Bacto	Group Factor
	101365	101254	54
	101366	101255	55
	101367	101256	56
	101368	101257	57
	101369	101258	58
	101370	101259	59
	101371	101260	60
	101372	101261	61
	101375	101264	64

[NOTE 4: The items in the Table pertain to a statement in the following format:

101735: Organism reacts with Wellcome *Salmonella* flagellar (H) antisera a.]

Wellcome	CDC	Bacto	H Ag
101735	101701	101601	a
101736	101702	101602	b
101737	101703	101603	c
101738	101704	101604	d
		101605	f
		101606	g
		101607	h
101739	101705	101608	i
101740	101706	101609	k
		101610	m
		101611	n
		101612	p
		101613	q
101741	101707	101614	r
		101615	s
		101616	t
		101617	u
		101618	v
		101619	w
		101620	x
101742	101708	101621	y
101743	101709	101622	z
101744	101710	101623	Z6
101745	101711	101624	Z10
		101625	Z13
		101626	Z15
		101627	Z23
		101628	Z24
101746	101712	101629	Z27
		101630	Z28
101747	101713	101631	Z29
		101632	Z32
	101714	101633	Z35

Section 101: *Salmonella*

Wellcome	CDC	Bacto	H Ag
101748	101715	101634	Z36
	101716	101635	Z37
101749	101717	101636	Z38
	101718	101637	Z39
		101638	Z40
	101719	101639	Z41
	101720	101640	Z42
	101721	101641	Z43
	101722	101642	Z44
	101723		Z45
	101724	101643	Z46
	101725	101644	Z47
	101726	101645	Z48
	101727	101646	Z49
		101647	Z50
		101648	Z51
	101728	101649	Z52
	101729	101650	Z53
		101651	Z54
		101652	Z55
		101653	Z56
		101654	Z57
		101655	2
		101656	5
		101657	6
		101658	7

[NOTE 5: The items in the Table pertain to a statement in the following format:

101557: Organism reacts with Wellcome *Salmonella* H Pooled Antisera e,h.]

Wellcome	CDC	H Pooled Antisera
101557	101570	e,h
101558	101571	e,n,x
101559	101572	e,n,z15
101560	101573	f,g
101561	101574	g,m
	101575	g,m,s
101562	101576	g,p
	101577	g,p,u
101563	101578	g,q
101564	101579	g,s,t
	101580	g,z51
101567	101581	m,t
101565	101582	l,v
101566	101583	l,w
	101584	l,z13
	101585	l,z28
	101586	l,z40
101568	101587	z4,z23
	101588	z4,z24
	101589	z4,z32
101595	101590	1,2
101596	101591	1,5
101597	101592	1,6
101598	101593	1,7

SECTION 103: *STREPTOCOCCUS*

NEUFELD QUELLUNG REACTION

103601: Neufeld Quellung reaction is type 1.
103602: Neufeld Quellung reaction is type 2.
103603: Neufeld Quellung reaction is type 3.
103604: Neufeld Quellung reaction is type 4.
103605: Neufeld Quellung reaction is type 5.
103606: Neufeld Quellung reaction is type 6.
103607: Neufeld Quellung reaction is type 7.
103608: Neufeld Quellung reaction is type 8.
103609: Neufeld Quellung reaction is type 9.
103610: Neufeld Quellung reaction is type 10.
103611: Neufeld Quellung reaction is type 11.
103612: Neufeld Quellung reaction is type 12.
103613: Neufeld Quellung reaction is type 13.
103614: Neufeld Quellung reaction is type 14.
103615: Neufeld Quellung reaction is type 15.
103616: Neufeld Quellung reaction is type 16.
103617: Neufeld Quellung reaction is type 17.
103618: Neufeld Quellung reaction is type 18.
103619: Neufeld Quellung reaction is type 19.
103620: Neufeld Quellung reaction is type 20.
103621: Neufeld Quellung reaction is type 21.
103622: Neufeld Quellung reaction is type 22.
103623: Neufeld Quellung reaction is type 23.
103624: Neufeld Quellung reaction is type 24.
103625: Neufeld Quellung reaction is type 25.
103626: Neufeld Quellung reaction is type 26.
103627: Neufeld Quellung reaction is type 27.
103628: Neufeld Quellung reaction is type 28.
103629: Neufeld Quellung reaction is type 29.
103630: Neufeld Quellung reaction is type 30.
103631: Neufeld Quellung reaction is type 31.
103632: Neufeld Quellung reaction is type 32.
103633: Neufeld Quellung reaction is type 33.

LANCEFIELD GROUPS

103701: Lancefield precipitin reaction is positive for Group A.
103702: Lancefield precipitin reaction is positive for Group B.
103703: Lancefield precipitin reaction is positive for Group C.
103704: Lancefield precipitin reaction is positive for Group D.
103705: Lancefield precipitin reaction is positive for Group E.
103706: Lancefield precipitin reaction is positive for Group F.
103707: Lancefield precipitin reaction is positive for Group G.
103708: Lancefield precipitin reaction is positive for Group H.

103709: Lancefield precipitin reaction is positive for Group J.
103710: Lancefield precipitin reaction is positive for Group K.
103711: Lancefield precipitin reaction is positive for Group L.
103712: Lancefield precipitin reaction is positive for Group M.
103713: Lancefield precipitin reaction is positive for Group N.
103714: Lancefield precipitin reaction is positive for Group O.
103715: Lancefield precipitin reaction is positive for Group P.
103716: Lancefield precipitin reaction is positive for Group Q.
103717: Lancefield precipitin reaction is positive for Group R.
103718: Lancefield precipitin reaction is positive for Group S.
103719: Lancefield precipitin reaction is positive for Group T.
103720: Lancefield precipitin reaction is positive for Group A Var.

SECTION 105: *STAPHYLOCOCCUS*

STAPHYLOCOCCAL TOXIN

105050: Organism reacts with staphylococcal antitoxin A.
105051: Organism reacts with staphylococcal antitoxin B.
105052: Organism reacts with staphylococcal antitoxin C.
105053: Organism reacts with staphylococcal antitoxin D.
105054: Organism reacts with staphylococcal antitoxin E.

IMMUNODIFFUSION

105001: Antigen yields a precipitation band of identity with *Staphylococcus* Antitoxin A.
105002: Antigen yields a precipitation band of partial identity (spur) with *Staphylococcus* Antitoxin A.
105003: Antigen yields precipitation bands of non-identity (crossed) with *Staphylococcus* Antitoxin A.
105004: Precipitation band is absent when antigen is tested with *Staphylococcus* Antitoxin A.
105005: Antigen yields a precipitation band of identity with *Staphylococcus* Antitoxin B.
105006: Antigen yields a precipitation band of partial identity (spur) with *Staphylococcus* Antitoxin B.
105007: Antigen yields precipitation bands of non-identity (crossed) with *Staphylococcus* Antitoxin B.
105008: Precipitation band is absent when antigen is tested with *Staphylococcus* Antitoxin B.
105009: Antigen yields a precipitation band of identity with *Staphylococcus* Antitoxin C.
105010: Antigen yields a precipitation band of partial identity (spur) with *Staphylococcus* Antitoxin C.
105011: Antigen yields precipitation bands of non-identity (crossed) with *Staphylococcus* Antitoxin C.
105012: Precipitation band is absent when antigen is tested with *Staphylococcus* Antitoxin C.
105013: Antigen yields a precipitation band of identity with *Staphylococcus* Antitoxin D.
105014: Antigen yields a precipitation band of partial identity (spur) with *Staphylococcus* Antitoxin D.
105015: Antigen yields precipitation bands of non-identity (crossed) with *Staphylococcus* Antitoxin D.
105016: Precipitation band is absent when antigen is tested with *Staphylococcus* Antitoxin D.
105017: Antigen yields a precipitation band of identity with *Staphylococcus* Antitoxin E.
105018: Antigen yields a precipitation band of partial identity (spur) with *Staphylococcus* Antitoxin E.
105019: Antigen yields precipitation bands of non-identity (crossed) with *Staphylococcus* Antitoxin E.
105020: Precipitation band is absent when antigen is tested with staphylococcus Antitoxin E.

PHAGE TYPING - LYTIC GROUPS

105021: Organism is lysed by staphylococcal Lytic Group I.
105022: Organism is lysed by staphylococcal Lytic Group II.
105023: Organism is lysed by staphylococcal Lytic Group III.
105024: Organism is lysed by staphylococcal Lytic Group IV.

PHAGE TYPES

105026: Organism is lysed by staphylococcal Phage 3A.
105027: Organism is lysed by staphylococcal Phage 3C.
105028: Organism is lysed by staphylococcal Phage 6.
105029: Organism is lysed by staphylococcal Phage 29.
105030: Organism is lysed by staphylococcal Phage 42D.
105031: Organism is lysed by staphylococcal Phage 42E.
105032: Organism is lysed by staphylococcal Phage 47.
105033: Organism is lysed by staphylococcal Phage 52.
105034: Organism is lysed by staphylococcal Phage 52A.
105035: Organism is lysed by staphylococcal Phage 53.
105036: Organism is lysed by staphylococcal Phage 54.
105037: Organism is lysed by staphylococcal Phage 55.
105038: Organism is lysed by staphylococcal Phage 71.
105039: Organism is lysed by staphylococcal Phage 75.
105040: Organism is lysed by staphylococcal Phage 77.
105041: Organism is lysed by staphylococcal Phage 79.
105042: Organism is lysed by staphylococcal Phage 80.
105043: Organism is lysed by staphylococcal Phage 81.
105044: Organism is lysed by staphylococcal Phage 83A.
105045: Organism is lysed by staphylococcal Phage 84.
105046: Organism is lysed by staphylococcal Phage 85.
105047: Organism is lysed by staphylococcal Phage 187.

SECTION 106: *ESCHERICHIA*

[NOTE 1: The information on the *E. coli* antisera was obtained from the list of Bacto products. This is not intended as an endorsement. Other manufacturers may make equivalent products.]

O ANTIGENS - BACTO

106001: Organism reacts with Bacto *E. coli* O Antisera 01.
106002: Organism reacts with Bacto *E. coli* O Antisera 02.
106003: Organism reacts with Bacto *E. coli* O Antisera 03.
106004: Organism reacts with Bacto *E. coli* O Antisera 04.
106005: Organism reacts with Bacto *E. coli* O Antisera 05.
106006: Organism reacts with Bacto *E. coli* O Antisera 06.
106007: Organism reacts with Bacto *E. coli* O Antisera 07.
106008: Organism reacts with Bacto *E. coli* O Antisera 08.
106009: Organism reacts with Bacto *E. coli* O Antisera 09.
106010: Organism reacts with Bacto *E. coli* O Antisera 010.
106011: Organism reacts with Bacto *E. coli* O Antisera 011.
106012: Organism reacts with Bacto *E. coli* O Antisera 012.
106013: Organism reacts with Bacto *E. coli* O Antisera 013.
106014: Organism reacts with Bacto *E. coli* O Antisera 014.
106015: Organism reacts with Bacto *E. coli* O Antisera 015.
106016: Organism reacts with Bacto *E. coli* O Antisera 016.
106017: Organism reacts with Bacto *E. coli* O Antisera 017.
106018: Organism reacts with Bacto *E. coli* O Antisera 018.
106019: Organism reacts with Bacto *E. coli* O Antisera 019.
106020: Organism reacts with Bacto *E. coli* O Antisera 020.
106021: Organism reacts with Bacto *E. coli* O Antisera 021.
106022: Organism reacts with Bacto *E. coli* O Antisera 022.
106023: Organism reacts with Bacto *E. coli* O Antisera 023.
106024: Organism reacts with Bacto *E. coli* O Antisera 024.
106025: Organism reacts with Bacto *E. coli* O Antisera 025.
106026: Organism reacts with Bacto *E. coli* O Antisera 026.
106027: Organism reacts with Bacto *E. coli* O Antisera 028.
106028: Organism reacts with Bacto *E. coli* O Antisera 034.
106029: Organism reacts with Bacto *E. coli* O Antisera 036.
106030: Organism reacts with Bacto *E. coli* O Antisera 040.
106031: Organism reacts with Bacto *E. coli* O Antisera 044.
106032: Organism reacts with Bacto *E. coli* O Antisera 045.
106033: Organism reacts with Bacto *E. coli* O Antisera 050.
106034: Organism reacts with Bacto *E. coli* O Antisera 055.
106035: Organism reacts with Bacto *E. coli* O Antisera 060.
106036: Organism reacts with Bacto *E. coli* O Antisera 068.
106037: Organism reacts with Bacto *E. coli* O Antisera 069.
106038: Organism reacts with Bacto *E. coli* O Antisera 071.
106039: Organism reacts with Bacto *E. coli* O Antisera 075.
106040: Organism reacts with Bacto *E. coli* O Antisera 077.
106041: Organism reacts with Bacto *E. coli* O Antisera 078.
106042: Organism reacts with Bacto *E. coli* O Antisera 080.
106043: Organism reacts with Bacto *E. coli* O Antisera 083.
106044: Organism reacts with Bacto *E. coli* O Antisera 086.
106045: Organism reacts with Bacto *E. coli* O Antisera 0102.
106046: Organism reacts with Bacto *E. coli* O Antisera 0111.
106047: Organism reacts with Bacto *E. coli* O Antisera 0112.
106048: Organism reacts with Bacto *E. coli* O Antisera 0113.

106049: Organism reacts with Bacto *E. coli* O Antisera 0114.
106050: Organism reacts with Bacto *E. coli* O Antisera 0117.
106051: Organism reacts with Bacto *E. coli* O Antisera 0119.
106052: Organism reacts with Bacto *E. coli* O Antisera 0120.
106053: Organism reacts with Bacto *E. coli* O Antisera 0124.
106054: Organism reacts with Bacto *E. coli* O Antisera 0125.
106055: Organism reacts with Bacto *E. coli* O Antisera 0126.
106056: Organism reacts with Bacto *E. coli* O Antisera 0127.
106057: Organism reacts with Bacto *E. coli* O Antisera 0128.
106058: Organism reacts with Bacto *E. coli* O Antisera 0136.
106059: Organism reacts with Bacto *E. coli* O Antisera 0140.
106060: Organism reacts with Bacto *E. coli* O Antisera 0145.
106061: Organism reacts with Bacto *E. coli* O Antisera 0148.
106062: Organism reacts with Bacto *E. coli* O Antisera 0150.

H ANTIGENS - POLYVALENT ANTISERA - BACTO

106101: Organism reacts with Bacto *E. coli* H Antisera Poly 1
(Agglutinins for Serogroups H1,H2,H3,H4,AND H12).
106102: Organism reacts with Bacto *E. coli* H Antisera Poly 2
(Agglutinins for Serogroups H5,H6,H7,H8,AND H40).
106103: Organism reacts with Bacto *E. coli* H Antisera Poly 3
(Agglutinins for Serogroups H9,H10,H11,H14,AND H21).
106104: Organism reacts with Bacto *E. coli* H Antisera Poly 4
(Agglutinins for Serogroups H15,H16,H17,H18,AND H19).
106105: Organism reacts with Bacto *E. coli* H Antisera Poly 5
(Agglutinins for Serogroups H20,H23,H24,H25,AND H26).
106106: Organism reacts with Bacto *E. coli* H Antisera Poly 6
(Agglutinins for Serogroups H27,H28,H29,H30,AND H32).
106107: Organism reacts with Bacto *E. coli* H Antisera Poly 7
(Agglutinins for Serogroups H31,H33,H34,H35,AND H36).
106108: Organism reacts with Bacto *E. coli* H Antisera Poly 8
(Agglutinins for Serogroups H37,H38,H39,H41,AND H42).
106109: Organism reacts with Bacto *E. coli* H Antisera Poly 9
(Agglutinins for Serogroups H43,H44,H45,H46,AND H47).
106110: Organism reacts with Bacto *E. coli* H Antisera Poly 10
(Agglutinins for Serogroups H48 AND 49).

H ANTIGENS - SINGLE FACTORS - BACTO

106150: Organism reacts with Bacto *E. coli* Antisera H1.
106151: Organism reacts with Bacto *E. coli* Antisera H2.
106152: Organism reacts with Bacto *E. coli* Antisera H3.
106153: Organism reacts with Bacto *E. coli* Antisera H4.
106154: Organism reacts with Bacto *E. coli* Antisera H5.
106155: Organism reacts with Bacto *E. coli* Antisera H6.
106156: Organism reacts with Bacto *E. coli* Antisera H7.
106157: Organism reacts with Bacto *E. coli* Antisera H8.
106158: Organism reacts with Bacto *E. coli* Antisera H9.
106159: Organism reacts with Bacto *E. coli* Antisera H10.
106160: Organism reacts with Bacto *E. coli* Antisera H11.
106161: Organism reacts with Bacto *E. coli* Antisera H12.
106162: Organism reacts with Bacto *E. coli* Antisera H14.
106163: Organism reacts with Bacto *E. coli* Antisera H15.
106164: Organism reacts with Bacto *E. coli* Antisera H16.
106165: Organism reacts with Bacto *E. coli* Antisera H17.
106166: Organism reacts with Bacto *E. coli* Antisera H18.

Section 106: *Escherichia* 267

106167: Organism reacts with Bacto *E. coli* Antisera H19.
106168: Organism reacts with Bacto *E. coli* Antisera H20.
106169: Organism reacts with Bacto *E. coli* Antisera H21.
106170: Organism reacts with Bacto *E. coli* Antisera H23.
106171: Organism reacts with Bacto *E. coli* Antisera H24.
106172: Organism reacts with Bacto *E. coli* Antisera H25.
106173: Organism reacts with Bacto *E. coli* Antisera H26.
106174: Organism reacts with Bacto *E. coli* Antisera H27.
106175: Organism reacts with Bacto *E. coli* Antisera H28.
106176: Organism reacts with Bacto *E. coli* Antisera H29.
106177: Organism reacts with Bacto *E. coli* Antisera H30.
106178: Organism reacts with Bacto *E. coli* Antisera H31.
106179: Organism reacts with Bacto *E. coli* Antisera H32.
106180: Organism reacts with Bacto *E. coli* Antisera H33.
106181: Organism reacts with Bacto *E. coli* Antisera H34.
106182: Organism reacts with Bacto *E. coli* Antisera H35.
106183: Organism reacts with Bacto *E. coli* Antisera H36.
106184: Organism reacts with Bacto *E. coli* Antisera H37.
106185: Organism reacts with Bacto *E. coli* Antisera H38.
106186: Organism reacts with Bacto *E. coli* Antisera H39.
106187: Organism reacts with Bacto *E. coli* Antisera H40.
106188: Organism reacts with Bacto *E. coli* Antisera H41.
106189: Organism reacts with Bacto *E. coli* Antisera H42.
106190: Organism reacts with Bacto *E. coli* Antisera H43.
106191: Organism reacts with Bacto *E. coli* Antisera H44.
106192: Organism reacts with Bacto *E. coli* Antisera H45.
106193: Organism reacts with Bacto *E. coli* Antisera H46.
106194: Organism reacts with Bacto *E. coli* Antisera H47.
106195: Organism reacts with Bacto *E. coli* Antisera H48.
106196: Organism reacts with Bacto *E. coli* Antisera H49.

OK ANTIGENS - POLYVALENT ANTISERA - BACTO

106300: Organism reacts with Bacto *E. coli* OK(OB) Antisera Poly A (Agglutinins for Serogroups O26:K60(B6), O55:K59(B5), O111:K58(B4), O127aK63(B8)).
106301: Organism reacts with Bacto *E. coli* OK(OB) Antisera Poly B (Agglutinins for Serogroups O86a:K61(B7), O119:K69(B14), O124:K72(B17), O125:K70(B15), O126:K71(B16), O128:K67(B12)).
106302: Organism reacts with Bacto *E. coli* OK(OB) Antisera Poly C (Agglutinins for Serogroups O18aO18c:K77(B21), O20aO20c:K61(B7), O20aO20b:K84(B), O28:K73(B18), O44:K74, O112aO112c:K66(B11)).
106303: Organism reacts with Bacto *E. coli* OK(OB) Antisera Poly D (Agglutinins for Serogroups O2:K56(B1), O8:K25(B2), O9:K57(B3), O18aO18b:K76(B20)).
106304: Organism reacts with Bacto *E. coli* OK(OB) Antisera Poly E (Agglutinins for Serogroups O112aO112b:K68(B13), O113:K75(B19), O127aO127B:K65(B10), O136:K78(B22)).

SPECIFIC OK ANTIGENS - BACTO

106350: Organism reacts with Bacto *E. coli* OK(OB) Antisera O2:K56(B1).

106351: Organism reacts with Bacto *E. coli* OK(OB) Antisera O8:K25(B2).
106352: Organism reacts with Bacto *E. coli* OK(OB) Antisera O9:K57(B3).
106353: Organism reacts with Bacto *E. coli* OK(OB) Antisera O18aO18b:K76(B20).
106354: Organism reacts with Bacto *E. coli* OK(OB) Antisera O18aO18c:K77(B21).
106355: Organism reacts with Bacto *E. coli* OK(OB) Antisera O20aO20c:K61(B7).
106356: Organism reacts with Bacto *E. coli* OK(OB) Antisera O20aO20b:K84(B).
106357: Organism reacts with Bacto *E. coli* OK(OB) Antisera O26:K60(B6).
106358: Organism reacts with Bacto *E. coli* OK(OB) Antisera O28:K73(B18).
106359: Organism reacts with Bacto *E. coli* OK(OB) Antisera O44:K74.
106360: Organism reacts with Bacto *E. coli* OK(OB) Antisera O55:K59(B5).
106361: Organism reacts with Bacto *E. coli* OK(OB) Antisera O86a:K61(B7).
106362: Organism reacts with Bacto *E. coli* OK(OB) Antisera O86aO86b:K64(B9).
106363: Organism reacts with Bacto *E. coli* OK(OB) Antisera O111:K58(B4).
106364: Organism reacts with Bacto *E. coli* OK(OB) Antisera O112aO112c:K66(B11).
106365: Organism reacts with Bacto *E. coli* OK(OB) Antisera O112aO112b:K68(B13).
106366: Organism reacts with Bacto *E. coli* OK(OB) Antisera O113:K75(B19).
106367: Organism reacts with Bacto *E. coli* OK(OB) Antisera O119:K69(B14).
106368: Organism reacts with Bacto *E. coli* OK(OB) Antisera O124:K72(B17).
106369: Organism reacts with Bacto *E. coli* OK(OB) Antisera O125:K72(B17).
106370: Organism reacts with Bacto *E. coli* OK(OB) Antisera O125:K70(B15).
106371: Organism reacts with Bacto *E. coli* OK(OB) Antisera O126:K71(B16).
106372: Organism reacts with Bacto *E. coli* OK(OB) Antisera O127a:K63(B8).
106373: Organism reacts with Bacto *E. coli* OK(OB) Antisera O127aO127b:K65(B10).
106374: Organism reacts with Bacto *E. coli* OK(OB) Antisera O128:K67(B12).
106375: Organism reacts with Bacto *E. coli* OK(OB) Antisera O136:K78(B22).
106376: Organism reacts with Bacto *E. coli* OK(OB) Antisera O138:K81.
106377: Organism reacts with Bacto *E. coli* OK(OB) Antisera O139:K82.

Section 106: *Escherichia*

106378: Organism reacts with Bacto *E. coli* OK(OB) Antisera O141:K85.
106379: Organism reacts with Bacto *E. coli* OK(OB) Antisera O141:K87.
106380: Organism reacts with Bacto *E. coli* OK(OB) Antisera O141:K88.

SECTION 107: *CAMPYLOBACTER*

PENNER METHOD - HEMAGGLUTINATION - CDC

107001: Heat-stable antigen reacts with CDC *Campylobacter* Antiserum 1.
107002: Heat-stable antigen reacts with CDC *Campylobacter* Antiserum 2.
107003: Heat-stable antigen reacts with CDC *Campylobacter* Antiserum 3.
107004: Heat-stable antigen reacts with CDC *Campylobacter* Antiserum 4.
107005: Heat-stable antigen reacts with CDC *Campylobacter* Antiserum 5.
107006: Heat-stable antigen reacts with CDC *Campylobacter* Antiserum 6.
107007: Heat-stable antigen reacts with CDC *Campylobacter* Antiserum 7.
107008: Heat-stable antigen reacts with CDC *Campylobacter* Antiserum 8.
107009: Heat-stable antigen reacts with CDC *Campylobacter* Antiserum 9.
107010: Heat-stable antigen reacts with CDC *Campylobacter* Antiserum 10.
107011: Heat-stable antigen reacts with CDC *Campylobacter* Antiserum 11.
107012: Heat-stable antigen reacts with CDC *Campylobacter* Antiserum 12.
107013: Heat-stable antigen reacts with CDC *Campylobacter* Antiserum 13.
107014: Heat-stable antigen reacts with CDC *Campylobacter* Antiserum 14.
107015: Heat-stable antigen reacts with CDC *Campylobacter* Antiserum 15.
107016: Heat-stable antigen reacts with CDC *Campylobacter* Antiserum 16.
107017: Heat-stable antigen reacts with CDC *Campylobacter* Antiserum 17.
107018: Heat-stable antigen reacts with CDC *Campylobacter* Antiserum 18.
107019: Heat-stable antigen reacts with CDC *Campylobacter* Antiserum 19.
107020: Heat-stable antigen reacts with CDC *Campylobacter* Antiserum 20.
107021: Heat-stable antigen reacts with CDC *Campylobacter* Antiserum 21.
107022: Heat-stable antigen reacts with CDC *Campylobacter* Antiserum 22.
107023: Heat-stable antigen reacts with CDC *Campylobacter* Antiserum 23.
107024: Heat-stable antigen reacts with CDC *Campylobacter* Antiserum 24.
107025: Heat-stable antigen reacts with CDC *Campylobacter* Antiserum 25.
107026: Heat-stable antigen reacts with CDC *Campylobacter* Antiserum 26.

Section 107: *Campylobacter* 271

107027: Heat-stable antigen reacts with CDC *Campylobacter* Antiserum 27.
107028: Heat-stable antigen reacts with CDC *Campylobacter* Antiserum 28.
107029: Heat-stable antigen reacts with CDC *Campylobacter* Antiserum 29.
107030: Heat-stable antigen reacts with CDC *Campylobacter* Antiserum 30.
107031: Heat-stable antigen reacts with CDC *Campylobacter* Antiserum 31.
107032: Heat-stable antigen reacts with CDC *Campylobacter* Antiserum 32.
107033: Heat-stable antigen reacts with CDC *Campylobacter* Antiserum 33.
107034: Heat-stable antigen reacts with CDC *Campylobacter* Antiserum 34.
107035: Heat-stable antigen reacts with CDC *Campylobacter* Antiserum 35.
107036: Heat-stable antigen reacts with CDC *Campylobacter* Antiserum 36.
107037: Heat-stable antigen reacts with CDC *Campylobacter* Antiserum 37.
107038: Heat-stable antigen reacts with CDC *Campylobacter* Antiserum 38.
107039: Heat-stable antigen reacts with CDC *Campylobacter* Antiserum 39.
107040: Heat-stable antigen reacts with CDC *Campylobacter* Antiserum 40.
107041: Heat-stable antigen reacts with CDC *Campylobacter* Antiserum 41.
107042: Heat-stable antigen reacts with CDC *Campylobacter* Antiserum 42.
107043: Heat-stable antigen reacts with CDC *Campylobacter* Antiserum 43.
107044: Heat-stable antigen reacts with CDC *Campylobacter* Antiserum 44.
107045: Heat-stable antigen reacts with CDC *Campylobacter* Antiserum 45.
107046: Heat-stable antigen reacts with CDC *Campylobacter* Antiserum 46.
107047: Heat-stable antigen reacts with CDC *Campylobacter* Antiserum 47.
107048: Heat-stable antigen reacts with CDC *Campylobacter* Antiserum 48.
107049: Heat-stable antigen reacts with CDC *Campylobacter* Antiserum 49.
107050: Heat-stable antigen reacts with CDC *Campylobacter* Antiserum 50.
107051: Heat-stable antigen reacts with CDC *Campylobacter* Antiserum 51.
107052: Heat-stable antigen reacts with CDC *Campylobacter* Antiserum 52.
107053: Heat-stable antigen reacts with CDC *Campylobacter* Antiserum 53.
107054: Heat-stable antigen reacts with CDC *Campylobacter* Antiserum 54.

107055: Heat-stable antigen reacts with CDC *Campylobacter* Antiserum 55.
107056: Heat-stable antigen reacts with CDC *Campylobacter* Antiserum 56.

LIOR METHOD - POLYVALENT ANTISERA - HEAT LABILE ANTIGENS - SLIDE AGGLUTINATION

107062: Organism agglutinates in CDC *Campylobacter* unabsorbed polyvalent Pool 1 (contains antisera 4,5,6,12,17,20).
107063: Organism agglutinates in CDC *Campylobacter* unabsorbed polyvalent Pool 2 (contains antisera 1,2,7,22,36).
107064: Organism agglutinates in CDC *Campylobacter* unabsorbed polyvalent Pool 3 (contains antisera 8,9,11,16,21).
107065: Organism agglutinates in CDC *Campylobacter* unabsorbed polyvalent Pool 4 (contains antisera 18,19,23,24,25,26,27,30).
107066: Organism agglutinates in CDC *Campylobacter* unabsorbed polyvalent Pool 5 (contains antisera 28,29,31,32,33,34,35).
107067: Organism agglutinates in CDC *Campylobacter* unabsorbed polyvalent Pool 6 (contains antisera 10,13,14,15,39).
107068: Organism agglutinates in CDC *Campylobacter* unabsorbed polyvalent Pool 7 (see note 2))includes antisera 38,40,42,43,44,45).
107069: Organism agglutinates in CDC *Campylobacter* unabsorbed polyvalent Pool 8 (see note 2) (includes antisera 46,47,48,49,50,51).
107070: Organism agglutinates in CDC *Campylobacter* unabsorbed polyvalent Pool 9 (see note 2) (includes antisera 52,53,54,55,56,57,58,60).

LIOR METHOD - SINGLE FACTOR ANTISERA - HEAT LABILE ANTIGENS - SLIDE AGGLUTINATION

107071: Heat labile antigens agglutinate in CDC *Campylobacter* absorbed antisera type 1.
107072: Heat labile antigens agglutinate in CDC *Campylobacter* absorbed antisera type 2.
107074: Heat labile antigens agglutinate in CDC *Campylobacter* absorbed antisera type 4.
107075: Heat labile antigens agglutinate in CDC *Campylobacter* absorbed antisera type 5.
107076: Heat labile antigens agglutinate in CDC *Campylobacter* absorbed antisera type 6.
107077: Heat labile antigens agglutinate in CDC *Campylobacter* absorbed antisera type 7.
107078: Heat labile antigens agglutinate in CDC *Campylobacter* absorbed antisera type 8.
107079: Heat labile antigens agglutinate in CDC *Campylobacter* absorbed antisera type 9.

Section 107: *Campylobacter* 273

107080: Heat labile antigens agglutinate in CDC *Campylobacter* absorbed antisera type 10.
107081: Heat labile antigens agglutinate in CDC *Campylobacter* absorbed antisera type 11.
107082: Heat labile antigens agglutinate in CDC *Campylobacter* absorbed antisera type 12.
107083: Heat labile antigens agglutinate in CDC *Campylobacter* absorbed antisera type 13.
107084: Heat labile antigens agglutinate in CDC *Campylobacter* absorbed antisera type 14.
107085: Heat labile antigens agglutinate in CDC *Campylobacter* absorbed antisera type 15.
107086: Heat labile antigens agglutinate in CDC *Campylobacter* absorbed antisera type 16.
107087: Heat labile antigens agglutinate in CDC *Campylobacter* absorbed antisera type 17.
107088: Heat labile antigens agglutinate in CDC *Campylobacter* absorbed antisera type 18.
107089: Heat labile antigens agglutinate in CDC *Campylobacter* absorbed antisera type 19.
107090: Heat labile antigens agglutinate in CDC *Campylobacter* absorbed antisera type 20.
107091: Heat labile antigens agglutinate in CDC *Campylobacter* absorbed antisera type 21.
107092: Heat labile antigens agglutinate in CDC *Campylobacter* absorbed antisera type 22.
107093: Heat labile antigens agglutinate in CDC *Campylobacter* absorbed antisera type 23.
107094: Heat labile antigens agglutinate in CDC *Campylobacter* absorbed antisera type 24.
107095: Heat labile antigens agglutinate in CDC *Campylobacter* absorbed antisera type 25.
107096: Heat labile antigens agglutinate in CDC *Campylobacter* absorbed antisera type 26.
107097: Heat labile antigens agglutinate in CDC *Campylobacter* absorbed antisera type 27.
107098: Heat labile antigens agglutinate in CDC *Campylobacter* absorbed antisera type 28.
107099: Heat labile antigens agglutinate in CDC *Campylobacter* absorbed antisera type 29.
107100: Heat labile antigens agglutinate in CDC *Campylobacter* absorbed antisera type 30.
107101: Heat labile antigens agglutinate in CDC *Campylobacter* absorbed antisera type 31.
107102: Heat labile antigens agglutinate in CDC *Campylobacter* absorbed antisera type 32.
107103: Heat labile antigens agglutinate in CDC *Campylobacter* absorbed antisera type 33.
107104: Heat labile antigens agglutinate in CDC *Campylobacter* absorbed antisera type 34.
107105: Heat labile antigens agglutinate in CDC *Campylobacter* absorbed antisera type 35.
107106: Heat labile antigens agglutinate in CDC *Campylobacter* absorbed antisera type 36.
107107: Heat labile antigens agglutinate in CDC *Campylobacter* absorbed antisera type 37.

107108: Heat labile antigens agglutinate in CDC *Campylobacter* absorbed antisera type 38.
107109: Heat labile antigens agglutinate in CDC *Campylobacter* absorbed antisera type 39.
107110: Heat labile antigens agglutinate in CDC *Campylobacter* absorbed antisera type 40.
107111: Heat labile antigens agglutinate in CDC *Campylobacter* absorbed antisera type 41.
107112: Heat labile antigens agglutinate in CDC *Campylobacter* absorbed antisera type 42.
107113: Heat labile antigens agglutinate in CDC *Campylobacter* absorbed antisera type 43.
107114: Heat labile antigens agglutinate in CDC *Campylobacter* absorbed antisera type 44.
107115: Heat labile antigens agglutinate in CDC *Campylobacter* absorbed antisera type 45.
107116: Heat labile antigens agglutinate in CDC *Campylobacter* absorbed antisera type 46.
107117: Heat labile antigens agglutinate in CDC *Campylobacter* absorbed antisera type 47.
107118: Heat labile antigens agglutinate in CDC *Campylobacter* absorbed antisera type 48.
107119: Heat labile antigens agglutinate in CDC *Campylobacter* absorbed antisera type 49.
107120: Heat labile antigens agglutinate in CDC *Campylobacter* absorbed antisera type 50.
107121: Heat labile antigens agglutinate in CDC *Campylobacter* absorbed antisera type 51.
107122: Heat labile antigens agglutinate in CDC *Campylobacter* absorbed antisera type 52.
107123: Heat labile antigens agglutinate in CDC *Campylobacter* absorbed antisera type 53.
107124: Heat labile antigens agglutinate in CDC *Campylobacter* absorbed antisera type 54.
107125: Heat labile antigens agglutinate in CDC *Campylobacter* absorbed antisera type 55.
107126: Heat labile antigens agglutinate in CDC *Campylobacter* absorbed antisera type 56.
107127: Heat labile antigens agglutinate in CDC *Campylobacter* absorbed antisera type 57.
107128: Heat labile antigens agglutinate in CDC *Campylobacter* absorbed antisera type 58.
107129: Heat labile antigens agglutinate in CDC *Campylobacter* absorbed antisera type 59.
107130: Heat labile antigens agglutinate in CDC *Campylobacter* absorbed antisera type 60.
107131: Heat labile antigens agglutinate in CDC *Campylobacter* unabsorbed antisera type 46.
107132: Heat labile antigens agglutinate in CDC *Campylobacter* unabsorbed antisera type 50.
107133: Heat labile antigens agglutinate in CDC *Campylobacter* unabsorbed antisera type 51.

APPENDIX

CROSS-REFERENCED SYNONYMS OF COMMON NAMES OF COMPOUNDS
IN THE RKC FORMAT

A large number of the compounds or substances listed herein have been known by different names in different laboratories and countries. Often these historically acquired synonyms are multiple for a given substance. In order to avoid misunderstandings and errors, we have compiled the following cross-referenced list of names of materials. The list is meant to be practical with respect to microbiology. Therefore, we have often omitted the precise standard chemical and many rarely used synonyms. Although the use of terminology descriptive of the structure of compounds would have precise connotations, for the present purposes we need not describe sucrose as:

β-D-Fructofuranosyl-α-D-Glucopyranoside
or
α-D-Glucopyranosyl-β-D-Fructofuranoside.

Starred names (*) are those encoded in the RKC coding system. Non-starred names are considered to be synonyms and are generally not mentioned in the RKC format.

The starred names are listed on one line followed by all the common synonyms on the next lines indented as follows:

*Coniferin
    Abietin
    Laricin

The non-starred names are always cross-referenced to the pertinent starred synonyms as the following example shows:

Abietin   *Coniferin

Although Laricin is another synonym of Abietin, it is treated separately in an alphabetic sequence in the same way:

Laricin   Coniferin

(In a few cases, the line length was too long. The synonym is placed on the line below and indented ten spaces.)

We believe we have been quite successful in avoiding the assignation of different code numbers to synonyms. We know of only one instance, i.e., the synonym pair:

Valeric-Valerianic Acids

in which coding of this synonym pair by different numbers was not detected in the original editing process.

The principle references consulted in compiling this list of synonyms were Gardner and Cooke (1971), Stecher (1968), and Weast, (1974).

```
Abietin      *Coniferin
Acacia       *Gum Arabic
*Acetaldehyde
    Ethanal
    Ethylaldehyde
*Acetamide
    Acetic Acid Amide
Acetic Acid Amide      *Acetamide
*Acetoacetic Acid
    Diacetic Acid
    *β-Oxobutyric Acid
*Acetoin
    Acetyl Methyl Carbinol
    3-Hydroxy-2-Butanone
*Acetone
    Dimethylketone
    2-Propanone
*Acetylene
    Ethine
Acetylformic Acid      *Pyruvic Acid
Acetyl Methyl Carbinol      *Acetoin
β-Acetylpropionic Acid      *2-Ketogluconic Acid
Acid Lactone      *Coumarin
Acid Violet 19      *Fuchsin
*Aconitic Acid
    Equisetic Acid
    1,2,3-Propenetricarboxylic Acid
Acorn Sugar      *Quercitol
*Acrylamide
    Propenamide
Adeps      *Lard
*Adipic Acid
    1,4-Butanedicarboxylic Acid
    Hexanedioic Acid
Adonite      *Adonitol
*Adonitol
    Adonite
    Ribitol
D-Adonose      *D-Ribulose
*Agar
    Gelose
Agrozyme      *α-Amylase
Alabaster      *Selenite
Alantin      *Inulin
Alcohol      *Ethanol
*Alginic Acid
    Polymannuronic Acid
*Allantoin
    Glyoxyldiureide
    5-Ureidohydantoin
```

Appendix 277

```
  Allomaleic Acid      *Fumaric Acid
  β-D-Allopyranose     *D-Allose
*D-Allose
      β-D-Allopyranose
*Allyl Alcohol
      2-Propen-1-ol
*Allylamine
      3-Aminopropylene
  Amidazine        *Ethionamide
  Aminobenzene     *Phenylamine
  1-Aminobutane    *Butylamine
  Aminocaproic Lactam    *ε-Caprolactam
  2-Amino-2-Deoxyglucose    *Glucosamine
  2-Aminoethanesulfonic Acid    *Taurine
  2-Aminoethanol       *Ethanolamine
  α-Aminoglutaric Acid Lactam     *L-Pyroglutamic Acid
  Aminomethane     *Methylamine
  Aminopentamide   *Valeramide
  3-Aminopropylene     *Allylamine
  Aminotoluene     *Benzylamine
  α-Amino-δ-Ureidovaleric Acid    *Citrulline
  Ammonium Rhodamide       *Ammonium Thiocyanate
  Ammonium Sulfocyanate    *Ammonium Thiocyanate
*Ammonium Thiocyanate
      Ammonium Rhodamide
      Ammonium Sulfocyanate
*Amygdalin
      Amygdaloside
      Mandelonitrile-β-Gentiobioside
  Amygdaloside     *Amygdalin
*α-Amylase
      Agrozyme
      Amylo-Liquifase
      Biolase
  Amyl Carbinol    *1-Hexanol
  Amylo-Liquifase      *α-Amylase
  Amylum     *Starch
  Aniline    *Phenylamine
  Animal Coniine       *Cadaverine
  Animal Starch    *Glycogen
  Anise Alcohol    *Anisyl Alcohol
*Anisyl Alcohol
      Anise Alcohol
      P-Methoxybenzyl Alcohol
  Arabite    *Arabitol
*Arabitol
      Arabite
      1,2,3,4,5-Pentanepentol
*Arbutin
      Arbutoside
      Ursin
  Arbutoside     *Arbutin
*Ascorbic Acid
      Vitamin C
*Asphalt
      Bitumen
  Axungia Porci      *Lard
```

*Azelaic Acid
    1,7-Heptanedicarboxylic Acid
    Lepargylic Acid
    Nonanedioic Acid
*Behenic Acid
    Docosanoic Acid
*Benzamide
    Benzoylamide
*Benzene
    Benzol
    Cyclohexatriene
 Benzenecarboxylic Acid   *Benzoic Acid
 p-Benzenecarboxylic Acid   *Terephthalic Acid
 m-Benzenedicarboxylic Acid   *m-Isophthalic Acid
 o-Benzenedicarboxylic Acid   *Phthalic Acid
 1,2-Benzenediol   *Catechol
 1,3-Benzenediol   *Resorcinol
*Benzoic Acid
    Benzenecarboxylic Acid
    Dracylic Acid
    Phenylformic Acid
 Benzol   *Benzene
 Benzo-Phenanthrene   *Pyrene
 1,2-Benzopyrone   *Coumarin
 2,3-Benzopyrrole   *Indole
 Benzoylamide   *Benzamide
 1,2-Benzphenanthrene   *Chrysene
 9,10-Benzphenanthrene   *Triphenylene
 Benzylacetic Acid   *β-Phenylpropionic Acid
*Benzyl Alcohol
    α-Hydroxytoluene
    Phenylcarbinol
    Phenylmethanol
*Benzylamine
    Aminotoluene
    Moringine
    Phenylmethylamine
 Benyzlbenzene   *Diphenylmethane
 Benzyl Carbinol   *2-Phenylethanol
*Benzylviologen
    Viologen
 Bilineurine   *Choline
 Biolase   *α-Amylase
 Bitumen   *Asphalt
 Boletic Acid   *Fumaric Acid
 Brain Sugar   *D-Galactose
 British Gum   *Dextrin
 1,4-Butanedicarboxylic Acid   *Adipic Acid
*meso-2,3-Butanediol
    Dimethylene Glycol
*DL-1,3-Butanediol
    1,3-Dihydroxybutane
*D(-)2,3-Butanediol
    2,3-Dihydroxybutane
 1,2,3,4-Butanetriol   *Erythritol
 Butanoic Acid   *n-Butyric Acid
*2-Butanol

Appendix

```
     Methyl Ethyl Carbinol
*1-Butanol
     n-Butyl Alcohol
     Propyl Carbinol
*2-Butanone
     Methylethyl Ketone
 cis-Butenedioic Acid      *Maleic Acid
 trans-Butenedioic Acid    *Fumaric Acid
 tert-Butyl Alcohol    *2-Methyl-2-Propanol
*Butylamine
     1-Aminobutane
*n-Butylbenzene
     1-Phenylbutane
*n-Butyric Acid
     Butanoic Acid
     Ethylacetic Acid
*Cadaverine
     Animal Coniine
     1,5-Pentanediamine
 Caffearine      *Trigonelline
 Calcium Sulfate Dihydrate      *Selenite
*Camphor
     Gum Camphor
*Capric Acid
     Decanoic Acid
*Caproic Acid
     Hexanoic Acid
*ε-Caprolactam
     Aminocaproic Lactam
     Hexahydro-2H-Azepin-2-one
*Caprylic Acid
     Octanoic Acid
 Caprylic Alcohol     *1-Octanol
 Carbinol      *Methanol
 Carbowax      *Polyethylene Glycol
 Carubinose      *D-Mannose
*Catechol
     1,2-Benzenediol
     Pyrocatechol
*Cellobiose
     Cellose
 Cellose      *Cellobiose
 Cellosolve Acetate     *2-Ethoxy Ethyl Acetic Acid
 Cerebrose     *D-Galactose
 Cetylic Acid     *Palmitic Acid
 Cetyl Alcohol     *1-Hexadecanol
 Chitosamine     *Glucosamine
*Choline
     Bilineurine
     Sincaline
*Chrysene
     1,2-Benzphenanthrene
*Cinnamic Acid
     β-Phenylacrylic Acid
     3-Phenylpropenoic Acid
*Cinnamyl Alcohol
```

γ-Phenylallyl Alcohol
3-Phenyl-2-Propen-1-ol
Styryl Carbinol
*Citraconic Acid
Methylmaleic Acid
*Citric Acid
2-Hydroxy-1,2,3-Propanetricarboxylic Acid
β-Hydroxytricarballylic Acid
*Citrulline
α-Amino-δ-Ureidovaleric Acid
δ-Ureidonorvaline
*Coconut Oil
Copra Oil
Coffearine   *Trigonelline
Colamine   *Ethanolamine
*Coniferin
Abietin
Laricin
*Coniferyl Alcohol
3-(4-Hydroxy-3-Methoxyphenyl)-2-Propen-1-ol
Copra Oil   *Coconut Oil
*Corn Oil
Maize Oil
*Coumarin
Acid Lactone
1,2-Benzopyrone
Tonka Bean Camphor
*Cresol (m-o-p)
Cresylic Acid
Cresylol
Tricresol
Cresylic Acid   *Cresol (m-o-p)
Cresylol   *Cresol (m-o-p)
Crude Oil   *Petroleum
Cyclohexane Carboxylic Acid   *Hexahydrobenzoic Acid
1,2,3,4,5-Cyclohexanepentol   *Quercitol
*Cyclohexanol
Hexahydrophenol
Hexalin
Cyclohexatriene   *Benzene
*Cyclopentanol
Cyclopentyl Alcohol
Hydroxycyclopentane
Cyclopentyl Alcohol   *Cyclopentanol
p-Cymene   *Isopropyltoluene
Dahlin   *Inulin
*Decahydronaphthalene   cis-Decalin
Naphthalane (naphthane)
cis-Decalin
*Decahydronaphthalene
Decanoic Acid   *Capric Acid
*Deoxyribose
Thyminose
Detergent Alkylate #5   *ω-Phenyldecane
*Dextran
Macrose

*Dextrin
　　British Gum
　Dextronic Acid　　*D-Gluconic Acid
　Dextrose　　*D-Glucose
　Dextrotartaric Acid　　*L(+)-Tartaric Acid
　Diacetic Acid　　*Acetoacetic Acid or
　　　　*β-Oxobutyric Acid
　2,5-Diaminopentanoic Acid　　*Ornithine
*Diethylene Glycol
　　2,2'-Oxydiethanol
　Diffusing Factor　　*Hyaluronidase
　Digitin　　*Digitonin
*Digitonin
　　Digitin
*Dihydroxyacetone
　　1,3-Dihydroxydimethyl Ketone
　m-Dihydroxybenzene　　*Resorcinol
　1,3-Dihydroxybutane　　*DL-1,3-Butanediol
　2,3-Dihydroxybutane　　*D(-)2,3-Butanediol
　1,3-Dihydroxydimethyl Ketone　　*Dihydroxyacetone
　3',6'-Dihydroxyfluoran　　*Fluorescein
　3,4-Dihydroxyhexanol　　*D(+)3,4-Hexanediol
　2,5-Dihydroxyhexanol　　*2,5-Hexanediol
　1,3-Dihydroxymethyl Ketone　　*Dihydroxyacetone
　3,5-Dihydroxy-3-Methylvaleric Acid　　*Mevalonic Acid
　1,3-Dihydroxypentane　　*1,3-Pentanediol
　1,5-Dihydroxypentane　　*Pentanediol
　2,3-Dihydroxypropanal　　*Glyceraldehyde
　Dihydroxypropionic Acid　　*Glyceric Acid
　Dimethyl Benzene　　*Xylene
　trans-3,7-Dimethyl-2,6-Octadiene-1-ol　　*Geraniol
　Dimethylene Glycol　　*meso-2,3-Butanediol
　Dimethylketone　　*Acetone
*Diphenylmethane
　　Benzylbenzene
　Docosanoic Acid　　*Behenic Acid
　cis-13-Docosenoic Acid　　*Erucic Acid
　Dodecanoic Acid　　*Lauric Acid
　Dracylic Acid　　*Benzoic Acid
　Dulcite　　*Dulcitol
*Dulcitol
　　Dulcite
　　Dulcose
　　Euonymit
　　Galactitol
　　Melampyrin
　　Melampyrum
　　Melampyrite
　Dulcose　　*Dulcitol
　Equisetic Acid　　*Aconitic Acid
*Erucic Acid
　　cis-13-Docosenoic Acid
*Erythritol
　　1,2,3,4-Butanetriol
　　Phycite
　　Tetrahydroxybutane

*L-Erythrulose
    L-Glycero-Tetrulose
Ethal   *1-Hexadecanol
Ethanal   *Acetaldehyde
Ethanediamide   *Oxamide
*1,2-Ethanediol
    Ethylene Glycol
*Ethanol
    Alcohol
*Ethanolamine
    2-Aminoethanol
    Colamine
    2-Hydroxyethylamine
Ethine   *Acetylene
*Ethionamide
    Amidazine
Ethol   *1-Hexadecanol
2-Ethoxy Ethanol   *Ethylene Glycol Monoethyl Ether
*2-Ethoxy Ethyl Acetic Acid
    Cellosolve Acetate
Ethylacetic Acid   *n-Butyric Acid
Ethylaldehyde   *Acetaldehyde
cis-1,2-Ethylenedicarboxylic Acid   *Maleic Acid
trans-1,2-Ethylenedicarboxylic Acid   *Fumaric Acid
Ethylene Glycol   *1,2-Ethanediol
*Ethylene Glycol Monoethyl Ether
    Cellosolve
    2-Ethoxy Ethanol
Euonymit   *Dulcitol
Fibrinolysin   *Plasmin
*Fluorescein
    3',6'-Dihydroxyfluoran
*Formaldehyde
    Methanal
    Methyl Aldehyde
    Methylene Oxide
*Formamide
    Methanamide
Formylformic Acid   *Glyoxylic Acid
*D-Fructose
    Levulose
*Fuchsin
    Acid Violet 19
*D-Fucose
    D-Galactomethylose
*L-Fucose
    L-Galactomethylose
*Fumaric Acid
    Allomaleic Acid
    Boletic Acid
    trans-Butenedioic Acid
    trans-1,2-Ethylenedicarboxylic Acid
Galactitol   *Dulcitol
D-Galactomethylose   *D-Fucose
L-Galactomethylose   *L-Fucose
Galactosaccharic Acid   *Mucic Acid

Appendix

*D-Galactose
    Brain Sugar
    Cerebrose
*Gelatin
    Puragel
 Gelose   *Agar
*Geraniol
    trans-3,7-Dimethyl-2,6-Octadiene-1-Ol
    Lemonol
 Gerontine   *Spermine
 D-Glucaric Acid   *Mucic Acid
*Glucitol
    D-Sorbitol
*D-Gluconic Acid
    Dextronic Acid
    Glyconic Acid
    Maltonic Acid
    Pentahydroxycaproic Acid
*Glucosamine
    2-Amino-2-Deoxyglucose
    Chitosamine
*D-Glucose
    Dextrose
*Glutaric Acid
    Pentanedioic Acid
    1,3-Propanedicarboxylic Acid
 Glutimic Acid   *L-Pyroglutamic Acid
*Glyceraldehyde
    2,3-Dihydroxypropanal
*Glyceric Acid
    Dihydroxypropionic Acid
 Glycerol   *1,2,3-Propanetriol
 L-Glycero-Tetrulose   *L-Erythrulose
 Glyceryl Tributyrate   *Tributyrin
 Glyceryl Trioleate   *Triolein
 Glyceryl Tripalmitate   *Tripalmitin
 Glyceryl Tristearin   *Tristearin
*Glycogen
    Animal Starch
    Liver Starch
*Glycolic Acid
    Hydroxyacetic Acid
    Hydroxyethanoic Acid
 Glyconic Acid   *D-Gluconic Acid
 Glyoxalic Acid   *Glyoxylic Acid
 Glyoxyldiureide   *Allantoin
*Glyoxylic Acid
    Formylformic Acid
    Glyoxalic Acid
    Oxoethanoic Acid
 Gossypose   *Raffinose
 Gum Camphor   *Camphor
*Gum Arabic
    Acacia
 Gynesine   *Trigonelline
 Gypsum   *Selenite
 1,7-Heptanedicarboxylic Acid   *Azelaic Acid

Heptanoic Acid   *Oenanthic Acid
Hexadecanoic Acid   *Palmitic Acid
*1-Hexadecanol
    Cetyl Alcohol
    Ethal
    Ethol
    Palmityl Alcohol
Hexahydro-2H-Azepin-2-One   *ε-Caprolactam
*Hexahydrobenzoic Acid
    Cyclohexane Carboxylic Acid
Hexahydrophenol   *Cyclohexanol
Hexahydroxycyclohexane   *meso-Inositol
Hexalin   *Cyclohexanol
Hexamethylene Glycol   *1,6-Hexanediol
Hexanedioic Acid   *Adipic Acid
*meso-3,4-Hexanediol
    3,4-Dihydroxyhexanol
*2,5-Hexanediol
    2,5-Dihydroxyhexanol
*D(+)3,4-Hexanediol
    3,4-Dihydroxyhexanol
*1,6 Hexanediol
    Hexamethylene Glycol
Hexanoic Acid   *Caproic Acid
*1-Hexanol
    Amyl Carbinol
    n-Hexyl Alcohol
    Pentyl Carbinol
n-Hexyl Alcohol   *1-Hexanol
Hiochic Acid   *Mevalonic Acid
*Histamine
    4-Imidazoleethylamine
*Hyaluronidase
    Spreading Factor
    Diffusing Factor
Hydrocinnamic Acid   *β-Phenylpropionic Acid
Hydroxyacetic Acid   *Glycolic Acid
*o-Hydroxybenzaldehyde
    Salicylaldehyde
o-Hydroxybenzamide   *Salicylamide
Hydroxybenzene   *Phenol
o-Hydroxybenzoic Acid   *Salicylic Acid
3-Hydroxybutanoic Acid   *β-Hydroxybutyric Acid
3-Hydroxy-2-Butanone   *Acetoin
*β-Hydroxybutyric Acid
    3-Hydroxybutanoic Acid
Hydroxycyclopentane   *Cyclopentanol
Hydroxyethanoic Acid   *Glycolic Acid
3-(4-Hydroxy-3-Methoxyphenyl)-2-Propen-1-ol
        *Coniferyl Alcohol
D-12-Hydroxyoleic Acid   *Ricinoleic Acid
4-Hydroxyphenethylamine   *Tyramine
7-Hydroxy-3H-Phenoxazin-3-One 10-Oxide   *Resazurin
2-Hydroxy-1,2,3-Propanetricarboxylic Acid
        *Citric Acid
Hydroxysuccinic Acid   *D-Malic Acid or *L-Malic Acid
α-Hydroxytoluene   *Benzyl Alcohol

β-Hydroxytricarballic Acid   *Citric Acid
8-Hydroxyxanthine   *Uric Acid
4-Imidazoleethylamine   *Histamine
*Indole
    2,3-Benzopyrrole
*meso-Inositol
    Hexahydroxycyclohexane
    Meat Sugar
*Inulin
    Alantin
    Dahlin
*Iodinin
    1,6-Phenazinediol 5,10-Dioxide
Isoamyl Alcohol   *3-Methyl-1-Butanol
Isoamylamine   *β-Methylbutylamine
Isobutyl Alcohol   *2-Methyl-1-Propanol
Isobutyl Carbinol   *3-Methyl-1-Butanol
Isodulcit   *L-Rhamnose
*Isoniazid
    Isonicotinic Acid Hydrazide
Isonicotinic Acid Hydrazide   *Isoniazid
Isopentylamine   *β-Methylbutylamine
*m-Isophthalic Acid
    m-Benzenedicarboxylic Acid
Isopropyl Alcohol   *2-Propanol
Isopropyl Carbinol   *2-Methyl-1-Propanol
*Isopropyltoluene
    p-Cymene
*Itaconic Acid
    Methylenesuccinic Acid
    Propylenedicarboxylic Acid
*2-Ketogluconic Acid
    β-Acetylpropionic Acid
*α-Ketoglutaric Acid
    2-Oxoglutaric Acid
1-Ketopropionaldehyde   *Pyruvaldehyde
Kinic Acid   *Quinic Acid
Kyanol   *Phenylamine
*L-Lactic Acid
    Sarcolactic Acid
*DL-Lactic Acid
    Racemic Lactic Acid
*Lard
    Adeps
    Axungia Porci
Laricin   *Coniferin
*Lauric Acid
    Dodecanoic Acid
    Laurostearic Acid
Laurostearic Acid   *Lauric Acid
Lemonol   *Geraniol
Lepargylic Acid   *Azelaic Acid
Levotartaric Acid   *D(-)-Tartaric Acid
Levulose   *D-Fructose
Lignin   *Lignocellulose

*Lignocellulose
    Lignin
9,12-Linoleic Acid   *Linoleic Acid
*Linoleic Acid
    9,12-Linoleic Acid
    Linolic
*Linolenic Acid
    9,12,15-Octadecatrienoic Acid
Linolic   *Linoleic Acid
Liver Starch   *Glycogen
Macrose   *Dextran
Maize Oil   *Corn Oil
*Maleic Acid
    cis-Butenedioic Acid
    cis-1,2-Ethylenedicarboxylic Acid
    Toxilic Acid
*L-Malic Acid
    Hydroxysuccinic Acid
*D-Malic Acid
    Hydroxysuccinic Acid
Maltobiose   *Maltose
Maltonic Acid   *D-Gluconic Acid
*Maltose
    Maltobiose
*Mandelic Acid
    Phenylhydroxyacetic Acid
Mandelonitrile-β-Gentiobioside   *Amygdalin
*D-Mannose
    Carubinose
    Seminose
Meat Sugar   *meso-Inositol
Melampyrite   *Dulcitol
Melampyrin   *Dulcitol
Melampyrum   *Dulcitol
Melitose   *Raffinose
Melitriose   *Raffinose
*Menadione
    2-Methyl-1,4-Naphthoquinone
    Vitamin K3
*Mesaconic Acid
    Methyl Fumaric Acid
*Mesitylene
    Sym-1,3,5-Trimethylbenzene
Methanal   *Formaldehyde
Methanamide   *Formamide
*Methanol
    Carbinol
P-Methoxybenzyl Alcohol   *Anisyl Alcohol
2-Methoxy-2-Nitrobutanol
        *2-Nitro-2-Ethyl-1,3-Propanediol
Methylacetaldehyde   *Propionaldehyde
Methyl Aldehyde   *Formaldehyde
*Methylamine
    Aminomethane
    Monomethylamine

Methylaminoethanoic Acid   *Sarcosine
Methylbenzene   *Toluene
Methylbenzylamine   *α-Phenylethylamine
*3-Methyl-1-Butanol
    Isoamyl Alcohol
    Isobutyl Carbinol
*β-Methylbutylamine
    Isoamylamine
    Isopentylamine
Methylene Oxide   *Formaldehyde
Methylenesuccinic Acid   *Itaconic Acid
Methyl Ethyl Carbinol   *2-Butanol
Methylethyl Ketone   *2-Butanone
Methyl Fumaric Acid   *Mesaconic Acid
N-Methylglycine   *Sarcosine
Methyl Glycol   *1,2-Propanediol
Methylglyoxal   *Pyruvaldehyde
Methylmaleic Acid   *Citraconic Acid
2-Methyl-1,4-Naphthoquinone   *Menadione
*2-Methyl-1-Propanol
    Isobutyl Alcohol
    Isopropyl Carbinol
*2-Methyl-2-Propanol
    tert-Butyl Alcohol
    Trimethyl Carbinol
*Mevalonic Acid
    3,5-Dihydroxy-3-Methylvaleric Acid
    Hiochic Acid
Monomethylamine   *Methylamine
Moringine   *Benzylamine
*Mucic Acid
    Galactosaccharic Acid
Musculamine   *Spermine
Mycose   *Trehalose
Naphthalane (naphthane)
    *Decahydronaphthalene
*Naphthalene
    Tar Camphor
Niacin   *Nicotinic Acid
Niacinamide   *Nicotinamide
*Nicotinamide
    Niacinamide
    Vitamin PP
*Nicotinic Acid
    Niacin
    Pyridine-3-Carboxylic Acid
*2-Nitro-2-Ethyl-1,3-Propanediol
    2-Methoxy-2-Nitrobutanol
Nonanedioic Acid   *Azelaic Acid
Nonanoic Acid   *Pelargonic Acid
Octadecanoic Acid   *Stearic Acid
9,12,15-Octadecatrienoic Acid   *Linolenic Acid
cis-6-Octadecenoic Acid   *Petroselenic Acid
cis-9-Octadecenoic Acid   *Oleic Acid
*9-Octadecynoic Acid
    Stearolic Acid
Octanoic Acid   *Caprylic Acid

*1-Octanol
    Caprylic Alcohol
*Oenanthic Acid
    Heptanoic Acid
*Oleic Acid
    cis-9-Octadecenoic Acid
 Olein   *Triolein
*Ornithine
    2,5-Diaminopentanoic Acid
*Orotic Acid
    Uracil-6-Carboxylic Acid
    Whey Factor
 Oxalic Acid Diamide   *Oxamide
*Oxamide
    Ethanediamide
    Oxalic Acid Diamide
 Oxobenzene   *Phenol
*β-Oxobutyric Acid
    *Acetoacetic Acid
    Diacetic Acid
 Oxoethanoic Acid   *Glyoxylic Acid
 2-Oxoglutaric Acid   *α-Ketoglutaric Acid
 2-Oxopropanal   *Pyruvaldehyde
 2-Oxopropanoic Acid   *Pyruvic Acid
 2,2'-Oxydiethanol   *Diethylene Glycol
*Palmitic Acid
    Cetylic Acid
    Hexadecanoic Acid
 Palmitin   *Tripalmitin
 Palmityl Alcohol   *1-Hexadecanol
*Paraffin Oil
    Petrolatum Liquid
*Pelargonic Acid
    Nonanoic Acid
*Pentaerythritol
    Tetramethylolmethane
 Pentahydroxycaproic Acid   *D-Gluconic Acid
 Pentamethylene Glycol   *1,5-Pentanediol
 1,5-Pentanediamine   *Cadaverine
 Pentanedioic Acid   *Glutaric Acid
*Pentanediol
    1,5-Dihydroxypentane
    Pentamethylene Glycol
*1,3-Pentanediol
    1,3-Dihydroxypentane
 1,2,3,4,5-Pentanepentol   *Arabitol
*1-Pentanol
    n-Amyl Alcohol
    n-Butyl Carbinol
    Pentyl Alcohol
 Pentyl Alcohol   *1-Pentanol
 Pentyl Carbinol   *1-Hexanol
 Petrolatum Liquid   *Paraffin Oil
*Petroleum
    Crude Oil
    Rock Oil

Appendix

*Petroselenic Acid
    cis-6-Octadecenoic Acid
1,6-Phenazinediol 5,10-Dioxide   *Iodinin
Phenethyl Alcohol   *2-Phenylethanol
*Phenol
    Hydroxybenzene
    Oxobenzene
    Phenylic Acid
*Phenylacetic Acid
    α-Toluic Acid
β-Phenylacrylic Acid   *Cinnamic Acid
γ-Phenylallyl Alcohol   *Cinnamyl Alcohol
*Phenylamine
    Aminobenzene
    Aniline
    Kyanol
1-Phenylbutane   *n-Butylbenzene
Phenylcarbinol   *Benzyl Alcohol
*ω-Phenyldecane
    Detergent Alkylate #5
    Tridecylbenzene
*1-Phenyl-1,2-Ethanediol
    Phenylglycol
    Styrene Glycol
*2-Phenylethanol
    Benzyl Carbinol
    Phenethyl Alcohol
*α-Phenylethylamine
    Methylbenzylamine
Phenylformic Acid   *Benzoic Acid
Phenylglycol   *1-Phenyl-1,2-Ethanediol
Phenylhydroxyacetic Acid   *Mandelic Acid
Phenylic Acid   *Phenol
Phenylmethanol   *Benzyl Alcohol
Phenylmethylamine   *Benzylamine
3-Phenyl-2-Propen-1-ol   *Cinnamyl Alcohol
*β-Phenylpropionic Acid
    Benzylacetic Acid
    Hydrocinnamic Acid
Phrenitene   *1,2,3,4-Tetramethylbenzene
*Phthalic Acid
    o-Benzenedicarboxylic Acid
Phycite   *Erythritol
*Plasmin
    Fibrinolysin
    Serum Tryptase
*Polyethylene Glycol
    Carbowax
Polymannuronic Acid   *Alginic Acid
Polysorbate 80   *Tween 80
Propanal   *Propionaldehyde
Propanamide   *Propionamide
1,3-Propanedicarboxylic Acid   *Glutaric Acid
*1,2-Propanediol
    Methyl Glycol
    Propylene Glycol

*1,2-3-Propanetriol
    Glycerol
*1-Propanol
    Propylic Acid
*2-Propanol
    Isopropyl Alcohol
 2-Propanone   *Acetone
 Propenamide   *Acrylamide
 3-Phenylpropenoic Acid   *Cinnamic Acid
 2-Propen-1-ol   *Allyl Alcohol
 Propene Oxide   *Propylene Oxide
 1,2,3-Propenetricarboxylic Acid   *Aconitic Acid
*Propionaldehyde
    Methylacetaldehyde
    Propanal
    Propylaldehyde
*Propionamide
    Propanamide
    Propionic Acid Amide
 Propionic Acid Amide   *Propionamide
 Propylacetic Acid   *Valerianic Acid or *Valeric Acid
 Propylaldehyde   *Propionaldehyde
 Propyl Carbinol   *1-Butanol
 Propylenedicarboxylic Acid   *Itaconic Acid
 Propylene Glycol   *1,2-Propanediol
*Propylene Oxide
    Propene Oxide
 Propylic Acid   *1-Propanol
*Pseudocumene
    1,2,4-Trimethylbenzene
 Puragel   *Gelatin
 2,6(1,3)-Purinedione   *Xanthine
 Purine-2,6,8-Triol   *Uric Acid
*Pyrene
    Benzo-Phenanthrene
 Pyridine-3-Carboxylic Acid   *Nicotinic Acid
 Pyrocatechol   *Catechol
*Pyrogallol
    1,2,3-Trihydroxybenzene
*L-Pyroglutamic Acid
    α-Aminoglutaric Acid Lactam
    Glutimic Acid
 Pyroracemic Acid   *Pyruvic Acid
*Pyruvaldehyde
    1-Ketopropionaldehyde
    Methylglyoxal
    2-Oxopropanal
*Pyruvic Acid
    Acetylformic Acid
    2-Oxopropanoic Acid
    Pyroracemic Acid
*Quercitol
    Acorn Sugar
    1,2,3,4,5-Cyclohexanepentol
*Quinic Acid
    Kinic Acid
 Racemic Lactic Acid   *DL-Lactic Acid

Appendix

```
*Raffinose
    Gossypose
    Melitose
    Melitriose
*Resazurin
    7-Hydroxy-3H-Phenoxazin-3-One 10-Oxide
*Resorcinol
    1,3-Benzenediol
    M-Dihydroxybenzene
*L-Rhamnose
    Isodulcit
 Rhodoviolascin     *Spirilloxanthin
 Ribitol    *Adonitol
 α-D-Ribofuranose    *D-Ribose
*D-Ribose
    α-D-Ribofuranose
*D-Ribulose
    D-Adonose
*Ricinoleic Acid
    D-12-Hydroxyoleic Acid
 Rock Oil    *Petroleum
*Saccharic Acid
    D-Glucaric Acid
 Saccharose    *Sucrose
*Salicin
    Salicyl Alcohol Glucoside
 Salicyl Alcohol    *Saligenin
 Salicyl Alcohol Glucoside    *Salicin
 Salicylaldehyde    *o-Hydroxybenzaldehyde
*Salicylamide
    o-Hydroxybenzamide
*Salicylic Acid
    o-Hydroxybenzoic Acid
*Saligenin
    Salicyl Alcohol
    Saligenol
 Saligenol    *Saligenin
 Sarcolactic Acid    *L-Lactic Acid
*Sarcosine
    Methylaminoethanoic Acid
    N-Methylglycine
*Selenite
    Alabaster
    Calcium Sulfate Dihydrate
    Gypsum
 Seminose    *D-Mannose
 Serum Tryptase    *Plasmin
*Shikimic Acid
    3,4,5-Trihydroxy-1-Cyclohexene-1-Carboxylic Acid
 Sincaline    *Choline
 Sodium Tellurate IV    *Tellurite
 Sodium Tetrathionate    *Tetrathionate
 Sodium Thiosulfate    *Thiosulfate
 Sorbin    *L-Sorbose
 Sorbinose    *L-Sorbose
 D-Sorbitol    *Glucitol
```

*L-Sorbose
    Sorbin
    Sorbinose
*Spermine
    Gerontine
    Musculamine
*Spirilloxanthin
    Rhodoviolascin
 Spreading Factor   *Hyaluronidase
*Starch
    Amylum
*Stearic Acid
    Octadecanoic Acid
 Stearin   *Tristearin
 Stearolic Acid   *9-Octadecynoic Acid
 Styrene Glycol   *1-Phenyl-1,2-Ethanediol
 Styryl Carbinol   *Cinnamyl Alcohol
*Sucrose
    Saccharose
 Tar Camphor   *Naphthalene
*D(-)-Tartaric Acid
    Levotartaric Acid
*L(+)-Tartaric Acid
    Dextrotartaric Acid
*DL-Tartaric Acid
    Uvic Acid
*meso-Tartaric Acid
    Unresolvable Tartaric Acid
*Taurine
    2-Aminoethanesulfonic Acid
*Tellurite
    Sodium Tellurate IV
 Tephthol   *Terephthalic Acid
*Terephthalic Acid
    p-Benzenecarboxylic Acid
    Tephthol
 Tetrahydroxybutane   *Erythritol
*1,2,3,4-Tetramethylbenzene
    Phrenitene
 Tetramethylolmethane   *Pentaerythritol
*Tetrathionate
    Sodium Tetrathionate
 2,2'-Thiodiethanol   *Thiodiethylene Glycol
*Thiodiethylene Glycol
    2,2'-Thiodiethanol
    Thiodiglycol
 Thiodiglycol   *Thiodiethylene Glycol
*Thiosulfate
    Sodium Thiosulfate
 Thyminose   *Deoxyribose
*Toluene
    Methylbenzene
 α-Toluic Acid   *Phenylacetic Acid
 Tonka Bean Camphor   *Coumarin
 Toxilic Acid   *Maleic Acid

Appendix

*Trehalose
    Mycose
*Tributyrin
    Glyceryl Tributyrate
 Tricresol   *Cresol (m-o-p)
 Tridecylbenzene   *ω-Phenyldecane
*Trigonelline
    Caffearine
    Coffearine
    Gynesine
 1,2,3-Trihydroxybenzene   *Pyrogallol
 3,4,5-Trihydroxy-1-Cyclohexene-1-Carboxylic Acid
         *Shikimic Acid
 Sym-1,3,5-Trimethylbenzene   *Mesitylene
 1,2,4-Trimethylbenzene   *Pseudocumene
 Trimethyl Carbinol   *2-Methyl-2-Propanol
*Triolein
    Glyceryl Trioleate
    Olein
*Tripalmitin
    Glyceryl Tripalmitate
    Palmitin
*Triphenylene
    9,10-Benzphenanthrene
*Tristearin
    Glyceryl Tristearin
    Stearin
*Tween 80
    Polysorbate 80
*Tyramine
    4-Hydroxyphenethylamine
    Tyrosamine
 Tyrosamine   *Tyramine
 Unresolvable Tartaric Acid   *meso-Tartaric Acid
 Uracil-6-Carboxylic Acid   *Orotic Acid
 Urastrat   *Urease
*Urease
    Urastrat
 5-Ureidohydantoin   *Allantoin
 δ-Ureidonorvaline   *Citrulline
*Uric Acid
    8-Hydroxyxanthine
    Purine-2,6,8-Triol
 Ursin   *Arbutin
 Uvic Acid   *DL-Tartaric Acid
*Valeramide
    Aminopentamide
*Valerianic Acid
    Propylacetic Acid
    *Valeric Acid
*Valeric Acid
    Propylacetic Acid
    *Valerianic Acid
 Viologen   *Benzylviologen
 Vitamin C   *Ascorbic Acid
 Vitamin K3   *Menadione

Vitamin PP   *Nicotinamide
Whey Factor   *Orotic Acid
*Xanthine
    2,6(1,3)-Purinedione
*Xylene
    Dimethyl Benzene

REFERENCES

Bauer, A. W., M. M. Kirby, J. C. Sherris, and M. Turck. 1966. Antibiotic susceptibility and testing by a standardized single disk method. *Am. J. Clin. Pathol.* **45**, 493-496.

Beveridge, T. J. and J. A. Davies. 1983. Re-examination of the Gram reaction: Trichloro (n2-ethylene) platinum II makes the crystal violet complex visible by electron microscopy. *Abstr. Annual Meeting of the American Society for Microbiology*, p. 171.

Bold, H. C. and M. J. Wynne. 1978. *Introduction to the algae. Structure and reproduction*, Prentice-Hall, Inc. Englewood Cliffs, 700 pp.

Buchanan, R. E. and N. E. Gibbons, eds. 1974. *Bergey's Manual of Determinative Bacteriology*, 8th ed, Williams & Wilkins Co., Baltimore, 1268 pp.

Buchanan, R. E., J. G. Holt, and E. F. Lessel, Jr. 1966. *Index Bergeyana*, The Williams & Wilkins Co., Baltimore, 1472 pp.

Bussard, A., M. I. Krichevsky, and L. D. Blaine. 1985. In *Monoclonal antibodies against bacteria*, vol. I. A. J. L. Macario, and E. C. de Macario, eds. Academic Press, Orlando, pp. 288-312.

Carlone, G. M., M. J. Valadez, and M. J. Pickett. 1982. Methods for distinguishing gram-positive from gram-negative bacteria. *J. Clin. Microbiol.* **16**, 1157-1159.

Corliss, J. O. 1979. *The Ciliated Protozoa*, 2nd ed. Pergamon Press, Oxford, 455 pp.

Daggett, P.-M., M. I. Krichevsky, J. O. Corliss, and J. P. Girolami. 1980. Method for coding data on protozoan strains for computers. *J. Protozool.* **27(4)**, 353-361.

Darland, G. T., T. D. Brock, W. S. Samsanoff, and S. F. Conti. 1970. A thermophilic, acidophilic mycoplasma isolated from a coal refuse pile. *Science* **170**, 1416-1418.

Enzyme Nomenclature. 1979. *Recommendations (1978) of the Nomenclature Committee of the International Union of Biochemistry*. Academic Press, New York, 606 pp.

Fox, G. E., E. Stackebrandt, R. B. Hespell, J. Gibson, J. Maniloff, T. A. Dyer, R. S. Wolfe, W. E. Balch, R. S. Tanner, L. J. Magrum, L. B. Zablen, R. Blakemore, R. Gupta, L. Bonen, B. J. Lewis, D. A. Stahl, K. R. Luehrsen, K. N. Chen, C. R. Woese. 1980. The phylogeny of prokaryotes. *Science* **209**, 457-463.

Gardner, W. and E. I. Cooke, eds. 1971. *Chemical Synonyms and Trade Names - A Dictionary and Commercial Handbook*, 7th ed. CRC Press, Cleveland, 689 pp.

Halebian, S., B. Harris, S. M. Finegold, and R. D. Rolfe. 1981. Rapid method that aids in distinguishing gram-positive from gram-negative anaerobic bacteria. *J. Clin. Microbiol.* 13, 444-448.

Helson, H. 1943. Some factors and implications of color constancy. *J. Optical Soc. America* 33 (10), 555-567.

Hurvich, L. and D. Jameson. 1957. An opponent-process theory of color vision. *Psychol. Rev.* 64, 384-404.

Johnson, R. 1979. Computer-aided identification. *FDA (Food and Drug Administration) By-Lines.* 9, 235-250.

Jones, J. B., B. Bowers, and T. C. Stadtman. 1977. *Methanococcus vanielli*: Ultrastructure and sensitivity to detergents and antibiotics. *J. Bacteriol*, 130, 1404-1406.

Kandler, O. 1982. Cell wall structures and their phylogenetic implications. *Zentr. Bakteriol. Mikrobiol. Hyg. I Abt. Orig.* C3, 149-160.

Kandler, O. and H. Konig. 1978. Chemical composition of the peptidoglcan-free cell walls of methanogenic bacteria. *Arch. Mikrobiol.* 118, 141-152.

Katz, D. 1935. *The world of color.* Kegan, Paul, Trench, Trubner & Co, London, 300 p.

Kelly, K. L. 1965. A universal color language. *Color Engineering* 3, 16.

Kelly, K. L. and D. B. Judd. 1955. The ISCC-NBS method of designating colors and a dictionary of color names. *National Bureau of Standards Circular 553*, U.S. Government Printing Office, Washington, DC.

Kelly, K. L. and D. B. Judd. 1976. Color. Universal language and dictionary of names. *National Bureau of Standards Publication 440.* U.S. Government Printing Office, Washington, DC.

Land, E. H. 1959a. Color vision and the natural image. Part 1. *Proc. Nat. Acad. Sci.* 45(1), 115-129.

Land, E. H. 1959b. Color vision and the natural image. Part 2. *Proc. Nat. Acad. Sci.* 45(4), 636-645.

Land, E. H. 1959c. Experiments in color vision. In *Readings from Scientific American. Perception: mechanisms and models*, 1971. W. H. Freeman and Company, San Francisco, pp. 286-298.

Le Minor, L. 1984. Genus III. *Salmonella*. In *Bergey's Manual of Systematic Bacteriology*, vol. 1, N. R. Krieg and J. G. Holt, eds. Williams & Wilkins, Baltimore, pp. 427-458.

Linnaeus, C. 1753. Species Plantarum. Vols. 1,2 In Stearn, W. T. 1957. *Carl Linnaeus Species Plantarum*. Roy. Society London, 1200 pp.

Munsell, A. H. 1971. *A color notation*, 12th ed. Munsell Color Company, Inc. Baltimore, Md. (note: Previous editions date from 1929; a glossy edition is dated 1976; and an edition for soil samples is dated 1976).

National Committee for Clinical Laboratory Standards. 1981. *Performance Standards for Antimicrobial Disc Susceptibility Tests*, vol. 1. National Committee for Clinical Laboratory Standards, Villanova.

Philpot, C. M., M. I. Krichevsky, and M. Rogosa. 1982. Coding of phenotypic data descriptive of selected group of fungi for entry into computers. *Int. J. System. Bacteriol.* 32, 175-190.

Quadling, C. and S. M. Martin. 1968. Organization of information about Microorganisms. *Progress in Industrial Microbiology* 7, 125-148. D. J. D. Hockenhull, ed. J. and A. Churchill Ltd., London.

Ridgeway, R. 1912. *Standards and color nomenclature* (color plates by Alfred Hoen of Baltimore, MD), Washington, DC.

Rogosa, M., M. I. Krichevsky, and R. R. Colwell. 1971. Method for coding data on microbial strains for computers (edition AB). *Int. J. Syst. Bacteriol.* 21, 1A-184A.

Rogosa, M., C. A. Walczak, J. M. Walat, D. McVey, and M. I. Krichevsky. 1986. Codes and abbreviations for approved or effectively published names of genera of bacteria published from January 1980 to October 1985 and of generally recognized yeast genera. *Int. J. Syst. Bacteriol.* 36, in press.

Simpson, F. J. 1968. A storage and retrieval system employing the digital computer for culture collections. In *Proceedings of the First International Conference on Culture Collections*, H. Iizuka and T. Hasegawa, eds. University of Tokyo Press, Tokyo, pp. 159-165.

Skerman, V. B. D. 1967. *A Guide to the Identification of the Genera of Bacteria*, 2nd ed., Williams & Wilkins Co., Baltimore, 303 pp.

Stecher, P. G., ed. 1968. *The Merck Index*. 8th ed. Merck and Co., Inc. Rahway, New Jersey 1713 pp.

Sturm, S., U. Schonefeld, W. Zillig, D. Janekovic, and K. O. Stetter. 1980. Structure and function of the DNA-dependent RNA polymerase of the archaebacterium *Thermoplasma acidophilus*. *Zentr. Bakteriol. Microbiol. Hyg. I. Abt. Orig.* **C1**, 12-25.

Systematics Association Committee for Descriptive Terminology, 1960. 1. Preliminary list of words relevant to Descriptive Biological Terminology, *Taxon* **99(8)**, 245-257.

Systematics Association Committee for Descriptive Biological Terminology. 1962a. II Terminology of simple symmetrical plane shapes (Chart 1). *Taxon* **11**, 145-162.

Systematics Association Committee for Descriptive Biological Terminology. 1962b. IIa. Terminology of simple symmetrical plane shapes. (Chart 1a). *Taxon* **11**, 245.

Van Valkenburg, S. D., E. P. Karlander, G. W. Patterson, and R. R. Colwell. 1977. Features for classifying photosynthetic aerobic nanoplankton by numerical taxonomy. *Taxon* **26(5/6)**, 497-505.

Wallach, H. 1948. Brightness constancy and the nature of achromatic colors. *J. Exper. Psychol.* **38(3)**, 310-324.

Wallach, H. 1963. The perception of neutral colors. In *Readings from Scientific American 1972. Perception: Mechanisms and Models.* W. Freeman and Company, San Francisco, pp. 280-285.

Weast, R. C., ed. 1974. *Handbook of Chemistry and Physics*, 55th ed. CRC Press, Cleveland, A1-I54 pp.

Weiss, R. L. 1974. Subunit cell wall of *Sulfolobus acidocaldarius*. *J. Bacteriol.* **118**, 275-284.

Wiegel, J. 1981. Distinction between the gram reaction and the gram type of bacteria. Int. J. System. **31**, 88.

Wiegel, J. and L. Quandt. 1982. Determination of the Gram type using the reaction between Polymyxin B and lipopolysaccharides of the outer cell wall of whole bacteria. *J. Gen. Microbiol.* **128**, 2261-2270.

Woese, C. R. 1982. Archaebacteria and cellular origins: an overview. *Zentr. Bakteriol. Mikrobiol. Hyg. I. Abt. Orig.* **C3**, 1-17.

Woese, C. R. and G. E. Fox. 1977. Phylogenetic structure of the procaryotic domain: The primary kingdoms. *Proc. Natl. Acad. Sci. U.S.A.* **74**, 5088-5090.

Wolfe, R. S. and I. J. Higgins. 1979. Biochemistry of methane - a study in contrasts. In *Microbial biochemistry*, J. R. Quayle, ed. M & P Press, Ltd., Lancaster, pp. 267-283.

Zillig, W., K. O. Stetter, and D. Janekovic. 1978. DNA-dependent RNA polymerase from *Halobacterium holobium*. *Europ. J. Biochem.* **91**, 193-199

RAYMOND H. FOGLER LIBRARY

DATE DUE